T0332883

Nonparametric Inference on Manifolds

This book introduces in a systematic manner a general nonparametric theory of statistics on manifolds, with emphasis on manifolds of shapes. The theory has important and varied applications in medical diagnostics, image analysis, and machine vision. An early chapter of examples establishes the effectiveness of the new methods and demonstrates how they outperform their parametric counterparts.

Inference is developed for both intrinsic and extrinsic Fréchet means of probability distributions on manifolds, and then applied to spaces of shapes defined as orbits of landmarks under a Lie group of transformations – in particular, similarity, reflection similarity, affine, and projective transformations. In addition, nonparametric Bayesian theory is adapted and extended to manifolds for the purposes of density estimation, regression, and classification. Ideal for statisticians who analyze manifold data and wish to develop their own methodology, this book is also of interest to probabilists, mathematicians, computer scientists, and morphometricians with mathematical training.

ABHISHEK BHATTACHARYA is Assistant Professor in the Theoretical Statistics and Mathematics Unit (SMU) at the Indian Statistical Institute, Kolkata.

RABI BHATTACHARYA is Professor in the Department of Mathematics at The University of Arizona, Tucson.

INSTITUTE OF MATHEMATICAL STATISTICS
MONOGRAPHS

Nonparametric Inference on Manifolds

With Applications to Shape Spaces

ABHISHEK BHATTACHARYA

Indian Statistical Institute, Kolkata

RABI BHATTACHARYA

University of Arizona

CAMBRIDGE
UNIVERSITY PRESS

CAMBRIDGE
UNIVERSITY PRESS

University Printing House, Cambridge CB2 8BS, United Kingdom

One Liberty Plaza, 20th Floor, New York, NY 10006, USA

477 Williamstown Road, Port Melbourne, VIC 3207, Australia

314-321, 3rd Floor, Plot 3, Splendor Forum, Jasola District Centre, New Delhi - 110025, India

103 Penang Road, #05-06/07, Visioncrest Commercial, Singapore 238467

Cambridge University Press is part of the University of Cambridge.

It furthers the University's mission by disseminating knowledge in the pursuit of education, learning and research at the highest international levels of excellence.

www.cambridge.org
Information on this title: www.cambridge.org/9781107019584

© A. Bhattacharya and R. Bhattacharya 2012

First published 2012
First paperback edition 2015

A catalogue record for this publication is available from the British Library

ISBN 978-1-107-01958-4 Hardback
ISBN 978-1-107-48431-3 Paperback

Contents

Commonly used notation

Manifolds

S^d: d-dimensional sphere

$\mathbb{R}P^k$, $\mathbb{C}P^k$: k-dimensional real, resp. complex, projective space; all lines in $\mathbb{R}^{k+1}$, resp. $\mathbb{C}^{k+1}$, passing through the origin

$V_{k,m}$: Stiefel manifold $\mathrm{St}_k(m)$; all k-frames (k mutually orthogonal directions) in $\mathbb{R}^m$; see §10.1

d_E, d_g: extrinsic distance, geodesic distance; see §4.1, §5.1

d_F, d_P: Procrustes distance: full, partial; see §9.5

$j: M \to E$: embedding of the manifold M into the Euclidean space E; $\tilde{M} = j(M)$; see §4.1

$T_u(M)$ or T_uM: tangent space of the manifold M at $u \in M$; see §4.2

$d_xP: T_xE \to T_{P(x)}\tilde{M}$: differential of the projection map $P: U \to \tilde{M}$, for U a neighborhood of nonfocal point μ; see §4.2

D_r, D: partial derivative with respect to rth coordinate, resp. vector of partial derivatives; see §3.4

$L = L_{\mu_E}$: orthogonal linear projection of vectors in $T_{\mu_E}E \equiv E$ onto $T_{\mu_E}\tilde{M}$; see §4.5.1

r_*: exponential map at p is injective on $\{v \in T_p(M) : |v| < r_*\}$; see §5.1

Shape spaces

Σ_m^k: similarity shape space of k-ads in m dimensions; see §7.1

Σ_{0m}^k: similarity shape space of "nonsingular" k-ads in $m > 2$ dimensions; see §7.2

S_m^k: preshape sphere of k-ads in m dimensions; see§7.1

S_m^k/G: preshape sphere of k-ads in m dimensions modulo transformation group G; see §7.1

$A\Sigma_m^k$: affine shape space of k-ads in m dimensions; see §11.1

$P\Sigma_m^k$: projective shape space of k-ads in m dimensions; see §12.3

$R\Sigma_m^k$: reflection shape space of k-ads in m dimensions; see §9.1

$S\Sigma_2^k$: planar size-and-shape space of k-ads; see §8.10

$SA\Sigma_2^k$: special affine shape space of k-ads in m dimensions; see §11.2

Transformation groups

A': transpose of the matrix A

U^*: conjugate transpose of the complex matrix U

GL(m): general linear group; $m \times m$ nonsingular matrices

$M(m, k)$: $m \times k$ real matrices

O(m): $m \times m$ orthogonal matrices

$S(k)$: $k \times k$ symmetric matrices

Skew(k): $k \times k$ skew-symmetric matrices

SO(m): special orthogonal group; $m \times m$ orthogonal matrices with determinant $+1$

SU(k): special unitary group; $k \times k$ unitary matrices with determinant $+1$

Statistics

$\xrightarrow{\mathcal{L}}$: convergence in distribution

$\mathcal{X}_d^2$: chi-squared distribution with d degrees of freedom

$N_d(0, \Sigma)$: mean zero d-dimensional Normal distribution

$Z_{1-\frac{\alpha}{2}}$: upper $(1 - \frac{\alpha}{2})$-quantile of $N(0, 1)$ distribution; see §5.4.1

μ_E, μ_{nE} (μ_I, μ_{nI}): extrinsic (intrinsic) mean of the distribution Q, resp. sample extrinsic (intrinsic) mean of the empirical distribution Q_n; see §4.1 (§5.1)

μ_F, μ_{nF}: (local) Fréchet mean of the distribution Q, resp. sample Fréchet mean of the empirical distribution Q_n; see §8.5

V, V_n: variation of the distribution Q, resp. sample variation of the empirical distribution Q_n; see §3.3

$K(m; \mu, \kappa)$: probability density (kernel) on a metric space M with variable $m \in M$ and parameters $\mu \in M$ and $\kappa \in N$, a Polish space; see §13.2

Preface

This book presents in a systematic manner a general nonparametric theory of statistics on manifolds with emphasis on manifolds of shapes, and with applications to diverse fields of science and engineering. There are many areas of significant application of statistics on manifolds. For example, directional statistics (statistics on the sphere S^2) are used to study shifts in the Earth's magnetic poles over geological time, which have an important bearing on the subject of tectonics. Applications in morphometrics involve classification of biological species and subspecies. There are many important applications to medical diagnostics, image analysis (including scene recognition), and machine vision (e.g., robotics). We take a fresh look here in analyzing existing data pertaining to a number of such applications. It is our goal to lay the groundwork for other future applications of this exciting emerging field of nonparametric statistics.

Landmark-based shape spaces were first introduced by D. G. Kendall more than three decades ago, and pioneering statistical work on shapes with applications to morphometrics was carried out by F. Bookstein around the same time. Statistics on spheres, or directional statistics, arose even earlier, and a very substantial statistical literature on directional statistics exists, including a seminal 1953 paper by R. A. Fisher and books by Watson (1983), Mardia and Jupp (2000), Fisher et al. (1987), and others. For statistics on shape spaces, important parametric models have been developed by Kent, Dryden, Mardia, and others, and a comprehensive treatment of the literature may be found in the book by Dryden and Mardia (1998). In contrast, the present book concerns nonparametric statistical inference, much of which is of recent origin.

Although the past literature on manifolds, especially that on shape spaces, has generally focused on parametric models, there were a number of instances of the use of model-independent procedures in the 1990s and earlier. In particular, Hendriks and Landsman (1996, 1998) provided nonparametric procedures for statistics on submanifolds of Euclidean spaces, which are special cases of what is described as extrinsic analysis in this book. Independently of this, Vic Patrangenaru, in his 1998 dissertation, arrived at nonparametric extrinsic methods for statistics on general manifolds. Intrinsic statistical inference, as well as further development of general extrinsic inference, with particular emphasis on Kendall's shape spaces, appeared in two papers in the *Annals of Statistics* (2003, 2005) by Patrangenaru and the second author of this monograph. Our aim here is to present the current state of this general theory and advances, including many new results that provide adequate tools of inference on shape spaces.

A first draft of the book submitted to the IMS was the Ph.D. dissertation of the first author at the University of Arizona. The present monograph under joint authorship came about at the suggestion of the IMS editors. It has substantially more material and greater emphasis on exposition than the first draft.

For the greater part of the book (Chapters 1–12) we focus on the *Fréchet mean* of a probability distribution Q on a manifold, namely, the minimizer, if unique, of the expected squared distance from a point of a manifold-valued random variable having the distribution Q. If the distance chosen is the geodesic distance with respect to a natural Riemannian structure, such a mean is called *intrinsic*. If, on the other hand, the manifold is embedded in a Euclidean space, or a vector space, then the distance induced on the manifold by the Euclidean distance is called *extrinsic*, and the corresponding Fréchet mean is termed an extrinsic mean. One would generally prefer an equivariant embedding that preserves a substantial amount of the geometry of the manifold. An advantage of extrinsic means is that they are generally unique. By contrast, sufficiently broad conditions for uniqueness of the intrinsic mean are not known, thus impeding its use somewhat.

The last two chapters of the book represent a point of departure from the earlier part. Based on recent joint work of the first author and David Dunson, nonparametric Bayes procedures are derived for functional inference on shape spaces – nonparametric density estimation, classification, and regression.

The manifolds of shapes arising in applications are of fairly high dimension, and the Fréchet means capture important and distinguishing features of the distributions on them. In analyzing real data, the nonparametric methods developed in the monograph often seem to provide sharper inferences than do their parametric counterparts. The parametric models do, however, play a significant role in the construction of nonparametric Bayes priors for density estimation and shape classification in the last two chapters.

Readership This monograph is suitable for graduate students in mathematics, statistics, science, engineering, and computer science who have taken (1) a graduate course in asymptotic statistics and (2) a graduate course in differential geometry. For such students special topics courses may be based on it. The book is also meant to serve as a reference for researchers in the areas mentioned. For the benefit of readers, extrinsic analysis, whose geometric component does not involve much more than introductory differentiable manifolds, is separated from intrinsic inference for the most part. Appendix A, on differentiable manifolds, provides some background. Some basic notions from Riemannian geometry are collected in Appendix B.

For background in statistics, one may refer to Bickel and Doksum (2001, Chaps. 1–5), Bhattacharya and Patrangenaru (2012, Pt. II), Casella and Berger (2001, Chaps. 5–10), or Ferguson (1996). For geometry, Do Carmo (1992) and Gallot et al. (1990) are good references that contain more than what is needed. Other good references are Boothby (1986) (especially for differentiable manifolds) and Lee (1997) (for Riemannian geometry). A very readable upper division introduction to geometry is given in Millman and Parker (1977), especially Chapters 4 and 7.

For a basic course on the subject matter of the monograph, we suggest Chapters 1–4, 6, 8 (Sections 8.1–8.3, 8.6, 8.11), and 9, as well as Appendices A, B, and C. A more specialized course on the application of nonparametric Bayes theory to density estimation, regression, and classification on manifolds may be based on Chapters 1, 5, 13, and 14 and Appendices A–D. Chapters 10–12, in addition to the basics, would be of special interest to students and researchers in computer science.

MATLAB code and data sets MATLAB was used for computation in the examples presented in the book. The MATLAB code, with all data sets embedded, can be downloaded from http://www.isical.ac.in/~abhishek.

Acknowledgments The authors are indebted to the series editors Xiao-Li Meng and David Cox for their kind suggestions for improving the substance of the book as well as its presentation, and thank Lizhen Lin for her invaluable help with corrections and editing. Also appreciated are helpful suggestions from a reviewer and from the Cambridge University Press editor Diana Gillooly. The authors gratefully acknowledge support from National Science Foundation grants DMS 0806011 and 1107053, and National Institute of Environmental Health Sciences grant R01ES017240.

1

Introduction

Digital images today play a vital role in science and technology, and also in many aspects of our daily life. This book seeks to advance the analysis of images, especially digitized ones, through the statistical analysis of shapes. Its focus is on the analysis of landmark-based shapes in which a k-ad, that is, a set of k labeled points or landmarks on an object or a scene, is observed in two or three dimensions, usually with expert help, for purposes of identification, discrimination, and diagnostics.

In general, consider the k-ad to lie in $\mathbb{R}^m$ (usually, $m = 2$ or 3) and assume that not all the k points are the same. Then the appropriate shape of the object is taken to be the k-ad modulo a group of transformations.

For example, one may first center the k-ad, by subtracting the mean of the k-ad from each of the k landmarks, to remove the effect of location. The centered k-ad then lies in a hyperplane of dimension $mk - m$, because the sum of each of the m coordinates of the centered k points is zero. Next one may scale the centered k-ad to unit size to remove the effect of scale or size. The scaled, centered k-ad now lies on the unit sphere $S^{m(k-1)-1}$ in a Euclidean space (the hyperplane) of dimension $m(k - 1)$ and is now called the *preshape* of the k-ad. Further, to remove the effect of orientation, the scaled, centered k-ad is rotated by means of elements of the (special orthogonal) group $SO(m)$ of rotations in $\mathbb{R}^m$. The orbit of the preshape under all rotations may then be taken to be the shape of the k-ad. This shape is called a *similarity shape*, and the space of these shapes comprises Kendall's similarity shape space Σ_m^k. While this is a proper choice for many problems in biology and in medical imaging, other notions of shape, such as affine shape and projective shape, are important in machine vision and bioinformatics. The *affine shape* of a k-ad is invariant under all affine transformations; that is, it may be identified as the orbit of the k-ad under (the group of) all affine transformations. This is an appropriate notion of shape of a k-ad based on images taken from far away, for example, from an airplane or a satellite. Here a rectangle may be transformed to a parallelogram. Similarly, *projective shapes* are invariant under

1

all projective transformations, and these are the appropriate shapes of k-ads recorded, for example, by central projection by a camera, where a line in three dimensions is projected to a point on the image plane, and a three-dimensional object appears as a two-dimensional image. These shape spaces are differentiable manifolds (sometimes after the removal of a small singular set), often with a Riemannian structure allowing one to measure (geodesic) lengths, angles, curvature, and so on. These notions are important in intrinsic analysis as mentioned briefly below and explained in detail in Chapter 5.

For nonparametric analysis of shape distributions Q on a manifold M with a distance d, we focus on the *Fréchet mean* of Q, which is the minimizer (if unique) of the *Fréchet function*, namely, the average squared distance from a point. Sometimes the corresponding minimum average is also considered, called the *variation* of Q. If the distance is the geodesic distance, these parameters, and the corresponding statistical analysis, are said to be *intrinsic*. Often it is more convenient, mathematically as well as from a computational point of view, to embed the manifold in a Euclidean space E (generally of higher dimension than that of M) and use the induced (Euclidean) distance on the image under the embedding j, say. The Fréchet mean and variation, and the statistical analysis based on them, are then said to be *extrinsic*. For $M = S^d$, a d-dimensional unit sphere, the intrinsic, or geodesic, distance between two points is the arc length between them measured on the great circle joining the points. This distance is sometimes referred to as the arc distance. Also, the sphere has a natural embedding into $\mathbb{R}^{d+1}$ by the *inclusion map j*. That is, if we represent S^d as $\{|x| = 1 : x = (x_1, \ldots, x_{d+1}) \in \mathbb{R}^{d+1}\}$, then $j(x) = x$ for $x \in S^d$. The extrinsic distance between two points is the length of the line segment joining the two points, the so-called chord distance.

Suppose one has a sample of k-ads of size n. This yields a sample of n shapes in the appropriate shape space (manifold) M. The common distribution of these n shapes is denoted by Q. The statistical analysis begins by first (1) finding broad conditions for uniqueness of the Fréchet minimizer, and then (2) finding the asymptotic distribution of the corresponding sample Fréchet mean as its estimate. Similarly one estimates the variation. There is no general simplifying procedure for the construction of the intrinsic mean. However, for the extrinsic mean of Q under an embedding j, one first computes the mean, say μ, of Q as a distribution on the ambient Euclidean space $E = \mathbb{R}^D$. The extrinsic mean of Q on the image $j(M)$ of M is given by the point in $j(M)$, if unique, which is at the minimum Euclidean distance from μ, denoted by $P(\mu)$, with P denoting the *projection* operation.

For the statistical problems at hand, consider the particular case of distinguishing between two shape distributions on a manifold. To understand the general nature of a two-sample test for equality of the extrinsic means of two distributions Q_1, Q_2 on M, based on independent samples $\{X_{j_i} : j = 1, \ldots, n_i\}$ of sizes n_i ($i = 1, 2$), observe that

$$\sqrt{n_i}\,[P(\hat{\mu}_i) - P(\mu_i)] = \sqrt{n_i}\,[d_{\mu_i}P(\hat{\mu}_i - \mu_i)] + o_p(1) \quad (i = 1, 2), \quad (1.1)$$

where P is the projection on $j(M)$ and $\mu_i, \hat{\mu}_i$ are the Euclidean population and sample means of the ith group ($i = 1, 2$), under the embedding j (i.e., in the ambient Euclidean space $\mathbb{R}^D$). Here $d_{\mu_i}P$ is the differential of the projection map P at μ_i, which is a linear map from the tangent space of $\mathbb{R}^D$ at μ_i, namely, $T_{\mu_i}\mathbb{R}^D = \mathbb{R}^D$, to the tangent space of $\tilde{M}$ at $P(\mu_i)$, that is, $T_{P(\mu_i)}\tilde{M} \subset T_\mu\mathbb{R}^D$ (here $\tilde{M} := j(M)$ is the image of M under the embedding j). Note that the Jacobian of $d_{\mu_i}P$ is a singular $D \times D$ matrix whose rank is the same as the dimension d of $\tilde{M}$ (or M). Now, for each i, viewed as a random element of $\mathbb{R}^D$, equation (1.1) converges to a (singular) Normal distribution $N(0, \Sigma_i)$ in $\mathbb{R}^D$, where Σ_i is a $D \times D$ covariance matrix. Under the null hypothesis H_0 of equality of the two extrinsic means, $P(\mu_1) = P(\mu_2) = j(\mu_E) = \tilde{\mu}_E$, say, the tangent spaces $T_{P(\mu_i)}\tilde{M}, i = 1, 2$, are the same. Let $n = n_1 + n_2$ and assume $\frac{n_1}{n} \to p$, $0 < p < 1$. Taking the difference between the two quantities in equation (1.1), one has the convergence in distribution $\sqrt{n}[P(\hat{\mu}_1) - P(\hat{\mu}_2)] \to N(0, p^{-1}\Sigma_1 + (1 - p)^{-1}\Sigma_2)$. But the right side of equation (1.1), excluding the term $o_p(1)$, lies in the tangent space $T_{\tilde{\mu}_E}\tilde{M}$ of dimension d and converges to a d-dimensional Normal $N(0, \Gamma)$, where $\Gamma = p^{-1}\Sigma_1 + (1-p)^{-1}\Sigma_2$, with Σ_i being the covariance matrix of the coordinates of $d_{\mu_i}(\tilde{X}_{j_i} - \mu_i)$ ($i = 1, 2$) with respect to the basis of the tangent space $T_{\tilde{\mu}_E}\tilde{M}$ of the embedded manifold $\tilde{M}$ at the common extrinsic mean $\tilde{\mu}_E$ (under H_0). Here $\tilde{X} = j(X)$. This leads to a chi-squared test. As usual, one replaces Σ_i by the sample estimate $\hat{\Sigma}_i$, obtained as the sample covariance of the coordinates of $d_{\hat{\mu}_i}(\tilde{X}_{j_i} - \hat{\mu}_i)$ ($j = 1, \ldots, n_i$). For intrinsic analysis, the computation of the intrinsic sample mean begins with a theoretical derivation of the geodesics as well as the Fréchet minimization involving the geodesic distance. Analogous to the extrinsic embedding, one then transfers the population (and sample) distributions to the tangent space at the intrinsic mean by the so-called inverse exponential map described in Chapter 5. The rest of the procedure is similar to that for the extrinsic mean.

It is useful to remember that the larger the group of transformations applied to the k-ads, the larger the orbit under it defining the shape of a k-ad, and the fewer are the details of the numerics of the k-ads preserved in their shapes. In particular, statistical significance (at a given level of significance) in a two-sample test based on a notion of shape invariant under a larger group is in general a stronger statement than that based on shape invariant under smaller groups. In this context one should note the increasing order of groups of transformations defining Kendall's similarity shapes, reflection similarity shapes, affine shapes, and projective shapes.

Here is a brief outline of the book.

To motivate the reader, Chapter 2 provides an exposition of several data examples, which are analyzed in detail in later chapters.

In Chapter 3, the concepts of Fréchet mean and variation are introduced. The idea is to define the Fréchet function of a probability distribution Q on a metric space M as the integral (with respect to Q) of the squared distance on M, and define the Fréchet mean as the minimizer (if unique) of this Fréchet function and the Fréchet variation as the minimum value (if finite). Conditions are derived for the consistency and asymptotic Normality of the sample estimates of the Fréchet mean. Confidence regions for the population parameters are constructed both by using the asymptotic distribution of the sample estimates and by pivotal bootstrap methods.

Chapter 4 is devoted to extrinsic inference on a differentiable manifold M. Here one embeds M into some higher dimensional Euclidean space and uses the distance induced by this embedding. Because many embeddings are available, one chooses an embedding that is equivariant with respect to a group of transformations large enough to preserve a great deal of the geometry of M. As the results in later chapters show, the corresponding analysis becomes simpler both mathematically and computationally than its intrinsic counterpart. For example, the extrinsic mean is known to exist under fairly broad conditions and in most cases has a closed-form analytic expression (see Chapters 8–12). In particular, the extrinsic mean is the projection of the Euclidean mean of the image of Q on the image manifold under the embedding. Hence there is a unique mean if and only if there is a unique projection. For asymptotic Normality of the sample mean and variation, one requires that this projection map is smooth in a neighborhood of the population mean (which is assumed to exist). The chapter concludes with two-sample nonparametric tests to distinguish between two probability distributions by comparing the sample extrinsic means and variations. Appropriate tests are constructed for both mutually independent and matched pair samples. The numerical examples in Chapter 8 show that in these examples the extrinsic and intrinsic means are very close to each other and the two-sample extrinsic and intrinsic tests yield similar results.

Chapter 5 performs Fréchet analysis on a Riemannian manifold M by using the geodesic distance as the distance metric in the definition of the Fréchet function. The resulting Fréchet parameters are called intrinsic and the corresponding statistical analysis is called intrinsic analysis on a manifold M. In this chapter, sufficient conditions for existence of a unique intrinsic mean are used to derive the asymptotic Normality of the sample intrinsic mean. Two-sample nonparametric tests are constructed to compare the sample intrinsic means and variations, which can be used to distinguish between two underlying distributions.

Chapter 6 introduces the different notions of shapes treated in this book. They include (direct) similarity shapes, reflection similarity shapes, affine shapes, and projective shapes. For problems in biology such as classifications of species, disease detection, and so on, similarity shape analysis has many uses, while for problems in machine vision and image analysis, affine and projective shape analyses are more appropriate.

In Chapters 7–12, the geometry of each of the shape spaces introduced in Chapter 6 is discussed in detail, and explicit forms of estimates and tests are derived in each case using the methods of Chapters 4 and 5.

In particular, Chapter 7 provides an exposition of the geometry of the (direct) *similarity shape space of k-ads in m* dimensions, or Σ_m^k. The cases of interest include $m = 2$ and 3. This space can be represented as the quotient of the unit sphere with respect to all rotations (in m dimensions), that is, the space of orbits of k-ads under rotations of the preshape sphere, as described at the outset. It is shown that, after removing some singularities, Σ_m^k is a Riemannian manifold. There are no such singularities when $m = 2$. This chapter identifies the tangent space of Σ_m^k, the exponential map, and the geodesic distance on Σ_m^k.

Chapter 8 considers in detail the similarity shape space Σ_2^k obtained when $m = 2$, which is also called the *planar shape space*. This is a compact, connected manifold. This chapter presents the geometry of this space and applies the methods of Chapters 4 and 5 for intrinsic and extrinsic analyses. Analytic expressions for the parameters in the asymptotic distribution of the sample extrinsic mean are derived. This enables one to perform two-sample tests to compare the extrinsic means and variations of two underlying probability distributions. The results of extrinsic and intrinsic analyses on Σ_2^k are applied to two examples.

When $m > 2$, the similarity shape space Σ_m^k fails to be a manifold. After singularities are excluded, the remaining set is a manifold that is not complete. As a consequence, the results from Chapters 4 and 5 cannot be applied to carry out intrinsic and extrinsic analyses. If, instead, one considers the *reflection similarity shape*, which is invariant under all orthogonal transformations, not just rotations, then one can embed the resulting shape space into the vector (or Euclidean) space of symmetric matrices $S(k, \mathbb{R})$ and carry out extrinsic analysis. This is discussed in Chapter 9.

The methods here allow one to extend nonparametric inference on Kendall-type shape spaces from two to higher dimensions. This chapter concludes with an application to a matched pair example.

Chapter 11 focuses on methodologies for nonparametric inference on the affine shape spaces $A\Sigma_m^k$. The *affine shape* of a k-ad x with landmarks in $\mathbb{R}^m$ is defined as the orbit of x under all affine transformations, $x \mapsto Ax + b$ ($A \in \text{GL}(m, \mathbb{R}), b \in \mathbb{R}^m$). The space of affine shapes of all centered k-ads whose columns span $\mathbb{R}^m$ is $A\Sigma_m^k$. Here $\text{GL}(m, \mathbb{R})$ is the so-called general linear space of all $m \times m$ nonsingular real matrices. For extrinsic analysis on $A\Sigma_m^k$, one embeds it into the vector space $S(k, \mathbb{R})$ of all $k \times k$ real symmetric matrices via an equivariant embedding. Using this embedding, an expression for the extrinsic mean and a condition for its uniqueness are derived. The results from Chapter 4 are used to derive the asymptotic distribution for the sample extrinsic mean and variation, and are applied to construct two-sample nonparametric tests to compare two probability distributions.

Chapter 12 presents methodologies for statistical analyses of projective shapes, which are useful in axial analysis and machine vision. For $m = 2$, for example, the projective space $\mathbb{R}P^2$ is the space of all lines in $\mathbb{R}^3$ passing through the origin. The image of a line, or axis, passing through the center of a pin-hole–type camera is recorded as a point on the plane of the camera film. Thus a scene in three dimensions is pictured on the plane of the camera film. A set of k distinct lines originating from the three-dimensional scene then yields a k-ad in $\mathbb{R}P^2$, that is, a point in $(\mathbb{R}P^2)^k$. To define the projective shape of the k-ad, one applies affine transformations $A \in \text{GL}(3, \mathbb{R})$ to the k points in three dimensions. The equivalence class of lines generated from these (or the corresponding equivalence class of points on the camera film) is the *projective shape* of the k-ad.

The last two chapters, Chapters 13 and 14, represent a different theme than the rest of the book. Here we consider functional inference – (a) density estimation, (b) classification and (c) regression – by applying the nonparametric Bayes methodology.

It may be noted that there now exists a substantial literature, mostly in computer science, but also in statistics, on what may be termed continuous shapes. The greater part of this work is in two dimensions, where such shapes may be analyzed by a deformable template of gray levels on a grid of points representing a digitized approximation of the image. Here one uses high-dimensional (parametric) Bayes methodology (see Amit, 2002). The other alternative that has gained popularity in recent years is to consider a two-dimensional shape as given by the actual boundary contour of the object of interest. There is also some work on the three-dimensional shape provided by the boundary surface. Apart from geometric considerations, much of this work focuses on analytical and computational problems involving matching (or discriminating among) given shapes, or changes in shapes, and not on statistical inference for discrimination among distributions of shapes (see Krim and Yezzi, 2006). For a nonparametric analysis of the latter, one views these spaces of shapes as infinite-dimensional (Hilbert) manifolds. Intrinsic inference here is limited by the fact that establishing the uniqueness of such basic quantities as the intrinsic mean even under restrictive conditions is difficult. In contrast, extrinsic analysis based on equivariant embeddings in vector (Hilbert) spaces holds some promise (see, e.g., the recent work of Ellingson et al., 2011).

Finally, a note on the use of bootstrapping in the text. Efron's *bootstrap* (Efron, 1979) has had a profound impact on modern statistical methodology. Apart from the fulfillment of its original intended goals such as estimating standard errors of rather complicated statistics and avoiding such computations altogether in providing confidence regions, for continuous data a great benefit of the method lies in its substantial edge over central limit theorem (CLT)–based confidence regions in reducing coverage errors. This aspect of the bootstrap's efficacy is established by the method of asymptotic expansions of distributions of smooth statistics (Bhattacharya, 1977; Bhattacharya and Ghosh, 1978) and their application to bootstrapped versions of such statistics. For this theory, see Singh (1981), Babu and Singh (1984), Bhattacharya (1987), Beran (1987), Bhattacharya and Qumsiyeh (1989), Bhattacharya and Denker (1990), and Lahiri (1994). It follows from this, in particular, that a confidence region based on pivotal bootstrapping of an asymptotic chi-squared statistic yields a coverage error of order $O(n^{-2})$, as opposed to the order $O(n^{-1})$ error resulting from the classical chi-squared approximation. For a readable account of the bootstrap's coverage error for symmetric confidence regions and various other matters we refer to Hall (1992). Unfortunately, for the generally high-dimensional shape spaces considered in this monograph, the bootstrapped covariance tends to be singular if the sample size is not very large, thus limiting the usefulness of the bootstrap somewhat.

2

Examples

This chapter collects together, and describes in a simple manner, a number of applications of the theory presented in this book. The examples are based on real data, and, where possible, results of parametric inference in the literature are cited for comparison with the new nonparametric inference theory.

2.1 Data example on S^1: wind and ozone

The wind direction and ozone concentration were observed at a weather station for 19 days. Table 2.1 shows the wind directions in degrees. The data are taken from Johnson and Wehrly (1977). The data viewed on the unit circle S^1 are plotted in Figure 3.1. We compute the sample extrinsic and intrinsic mean directions, which come out to be 16.71 and 5.68 degrees, respectively. They are displayed in the figure. We use angular coordinates for the data in degrees lying between $[0°, 360°)$ as in Table 2.1. An asymptotic 95% confidence region for the intrinsic mean as obtained in Section 3.7, Chapter 3, turns out to be

$$\{(\cos\theta, \sin\theta) : -0.434 \le \theta \le 0.6324\}.$$

The corresponding end points of this arc are also displayed in Figure 3.1.

Johnson and Wehrly (1977) computed the so-called angular–linear correlation $\rho_{AL} = \max_\alpha\{\rho(\cos(\theta - \alpha), X)\}$, where X is the ozone concentration when the direction of wind is θ. Here ρ denotes the true coefficient of correlation. Based on the sample counterpart r_{AL}, the 95% confidence interval for ρ_{AL} was found to be $(0.32, 1.00)$.

2.2 Data examples on S^2: paleomagnetism

We consider here an application of *directional statistics*, that is, statistics on the unit sphere S^d, with $d = 2$ in the present case, which has an important bearing on a fundamental issue in paleomagnetism. *Paleomagnetism*

Table 2.1 *Wind directions in degrees*

327	91	88	305	344	270	67
21	281	8	204	86	333	18
57	6	11	27	84		

Table 2.2 *Data from Fisher (1953) on remanent magnetism*

D	343.2	62.0	36.9	27.0	359.0	5.7	50.4	357.6	44.0
I	66.1	68.7	70.1	82.1	79.5	73.0	69.3	58.8	51.4

D: Declination; I: Inclination

is the field of earth science that is devoted to the study of fossil magnetism as contained in fossilized rock samples, known as remanent magnetism. It has been theorized for many years that the Earth's magnetic poles have shifted over geological time. This idea is related to the older theory of continental drift, namely, that the continents have changed their relative positions over a period of several hundred million years. If rock samples in different continents dating back to the same period exhibit different magnetic polarities, that would be a confirmation of the theory of continental drift. As pointed out by the geophysicist Irving (1964) in the preface of his book, over the years such confirmations have been achieved with the help of rigorous statistical procedures. In Chapter 4, Section 4.7, a multi-sample nonparametric test for the hypothesis of equality is provided for such purposes. In a seminal paper, Fisher (1953) used a parametric model known as the *Fisher* or *von Mises–Fisher distribution* on the sphere S^2 with a density $f(x; \mu, \tau) = c(\tau) \exp\{\tau x' \mu\}$ with respect to the uniform distribution on the sphere (see Appendix D), where μ is the true direction (given by a point on the unit sphere S^2) and $\tau > 0$ is the concentration parameter. The maximum likelihood estimate (MLE) of the true position μ, based on i.i.d. observations $X_1, \ldots, X_n$ on S^2, is given by $\overline{X}/|\overline{X}|$, assuming $\overline{X} \neq 0$. Thus the MLE is the same as the extrinsic mean of the sample (empirical) distribution on S^2, where μ is the extrinsic mean, as well as the intrinsic mean, of Fisher's distribution.

From the Icelandic lava flow of 1947–1948, nine specimens of remanent magnetism were collected. The data can be viewed as an i.i.d. sample on the manifold S^2 and can be found in Fisher (1953; the data were supplied by J. Hospers). They are displayed in Table 2.2.

The sample extrinsic mean is $\hat{\mu}_E = (.1346, .2984, .9449)$. The sample extrinsic and intrinsic means are very close, namely, at a geodesic distance of 0.0007 from each other. This estimate of magnetic north is close to the Earth's geographic north pole (0, 0, 1).

Based on his distribution, Fisher obtained a 95% confidence region for the mean direction μ. This region may be expressed as

$$\{p \in S^2 : d_g(\hat{\mu}_E, p) \leq 0.1536\},$$

where d_g denotes the geodesic distance. Fisher's confidence region, and our asymptotic confidence region for the population extrinsic mean derived in Chapter 4, are plotted in Figure 2.1. The former confidence region nearly contains the latter and is considerably larger than it.

To study possible shifts in the positions of Earth's magnetic poles, Fisher also analyzed a second set of data, supplied by Hospers, of remanent magnetism from the early Quaternary period (between 10,000 and one million years ago). The sample estimate (MLE) from this sample of 45 observations turns out to be $\hat{\mu}_E = (.0172, -.2978, -.9545)$, which shows a near reversal of the magnetic poles between the two geological periods. The 95% confidence region for the true direction by Fisher's method is a geodesic ball of radius .1475 around the MLE. Since we were unable to access the original data from the second example in Fisher's paper,

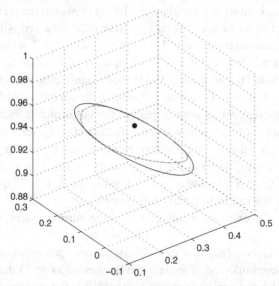

Figure 2.1 Boundaries of the confidence regions for the direction of Earth's magnetic pole, using Fisher's method (*solid*) and the nonparametric extrinsic method (*dashed*), based on data from Fisher (1953).

Table 2.3 *Data from Irving (1963)*

D	63	258	234	54	279	134	281	202	294	290	
I	−78	−77	−76	−74	−86	−85	−70	−81	−61	−76	
D	288	123	232	356	308	6	76	24	230	227	
I	−72	−78	−86	−84	−80	−85	−87	−82	−76	−76	
D	269	257	16	16	357	28	1	68	30	134	324
I	−75	−86	−80	−88	−71	−78	−80	−81	−86	−87	−69

D : Declination; I : Inclination

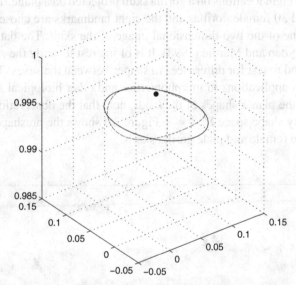

Figure 2.2 Boundaries of the confidence regions for the direction of earth's magnetic poles, using Fisher's method (*solid*) and the nonparametric extrinsic method (*dashed*), based on the Jurassic period data of Irving (1963).

the corresponding extrinsic (or intrinsic) nonparametric confidence region could not be computed.

We now consider another set of data from Irving (1963) from the Jurassic period (138–180 million years ago). Based on 31 observations from table 3 in the paper (each observation is the mean of two specimens from the same sample), the MLE of the von Mises–Fisher distribution, which is also the extrinsic sample mean, is (.1346, .2984, .9449). The data are presented in Table 2.3. Figure 2.2 shows Fisher's confidence region (*solid line*) covering an area of .0138 and the confidence region based on the nonparametric extrinsic analysis (*dashed line*) covering an area .0127.

The nonparametric methods, both extrinsic and intrinsic, seem to provide sharper confidence regions than those based on Fisher's parametric model.

2.3 Data example on Σ_2^k: shapes of gorilla skulls

In this example, we first discriminate between two planar shape distributions via their extrinsic (and intrinsic) means (Chapter 8). A classifier is then built and applied (Chapter 14).

Consider eight locations on a gorilla skull projected on a plane. There are 29 male and 30 female gorillas and the eight landmarks are chosen on the midline plane of the two-dimensional image of the skull. The data can be found in Dryden and Mardia (1998). It is of interest to study the shapes of the skulls and to test for differences in shapes between the sexes. This procedure finds applications in morphometrics and other biological sciences. To analyze the planar shapes of the k-ads, note that the observations lie in the similarity shape space Σ_2^k, $k = 8$. Figure 2.3 shows the preshapes of the k-ads for the female and male gorilla skulls.

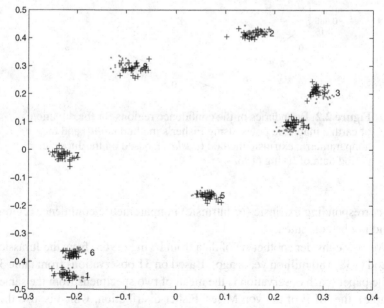

Figure 2.3 Eight landmarks from skulls of 30 females (*gray dot*) and 29 male gorillas (*black plus*)

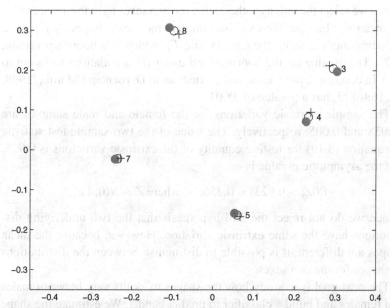

Figure 2.4 Landmarks from preshapes of extrinsic means for females (*filled gray circle*), males (*open black circle*) and pooled (*black plus*). From Bhattacharya and Dunson (2010a).

To distinguish between the distributions of shapes of skulls for the two sexes, one may compare the sample extrinsic mean shapes or variations in shape. Figure 2.4 plots the sample extrinsic means for the two sexes along with the pooled sample extrinsic mean. The sample intrinsic means are very close to their extrinsic counterparts; the geodesic distance between the intrinsic and extrinsic means is 5.54×10^{-7} for the female sample and 1.96×10^{-6} for the male sample.

The value of the two-sample test statistic, defined in equation (5.21) for comparing the intrinsic mean shapes, and the asymptotic p-value for the chi-squared test are

$$T_{n1} = 391.63, \quad \text{p-value} = P(X_{12}^2 > 391.63) < 10^{-16}.$$

Hence we reject the null hypothesis that the two sexes have the same intrinsic mean skull shape. The two-sample test statistics, defined in equations (4.15) and (4.18) for comparing the extrinsic mean shapes, and the corresponding asymptotic p-values are

$$T_1 = 392.6, \quad \text{p-value} = P(X_{12}^2 > 392.6) < 10^{-16},$$
$$T_2 = 392.0585, \quad \text{p-value} < 10^{-16}.$$

Hence we reject the null hypothesis that the two sexes have the same extrinsic mean skull shape. We can also compare the mean shapes by a pivotal bootstrap method using the test statistic T_2^*, which is a bootstrap version of T_2. The p-value for the bootstrap test using 10^5 simulations turns out to be 0. In contrast, a parametric test carried out in Dryden and Mardia (1998, pp. 168–172) has a p-value of .0001.

The sample extrinsic variations for the female and male samples are 0.0038 and 0.005, respectively. The value of the two-sample test statistic in equation (8.19) for testing equality of the extrinsic variations is 0.923, and the asymptotic p-value is

$$P(|Z| > 0.923) = 0.356, \quad \text{where } Z \sim N(0, 1).$$

Hence we do not reject the null hypothesis that the two underlying distributions have the same extrinsic variation. However, because the mean shapes are different, it is possible to distinguish between the distributions of shapes for the two sexes.

The next goal is to study how the shapes of skulls vary between males and females and build a classifier to predict gender. We estimate the shape densities for the two groups via nonparametric Bayesian methods and use them to derive the conditional distribution of gender given shape, as described in Chapter 13. Figure 13.1 shows the density estimates on a particular geodesic, along with 95% credible regions. We randomly pick 25 individuals of each gender as a training sample, with the remaining 9 used as test data. Table 13.1 presents the estimated posterior probability of being female for each of the gorillas in the test sample along with a 95% credible interval. For most of the gorillas, there is a high posterior probability of assigning the correct gender. There is misclassification only for the third female and the third male. For the third female, the credible interval includes 0.5, suggesting that there is insufficient information to be confident in the classification. However, for the third male, the credible interval suggests a high degree of confidence that this individual is female. In addition, we display the extrinsic distances between the shape for each gorilla and the female and male sample extrinsic means. Potentially we could define a distance-based classifier that allocates a test subject to the group having a mean shape closest to that subject's shape. The table suggests that such a classifier will yield results consistent with our nonparametric Bayes approach. Figure 13.2 shows the male and female training sample preshape clouds, along with the two misclassified test samples. There seems to be substantial deviation in the coordinates of these misclassified subjects from their respective gender training groups, especially for the male gorilla, even after rotating each training preshape separately to bring it as close as possible to the plotted test sample preshapes. It is expected that classification performance will improve in this application by also taking into account skull size (see Section 8.10 on planar *size-and-shape* manifolds). The proposed method can easily be extended to this case by using a Dirichlet process mixture density with the kernel being the product of a complex Watson kernel for the shape component and a log-Gaussian kernel for the size. Such a model induces a prior with support on the space of densities on the manifold $\Sigma_2^k \times \mathbb{R}^+$ (see Chapter 14).

2.4 Data example on Σ_2^k: brain scan shapes of schizophrenic and normal patients

In this example from Bookstein (1991), 13 landmarks are recorded on a midsagittal two-dimensional slice from magnetic resonance brain scans of each of 14 schizophrenic patients and 14 normal patients. It is of interest to study the differences in shapes of brains between the two groups, which can be used to detect schizophrenia. This is an application of disease detection by shape. The shapes of the sample k-ads lie in Σ_2^k, $k = 13$. Figures 8.2(a) and (b) show landmarks for the schizophrenic and normal patients, respectively. In this example, we have two independent random samples of size 14 each on Σ_2^k, $k = 13$. To distinguish between the underlying distributions, we compare the means and variations in shapes.

Figure 8.3 shows the preshapes of the sample extrinsic means for the two groups of patients along with a preshape for the pooled sample extrinsic mean. As in the case of the gorilla skull images from Section 2.3, the sample intrinsic means are very close to their extrinsic counterparts; the geodesic distances between the intrinsic and extrinsic means are 1.65×10^{-5} for the sample of normal patients and 4.29×10^{-5} for the sample of schizophrenic patients.

The value of the two-sample test statistic in equation (5.21) for testing the equality of the population intrinsic mean shapes and the asymptotic p-value are

$$T_{n1} = 95.4587, \quad \text{p-value} = P(X_{22}^2 > 95.4587) = 3.97 \times 10^{-11}.$$

The values of the two-sample test statistics defined through equations (4.15) and (4.18) for comparing the extrinsic mean shapes and the corresponding asymptotic p-values are

$$T_1 = 95.5476, \quad \text{p-value} = P(X_{22}^2 > 95.5476) = 3.8 \times 10^{-11},$$
$$T_2 = 95.2549, \quad \text{p-value} = 4.3048 \times 10^{-11}.$$

Table 2.4 *Percent of variation (P.V.) explained by different principal
components (P.C.) of* $\hat{\Sigma}$

P.C.	1	2	3	4	5	6	7	8	9	10	11
P.V.	21.6	18.4	12.1	10.0	9.9	6.3	5.3	3.6	3.0	2.5	2.1
P.C.	12	13	14	15	16	17	18	19	20	21	22
P.V.	1.5	1.0	0.7	0.5	0.5	0.3	0.2	0.2	0.1	0.1	0.0

Hence we reject the null hypothesis that the two groups have the same mean
shape (both extrinsic and intrinsic) at asymptotic levels as small as 10^{-10}.
The p-values above are smaller than their parametric counterparts (see
Dryden and Mardia, 1998, pp. 162–166) by many orders of magnitude.

Next we compare the extrinsic means by bootstrap methods. Since the
dimension 22 of the underlying shape space is much higher than the sample sizes, it becomes difficult to construct a bootstrap test statistic as in
the earlier section. That is because the bootstrap estimate of the standard
error $\hat{\Sigma}$ defined in equation (8.16) tends to be singular in most simulations.
Hence we compare only the first few principal scores of the coordinates of
the sample extrinsic means. Table 2.4 displays the percentage of variation
explained by each principal component of $\hat{\Sigma}$. The value of T_{21} from equation (8.17) for comparing the first five principal scores of $L[P(\hat{\mu}_1) - P(\hat{\mu}_2)]$
with zero and the asymptotic p-value are

$$T_{21} = 12.1872, \quad \text{p-value} = P(X_5^2 > 12.1872) = 0.0323.$$

The bootstrap p-value from 10^4 simulations equals 0.0168, which is fairly
small.

Remark 2.1   It may be remarked that there is no strong rationale in general to carry out tests for discrimination among mean values in a two-
sample (or a multi-sample) problem based on principal components (of
the pooled sample covariance) with the greatest variabilities. Indeed, components with small variability may reveal significant differences in mean
values that are subject to small sampling error. For example, in the gorilla
skulls problem, the test for the equality of the two means of the principal component with the twentieth smallest eigenvalue yields a p-value
of 4.84×10^{-6}. To illustrate that this small value cannot be ascribed
to large sampling errors in estimating small eigenvalues, we computed
the p-values .2663 and .2633 for the corresponding tests based on the
twenty-first and twenty-second principal components, respectively. Our
motivation for carrying out the bootstrap computation of the p-value displayed above is that its coverage error is small relative to the error of

the chi-square approximation based on the same components. Fortunately, this small p-value confirms the conclusion based on the prior chi-square approximations.

Finally, we test the null hypothesis of equality of the extrinsic variations of the two groups of children. The sample extrinsic variations for patient and normal samples turn out to be 0.0107 and 0.0093, respectively. The value of the two-sample test statistic in equation (8.19) for testing the equality of the population extrinsic variations is 0.9461 and the asymptotic p-value using a standard Normal approximation is 0.3441. The bootstrap p-value with 10^4 simulations equals 0.3564. Hence we conclude that the extrinsic variations in shapes for the two distributions are not significantly different.

Because the mean shapes are different, we conclude that the probability distributions of the shapes of brain scans of normal and schizophrenic children are distinct.

2.5 Data example on affine shape space $A\Sigma_2^k$: application to handwritten digit recognition

A random sample of 30 handwritten digits "3" were collected so as to devise a scheme to automatically classify handwritten characters. For this, 13 landmarks were recorded on each image by Anderson (1997). The landmark data can be found in Dryden and Mardia (1998).

We analyze the affine shape of the sample points and estimate the mean and variation in shape. This can be used as a prior model for digit recognition from images of handwritten codes. Our observations lie on the affine shape space $A\Sigma_2^k$, $k = 13$. Figure 11.1 shows the plot of the sample extrinsic mean. What is actually plotted is a representative of the mean in the Stiefel manifold $V_{2,13}$ (see Chapter 10). The sample extrinsic variation turns out to be 0.27, which is fairly large. There seems to be a lot of variability in the data. Following are the extrinsic squared distances of the sample points from the mean affine shape:

$$(\rho^2(X_j, \mu_E), \ j = 1, \ldots, n) = (1.64, 0.28, 1.00, 0.14, 0.13, 0.07, 0.20, 0.09,$$
$$0.17, 0.15, 0.26, 0.17, 0.14, 0.20, 0.42, 0.31, 0.14, 0.12, 0.51, 0.10, 0.06,$$
$$0.15, 0.05, 0.31, 0.08, 0.08, 0.11, 0.18, 0.64, 0.12).$$

Here $n = 30$ is the sample size. From these distances it is clear that observations 1 and 3 are outliers. We remove them and recompute the sample extrinsic mean and variation. The sample variation now turns out to be 0.19. An asymptotic 95% confidence region for the extrinsic mean μ_E is computed using equation (4.6). The two outliers are not in this region, not even in a 99% confidence region, thereby further justifying their status as outliers.

The dimension 20 of $A\Sigma_2^{13}$ is quite high compared to the sample size of 28. It is difficult to construct a bootstrap confidence region because the bootstrap covariance estimates Σ^* tend to be singular or close to singular in most simulations. Instead, we construct a nonpivotal bootstrap confidence region by replacing Σ^* with $\hat{\Sigma}$. Using 10^5 simulations, the bootstrap method yields a much smaller confidence region for the true mean shape compared to that obtained from the chi-squared approximation.

Remark 2.2   Note, however, that the bootstrapping here is of the chi-squared-like statistic in equation (4.6) with $\hat{\Sigma}$ replaced by Σ. Hence, although the bootstrap approximates the distribution of this modified statistic well, the bootstrap-based computation of the p-value is subject to the additional sampling error caused by using Σ in place of $\hat{\Sigma}$.

A 95% confidence interval for the extrinsic variation V by a Normal approximation, as described in equation (4.13), is given by $V \in [0.140, 0.243]$, while a pivotal bootstrap confidence interval using 10^5 simulations turns out to be $[0.119, 0.264]$.

In Dryden and Mardia (1998), the two-dimensional similarity shapes (planar shapes) of the sample k-ads are analyzed. A multivariate Normal distribution is assumed for the so-called Procrustes coordinates of the planar shapes of the sample points. Using these points, an F-test is carried out to test whether the population mean shape corresponds to that of an idealized template. The test yields a p-value of 0.0002 (see example 7.1 in Dryden and Mardia, 1998).

2.6 Data example on reflection similarity shape space $R\Sigma_3^k$: glaucoma detection

In this section, we see an application of three-dimensional similarity shape analysis in disease detection. Glaucoma is a leading cause of blindness. To detect any shape change due to glaucoma, three-dimensional images of the optic nerve head (ONH) of both eyes of 12 mature rhesus monkeys were collected. One of the eyes was treated to increase the intraocular pressure (IOP), which is often the cause of glaucoma onset, while the other was left untreated. Five landmarks were recorded on each eye. The landmark coordinates can be found in Bhattacharya and Patrangenaru (2005). In this section, we consider the reflection shape of the k-ads in $R\Sigma_3^k$ with $k = 5$. We want to test whether there is any significant difference between the shapes of the treated and untreated eyes by comparing the extrinsic means and variations.

Figures 9.1(a) and (b) show the preshapes of the untreated and treated eyes, respectively, along with a preshape of the corresponding sample extrinsic mean. Figure 9.2 shows the preshapes of the mean shapes for the two eyes along with a preshape of the pooled sample extrinsic mean. The sample extrinsic variations for the untreated and treated eyes are 0.041 and 0.038, respectively.

This is an example of a matched pair sample. To compare the extrinsic means and variations, we use the methodology of Section 4.5.2. The value of the matched pair test statistic T_{1p} in equation (4.22) is 36.29, and the asymptotic p-value for testing whether the shape distributions for the two eyes are the same is

$$P(X_8^2 > 36.29) = 1.55 \times 10^{-5}. \tag{2.1}$$

The value of the test statistic T_{2p} from equation (4.24) for testing whether the extrinsic means are the same is 36.56 and the p-value of the chi-squared test turns out to be 1.38×10^{-5}. Hence we conclude that the mean shapes of the two eyes are significantly different. Because the sample size is rather small and the dimension of the shape space is a little large, namely, 8, bootstrap calculations were carried out on differences in some smaller dimensional features, but they also yield very small p-values, although not as small as in equation (2.1). It may be noted that the p-values are much smaller than those obtained by preliminary methods in Bhattacharya and Patrangenaru (2005) and Bandulasiri et al. (2009). The inadequacy of principal components analysis (PCA) in two-sample problems has been pointed out in Remark 2.1.

Next we test whether the two eye shapes have the same extrinsic variation. The value of the test statistic T_{3p} from equation (4.27) equals -0.5572 and the asymptotic p-value equals

$$P(|Z| > 0.5572) = 0.577, \quad Z \sim N(0, 1).$$

The bootstrap p-value with 10^4 simulations equals 0.59. Hence we conclude that the extrinsic variations are not significantly different.

Because the mean shapes for the two eyes are found to be different, we conclude that the underlying probability distributions are distinct and hence glaucoma does indeed change the shape of the eyes, and may be diagnosed by such changes.

Remark 2.3 For data from a distribution with a density, the error of the chi-squared approximation is $O(n^{-1})$, with n being the sample size; the error of the bootstrap approximation of the asymptotic chi-squared statistic is $O(n^{-2})$ (see Bhattacharya and Ghosh, 1978; Chandra and Ghosh, 1979; Bhattacharya and Denker, 1990; Hall, 1992). Hence the accuracy of the very small p-values based on the chi-squared approximation in the examples may be questioned, especially because the sample sizes are not very large. However, the very large values of the asymptotic chi-squared statistic would indicate very small p-values. One way to think of this is in terms of the *nonuniform error bounds in the central limit theorem (CLT)* (see Bhattacharya and Ranga Rao, 2010, pp. 171–172). The true p-values may be larger or smaller than those computed, but they are exceedingly small nonetheless.

2.7 References

The data in Table 2.1 are from Johnson and Wehrly (1977), who used these to study the relationship between wind direction and ozone concentration. In Section 2.2, the seminal paper of R. A. Fisher (1953) is widely credited by geoscientists as having provided the first rigorous statistical confirmation of the wandering of the Earth's magnetic poles over geological time scales (see Irving, 1964). The second data example analyzed here is from Irving (1963). Intrinsic and extrinsic sample means of a set of magnetic pole data, along with a confidence region for the extrinsic mean, are computed in Bhattacharya and Patrangenaru (2005) by the nonparametric method developed there. Fisher et al. (1987) provide many examples of spherical data; their sample means generally correspond to what we term extrinsic sample means. Also see Mardia and Jupp (2000).

The gorilla skull data, as well as those for handwritten digits, are taken from Dryden and Mardia (1998), and the data on brain scans of normal and schizophrenic children are from Bookstein (1991). Various parametric precedents along with original references for these may be found in Dryden and Mardia (1998). The nonparametric tests and confidence regions presented here are, with some modifications, from Bhattacharya (2008a) and Bhattacharya and Bhattacharya (2008a, 2009).

The glaucoma data for Example 2.6 may be found in Burgoyne et al. (2000). This was first analyzed in Bhattacharya and Patrangenaru (2005) and later in Bandulasiri et al. (2009). The present analysis is largely due to Bhattacharya (2008b).

3

Location and spread on metric spaces

Consistency and asymptotic distributions of sample Fréchet means and variations are derived in this chapter for general distances and Fréchet functions.

3.1 Introduction

Much of this book is centered around the notion of the mean and variation of a probability measure Q on a manifold M. Generally, the mean is the minimizer of the expected squared distance of a point from an M-valued random variable X with distribution Q. Such an idea has a long history. Physicists have long considered the analogous notion of a center of mass on a general submanifold M of a Euclidean space, with the normalized volume measure as Q. The extension to general metric spaces M and arbitrary probability measures Q on M was made by Fréchet (1948). In this chapter we begin with a generalization of Fréchet's definition.

In general, we consider a loss function f that is an appropriate continuous increasing function on $[0, \infty)$ and define the expected loss F of a probability measure Q on a metric space (M, ρ) by

$$F(p) = \int_M f\{\rho(p, x)\} Q(dx), \quad p \in M,$$

where ρ is a distance on M. In the case of squared distance, $f(u) = u^2$. However, one may also consider minimization with respect to distance rather than squared distance, in which case $f(u) = u$. In a Euclidean space, for example, this would be like considering the median rather than the mean as a measure of location.

For the purposes of statistical inference that we pursue, it is important to have a unique minimizer of F, in which case the minimizer is called the mean of Q. The minimum value attained by F gives a measure of the spread of Q and is called the variation of Q.

21

In Section 3.2, the set of minimizers of F is shown to be nonempty and compact under some general assumptions on M and f, and the asymptotic behavior of the corresponding set for the empirical distribution Q_n, based on n i.i.d. observations, is derived (Theorems 3.2 and 3.3). It follows that if the mean of Q exists, that is, the minimizer of F is unique, then the sample mean (set) converges almost surely to this mean, as $n \to \infty$ (consistency). Consistency for the sample variation holds even when there is no unique mean, as is shown in Section 3.3.

As is usual, take f to be the squared loss function. That is, the mean of Q is the minimizer of the expected squared distance. On a differentiable manifold M, there are two classes of such means. If the distance ρ is induced on M from an embedding in a Euclidean space, it is called the *extrinsic distance* in this book, and the mean for the extrinsic distance is called the *extrinsic mean*. As we will see in the coming chapters, the extrinsic mean exists as a unique minimizer, and therefore consistency holds, under broad conditions. If, on the other hand, M has, or is given, a Riemannian structure, then we will take ρ to be the *geodesic distance* and term the corresponding mean as the *intrinsic mean*. Unfortunately, as attractive as this notion is, sufficiently broad conditions for the existence of a unique minimizer in intrinsic distance are not available.

Asymptotic inference, of course, is based on the asymptotic distributions of relevant statistics. For this, one needs M to be a differentiable manifold of dimension d, say. For the greater part of the book, a proper (equivariant) embedding of M into a Euclidean space of higher dimension is used, deriving the classical central limit theorem on this Euclidean space and lifting it to M (see Chapter 4). For example, the d-dimensional unit sphere S^d may be embedded in $\mathbb{R}^{d+1}$ by the inclusion map.

For the present chapter, however, the CLT is derived by a different route, which is more suitable for intrinsic analysis (Chapter 5), although it can be applied to the extrinsic case as well. Here we require that Q assign probability one to an open subset of the manifold, which is diffeomorphic to an open subset of $\mathbb{R}^d$. In that case, a central limit theorem on the image (under diffeomorphism) provides the required asymptotic distribution (Theorems 4.3 and 3.11). As restrictive as this hypothesis on Q may seem, it turns out that a natural diffeomorphism of this kind exists for general complete (Riemannian) manifolds when Q is absolutely continuous. (See Bhattacharya and Patrangenaru (2005) and Appendix B).

3.2 Location on metric spaces

Let (M, ρ) be a metric space, with ρ being the distance, and let $f \geq 0$ be a given continuous function on $[0, \infty)$. For a given probability measure Q

on (the Borel sigma-field of) M, define its *expected loss function*, or the *Fréchet function*, as

$$F(p) = \int_M f(\rho(p, x))Q(dx), \quad p \in M. \tag{3.1}$$

Definition 3.1 Suppose $F(p) < \infty$ for some $p \in M$. Then the set of all p for which $F(p)$ is the minimum value of F on M is called the *mean set* of Q, denoted by C_Q. If this set is a singleton, say $\{\mu\}$, then one says that the *Fréchet mean* of Q exists, and μ is called the *Fréchet mean* of Q. If $X_1, X_2, \ldots, X_n$ are i.i.d. M-valued random variables defined on some probability space $(\Omega, \mathcal{F}, P)$ with common distribution Q, and $Q_n := \frac{1}{n}\sum_{j=1}^{n} \delta_{X_j}$ is the corresponding empirical distribution, then the mean set of Q_n is called the *sample Fréchet mean set*, denoted by C_{Q_n}.

When M is compact, the sample mean set converges a.s. to the mean set of Q, or to a subset of it, as the sample size grows to infinity. This is established in Theorem 3.2.

Theorem 3.2 *Let M be a compact metric space and f a continuous loss function on $[0, \infty)$. Consider the expected loss function F of a probability measure Q given by equation (3.1). Given any $\epsilon > 0$, there exists a P-null set $\Omega(\epsilon)$ and an integer-valued random variable $N \equiv N(\omega, \epsilon)$ such that*

$$C_{Q_n} \subset C_Q^\epsilon := \{p \in M : \rho(p, C_Q) < \epsilon\}, \quad \forall n \geq N \tag{3.2}$$

outside of $\Omega(\epsilon)$.

Proof M being compact and f continuous implies that C_Q is nonempty and compact. Choose $\epsilon > 0$ arbitrarily. If $C_Q^\epsilon = M$, then equation (3.2) holds with $N = 1$. If $M_1 = M \setminus C_Q^\epsilon$ is nonempty, write

$$l = \min\{F(p) : p \in M\} = F(q), \quad \forall q \in C_Q,$$
$$l + \delta(\epsilon) = \min\{F(p) : p \in M_1\}, \quad \delta(\epsilon) > 0.$$

It is enough to show that

$$\max\{|F_n(p) - F(p)| : p \in M\} \longrightarrow 0 \text{ a.s.}, \quad \text{as } n \to \infty. \tag{3.3}$$

For if equation (3.3) holds, then there exists a positive integer-valued random variable N such that, outside a P-null set $\Omega(\epsilon)$,

$$\min\{F_n(p) : p \in C_Q^\epsilon\} \leq l + \frac{\delta(\epsilon)}{3},$$
$$\min\{F_n(p) : p \in M_1\} \geq l + \frac{\delta(\epsilon)}{2}, \quad \forall n \geq N. \tag{3.4}$$

Clearly equation (3.4) implies equation (3.2).

To prove equation (3.3), choose and fix $\epsilon' > 0$, however small. Note that $\forall p, p', x \in M, |\rho(p, x) - \rho(p', x)| \le \rho(p, p')$. Hence

$$|F(p) - F(p')| \le \max\{|f(\rho(p, x)) - f(\rho(p', x))| : x \in M\}$$
$$\le \max\{|f(u) - f(u')| : |u - u'| \le \rho(p, p')\},$$
$$|F_n(p) - F_n(p')| \le \max\{|f(u) - f(u')| : |u - u'| \le \rho(p, p')\}. \quad (3.5)$$

Because f is uniformly continuous on $[0, R]$, where R is the diameter of M, so are F and F_n on M, and there exists $\delta(\epsilon') > 0$ such that

$$|F(p) - F(p')| \le \frac{\epsilon'}{4}, \quad |F_n(p) - F_n(p')| \le \frac{\epsilon'}{4} \quad (3.6)$$

if $\rho(p, p') < \delta(\epsilon')$. Let $\{q_1, \dots, q_k\}$ be a $\delta(\epsilon')$-net of M, that is, $\forall p \in M$, there exists $q(p) \in \{q_1, \dots, q_k\}$ such that $\rho(p, q(p)) < \delta(\epsilon')$. By the strong law of large numbers, there exists a positive integer-valued random variable $N(\omega, \epsilon')$ such that, outside of a P-null set $\Omega(\epsilon')$, one has

$$|F_n(q_i) - F(q_i)| \le \frac{\epsilon'}{4}, \quad \forall i = 1, 2, \dots, k, \quad \text{if } n \ge N(\omega, \epsilon'). \quad (3.7)$$

From equations (3.6) and (3.7) we get

$$|F(p) - F_n(p)|$$
$$\le |F(p) - F(q(p))| + |F(q(p)) - F_n(q(p))| + |F_n(q(p)) - F_n(p)|$$
$$\le \frac{3\epsilon'}{4} < \epsilon', \quad \forall p \in M,$$

if $n \ge N(\omega, \epsilon')$ outside of $\Omega(\epsilon')$. This proves equation (3.3). $\qquad \square$

In view of Theorem 3.2, we define the *sample Fréchet mean* to be any measurable selection from the sample mean set, if the Fréchet mean of Q exists as the unique minimizer. Then, as stated in Corollary 3.4, the sample mean is a consistent estimator of the population mean.

Most of the manifolds in this book, including the shape spaces, are compact. Notable exceptions are the so-called size-and-shape spaces of Chapter 8, Section 8.10. We now turn to such noncompact spaces, taking the loss function $f(u) = u^\alpha$ $(\alpha \ge 1)$, $\alpha = 2$, as the most important.

Theorem 3.3 *Let M be a metric space such that every closed and bounded subset of M is compact. Suppose the expected loss function F corresponding to $f(u) = u^\alpha$ $(\alpha \ge 1)$ in equation (3.1) is finite for some p. Then (a) the Fréchet mean set C_Q is nonempty and compact, and (b) given any $\epsilon > 0$, there exists a positive integer-valued random variable $N \equiv N(\omega, \epsilon)$ and a P-null set $\Omega(\epsilon)$ such that*

$$C_{Q_n} \subseteq C_Q^{\epsilon} := \{p \in M : \rho(p, C_Q) < \epsilon\}, \quad \forall n \geq N \qquad (3.8)$$

outside of $\Omega(\epsilon)$.

Proof   (a) By the triangle inequality on ρ and by convexity of the function $u \mapsto u^{\alpha}$, $u \geq 0$, we obtain

$$\rho^{\alpha}(q, x) \leq \{\rho(p, q) + \rho(p, x)\}^{\alpha} \leq 2^{\alpha-1}\{\rho^{\alpha}(p, q) + \rho^{\alpha}(p, x)\},$$

which implies that

$$F(q) \leq 2^{\alpha-1}\rho^{\alpha}(p, q) + 2^{\alpha-1}F(p). \qquad (3.9)$$

Hence, if $F(p) < \infty$ for some p, then $F(q) < \infty$ $\forall q \in M$. When $\alpha = 1$, equation (3.9) also implies that F is continuous. When $\alpha > 1$, it is simple to check using Taylor expansions that

$$|\rho^{\alpha}(p, x) - \rho^{\alpha}(q, x)| \leq \alpha\rho(p, q)\{\rho^{\alpha-1}(p, x) + \rho^{\alpha-1}(q, x)\}.$$

This implies, by Lyapunov's inequality for moments, that

$$|F(p) - F(q)| \leq \alpha\rho(p, q)\{F^{\alpha/(\alpha-1)}(p) + F^{\alpha/(\alpha-1)}(q)\}.$$

This along with inequality (3.9) implies that F is continuous everywhere. Again, Lyapunov's inequality, together with the triangle inequality, implies that

$$\rho(p, q) \leq \int \rho(p, x)Q(dx) + \int \rho(q, x)Q(dx) \leq F^{1/\alpha}(p) + F^{1/\alpha}(q). \quad (3.10)$$

Because F is finite,

$$l = \inf\{F(q) : q \in M\} < \infty.$$

To show that this infimum is attained, let $\{p_n\}$ be a sequence such that $F(p_n) \to l$. Use equation (3.10) with $p = p_n$ and $q = p_1$ to obtain

$$\rho(p_n, p_1) \leq F^{1/\alpha}(p_n) + F^{1/\alpha}(p_1) \longrightarrow l^{1/\alpha} + F^{1/\alpha}(p_1).$$

Hence the sequence $\{p_n\}$ is bounded, so that its closure is compact by the hypothesis of the theorem. If $\{p_{n,k} : k = 1, 2, \ldots\}$ is a Cauchy subsequence of $\{p_n\}$, converging to p_{∞}, say, one has $F(p_{n,k}) \longrightarrow F(p_{\infty}) = l$. Thus $C_Q = \{p : F(p) = l\}$ is a nonempty closed set. Apply equation (3.10) again to arbitrary $p, q \in C_Q$ to get $\rho(p, q) \leq 2l^{1/\alpha}$. Thus C_Q is bounded and closed and, therefore, compact.

   (b) Given any $\epsilon > 0$, the task is to find a compact set M_1 containing C_Q and a positive integer-valued random variable $N_1 \equiv N_1(\omega, \epsilon)$ such that

$$\inf_{M\backslash M_1} F(p) \geq l + \epsilon, \quad \inf_{M\backslash M_1} F_n(p) \geq l + \epsilon \text{ a.s.}, \quad \forall n \geq N_1.$$

Then we can show as in the case of compact M (Theorem 3.2) that

$$\sup\{|F_n(p) - F(p)| : p \in M_1\} \longrightarrow 0 \text{ a.s.}, \quad \text{as } n \to \infty$$

and conclude that equation (3.8) holds. To get such a M_1, note that from equation (3.10) it follows that for any $p_1 \in C_Q$ and $p \in M$

$$F(p) \geq [\rho(p, p_1) - l^{1/\alpha}]^\alpha. \tag{3.11}$$

Let

$$M_1 = \{p : \rho(p, C_Q) \leq 2(l + \epsilon)^{1/\alpha} + l^{1/\alpha}\}.$$

Then, from equation (3.11), one can check that $F(p) \geq 2(l+\epsilon) \, \forall p \in M\backslash M_1$. Also from equation (3.10), we get for any $p \in M \backslash M_1$

$$F_n(p) \geq \{\rho(p, p_1) - F_n^{1/\alpha}(p_1)\}^\alpha.$$

From the definition of M_1,

$$\rho(p, p_1) - F_n^{1/\alpha}(p_1) > 2(l + \epsilon)^{1/\alpha} + l^{1/\alpha} - F_n^{1/\alpha}(p_1),$$

so that

$$\inf_{p \in M\backslash M_1} F_n(p) > \{2(l + \epsilon)^{1/\alpha} + l^{1/\alpha} - F_n^{1/\alpha}(p_1)\}^\alpha.$$

Because $F_n(p_1) \to l$ a.s., it follows that there exists a positive integer-valued random variable $N_1(\epsilon)$ and a null set $\Omega(\epsilon)$ such that $\forall n \geq N_1$:

$$\inf_{p \in M\backslash M_1} F_n(p) > l + \epsilon$$

outside of $\Omega(\epsilon)$. This completes the proof.   □

When M is compact, the hypothesis of Theorem 3.3 holds using any continuous loss function f and the conclusion that the Fréchet mean set is nonempty and compactness easily follows.

Corollary 3.4   *Under the hypothesis of Theorem 3.2 or that of Theorem 3.3, if C_Q is a singleton $\{\mu\}$, then the sample mean is a strongly consistent estimator of μ.*

Remark 3.5   Corollary 3.4 generalizes Theorem 2.3 in Bhattacharya and Patrangenaru (2003), where f is the squared loss function. In this case, consistency also follows from Ziezold (1977) when the metric space is compact. We will be working mainly with this loss function but will consider other extensions as well (see Section 3.6 and Chapter 8, Section 8.5).

Remark 3.6 From the Hopf–Rinow theorem (see Hopf and Rinow, 1931, or Do Carmo, 1992, pp. 146, 147), it follows that a complete and connected Riemannian manifold M satisfies the topological hypothesis of Theorem 3.3, that every closed and bounded subset of M is compact.

3.3 Variation on metric spaces

A notion of mean of a probability gives rise to a natural notion of spread or variation. In this section we study its properties. Consider the expected loss function F of a probability Q on a metric space M as defined in equation (3.1).

Definition 3.7 The infimum of F on M is called the *variation* of Q, denoted by V. Given an i.i.d. sample from Q, the variation of the empirical distribution Q_n is called the *sample variation*, denoted by V_n.

Proposition 3.8 proves the sample variation to be a consistent estimator of the variation of Q.

Proposition 3.8 *Under the hypothesis of Theorem 3.2 or that of Theorem 3.3, V_n is a strongly consistent estimator of V.*

Proof In view of Theorem 3.2 or Theorem 3.3, for any $\epsilon > 0$, there exists $N \equiv N(\omega, \epsilon)$ such that

$$|V_n - V| = |\inf_{p \in C_Q^\epsilon} F_n(p) - \inf_{p \in C_Q^\epsilon} F(p)| \leq \sup_{p \in C_Q^\epsilon} |F_n(p) - F(p)| \qquad (3.12)$$

for all $n \geq N$ a.s. Also from the theorems' proofs, it follows that

$$\sup_{p \in M_1} |F_n(p) - F(p)| \longrightarrow 0 \text{ a.s.,} \qquad \text{as } n \to \infty$$

whenever M_1 is compact. Also shown is that C_Q^ϵ is bounded and hence its closure is compact. Hence, from equation (3.12) it follows that

$$|V_n - V| \longrightarrow 0 \text{ a.s.,} \qquad \text{as } n \to \infty.$$

$\square$

Remark 3.9 In view of Proposition 3.8, the sample variation is consistent even when the expected loss function of Q does not have a unique minimizer, that is, even when Q does not have a mean.

3.4 Asymptotic distribution of the sample mean

In this section, we consider the asymptotic distribution of the sample mean μ_n. From now on, we assume M is a differentiable manifold of dimension d. Let ρ be a distance metrizing the topology of M. Theorem 3.10 below proves that under appropriate assumptions, the coordinates of μ_n are asymptotically normal. Here we denote by D_r the partial derivative with respect to the rth coordinate ($r = 1, \ldots, d$) and by D the vector of partial derivatives.

Theorem 3.10 *Suppose the following assumptions hold:*

A1 *Q has support in a single coordinate patch, (U, ϕ) ($\phi: U \longrightarrow \mathbb{R}^d$). Let $\tilde{X}_j = \phi(X_j)$, $j = 1, \ldots, n$.*

A2 *Q has a unique mean μ.*

A3 *For all x, the mapping $y \mapsto h(x, y) = f(\rho(\phi^{-1}(x), \phi^{-1}(y)))$ is twice continuously differentiable in a neighborhood of $\phi(\mu)$.*

A4 *$E\{D_r h(\tilde{X}_1, \phi(\mu))\}^2 < \infty \ \forall r = 1, \ldots, d$.*

A5 *$E\{ \sup_{|u-v| \le \epsilon} |D_s D_r h(\tilde{X}_1, v) - D_s D_r h(\tilde{X}_1, u)|\} \to 0$ as $\epsilon \to 0 \ \forall r, s$.*

A6 *$\Lambda = ((E\{D_s D_r h(\tilde{X}_1, \phi(\mu))\})) $ is nonsingular.*

Let μ_n be a measurable selection from the sample mean set. Then, under assumptions A1–A6,

$$\sqrt{n}\,(\phi(\mu_n) - \phi(\mu)) \xrightarrow{\mathcal{L}} N(0, \Lambda^{-1}\Sigma(\Lambda')^{-1}), \qquad (3.13)$$

where $\Sigma = \mathrm{Cov}[Dh(\tilde{X}_1, \phi(\mu))]$.

Proof Write $\psi^{(r)}(x, y) = D_r h(x, y) \equiv \frac{\partial}{\partial y_r} h(x, y)$ for $x, y \in \mathbb{R}^d$. Let $Q^\phi = Q \circ \phi^{-1}$. Denote

$$\tilde{F}(y) = \int_{\mathbb{R}^d} \rho^\alpha(\phi^{-1}(x), \phi^{-1}(y)) Q^\phi(dx), \quad \tilde{F}_n(y) = \frac{1}{n} \sum_{j=1}^{n} \rho^\alpha(\phi^{-1}(\tilde{X}_j), \phi^{-1}(y))$$

for $y \in \mathbb{R}^d$. Then $\tilde{F}$ has the unique minimizer $\phi(\mu)$ while $\tilde{F}_n$ has the minimizer $\phi(\mu_n)$. Therefore,

$$0 = \frac{1}{\sqrt{n}} \sum_{j=1}^{n} \psi^{(r)}(\tilde{X}_j, \phi(\mu_n)) = \frac{1}{\sqrt{n}} \sum_{j=1}^{n} \psi^{(r)}(\tilde{X}_j, \phi(\mu))$$

$$+ \sum_{s=1}^{d} \sqrt{n} \, (\phi(\mu_n) - \phi(\mu))_s \frac{1}{n} \sum_{j=1}^{n} D_s \psi^{(r)}(\tilde{X}_j, \phi(\mu))$$

$$+ \sum_{s=1}^{d} \sqrt{n} \, (\phi(\mu_n) - \phi(\mu))_s (\epsilon_n)_{rs}, \quad 1 \le r \le d, \qquad (3.14)$$

where $(\epsilon_n)_{rs} = \dfrac{1}{n} \sum_{j=1}^{n} [D_s \psi^{(r)}(\tilde{X}_j, \theta_n) - D_s \psi^{(r)}(\tilde{X}_j, \phi(\mu))]$

for some θ_n lying on the line segment joining $\phi(\mu)$ and $\phi(\mu_n)$. Equation (3.14) implies that

$$\left[\left(\left(\frac{1}{n} \sum_{j=1}^{n} D_s D_r h(\tilde{X}_j, \phi(\mu)) + \epsilon_n \right) \right) \right] \sqrt{n} \, (\phi(\mu_n) - \phi(\mu))$$

$$= -\frac{1}{\sqrt{n}} \sum_{j=1}^{n} Dh(\tilde{X}_j, \phi(\mu)).$$

In view of assumptions A5 and A6, it follows that

$$\sqrt{n} \, (\phi(\mu_n) - \phi(\mu)) = -\Lambda^{-1} \left(\frac{1}{\sqrt{n}} \sum_{j=1}^{n} Dh(\tilde{X}_j, \phi(\mu)) \right) + o_P(1),$$

which implies that

$$\sqrt{n} \, (\phi(\mu_n) - \phi(\mu)) \xrightarrow{\mathcal{L}} -\Lambda^{-1} N(0, \Sigma) = N(0, \Lambda^{-1} \Sigma (\Lambda')^{-1}). \qquad \square$$

From Theorem 3.10, it follows that, under assumptions A1–A6 and assuming Σ to be nonsingular,

$$n(\phi(\mu_n) - \phi(\mu))' \Lambda' \Sigma^{-1} \Lambda(\phi(\mu_n) - \phi(\mu)) \xrightarrow{\mathcal{L}} \mathcal{X}_d^2 \quad \text{as } n \to \infty.$$

Here $\mathcal{X}_d^2$ denotes the chi-squared distribution with d degrees of freedom. This can be used to construct an asymptotic confidence set for μ, namely,

$$\{\mu : n(\phi(\mu_n) - \phi(\mu))' \hat{\Lambda}' \hat{\Sigma}^{-1} \hat{\Lambda}(\phi(\mu_n) - \phi(\mu)) \le \mathcal{X}_d^2(1 - \theta)\}. \qquad (3.15)$$

Here $\hat{\Lambda}$ and $\hat{\Sigma}$ are the sample estimates of Λ and Σ, respectively, and $\mathcal{X}_d^2(1 - \theta)$ is the upper $(1 - \theta)$-quantile of the $\mathcal{X}_d^2$ distribution. The corresponding *pivotal bootstrap confidence region* is given by

$$\{\mu : n(\phi(\mu_n) - \phi(\mu))'\hat{\Lambda}'\hat{\Sigma}^{-1}\hat{\Lambda}(\phi(\mu_n) - \phi(\mu)) \le c^*(1 - \theta)\}, \tag{3.16}$$

where $c^*(1 - \theta)$ is the upper $(1 - \theta)$-quantile of the bootstrapped values of the statistic in equation (3.15).

3.5 Asymptotic distribution of the sample variation

Next we derive the asymptotic distribution of V_n when Q has a unique mean.

Theorem 3.11   *Let M be a differentiable manifold. Using the notation of Theorem 3.10, under assumptions A1–A6 and assuming that $E[\rho^{2\alpha}(X_1, \mu)] < \infty$, one has*

$$\sqrt{n}\,(V_n - V) \xrightarrow{\mathcal{L}} N\,(0, \mathrm{Var}(\rho^\alpha(X_1, \mu)))\,. \tag{3.17}$$

Proof   Let

$$\bar{F}(x) = \int_M \rho^\alpha(\phi^{-1}(x), m)Q(dm), \quad \tilde{F}_n(x) = \frac{1}{n}\sum_{j=1}^n \rho^\alpha(\phi^{-1}(x), X_j)$$

for $x \in \mathbb{R}^d$. Let μ_n be a measurable selection from the sample mean set. Then

$$
\begin{aligned}
\sqrt{n}(V_n - V) &= \sqrt{n}\,(\tilde{F}_n(\phi(\mu_n)) - \bar{F}(\phi(\mu))) \\
&= \sqrt{n}\,(\tilde{F}_n(\phi(\mu_n)) - \tilde{F}_n(\phi(\mu))) + \sqrt{n}\,(\tilde{F}_n(\mu) - \bar{F}(\mu)),
\end{aligned}
\tag{3.18}
$$

$$
\begin{aligned}
\sqrt{n}\,(\tilde{F}_n(\mu_n) - \tilde{F}_n(\mu)) = {} & \frac{1}{\sqrt{n}}\sum_{j=1}^n \sum_{r=1}^d (\phi(\mu_n) - \phi(\mu))_r D_r h(\tilde{X}_j, \phi(\mu)) \\
& + \frac{1}{2\sqrt{n}}\sum_{j=1}^n \sum_{r=1}^d \sum_{s=1}^d (\phi(\mu_n) - \phi(\mu))_r \\
& \times (\phi(\mu_n) - \phi(\mu))_s D_s D_r h(\tilde{X}_j, \theta_n)
\end{aligned}
\tag{3.19}
$$

for some θ_n in the line segment joining $\phi(\mu)$ and $\phi(\mu_n)$. By assumption A5 of Theorem 3.10 and because $\sqrt{n}\,(\phi(\mu_n) - \phi(\mu))$ is asymptotically normal, the second term on the right side of equation (3.19) converges to 0 in probability. Also,

$$\frac{1}{n}\sum_{j=1}^n Dh(\tilde{X}_j, \phi(\mu)) \xrightarrow{P} E\left(Dh(\tilde{X}_1, \phi(\mu))\right) = 0,$$

so that the first term on the right side of equation (3.19) converges to 0 in probability. Hence equation (3.18) becomes

$$\sqrt{n}\,(V_n - V) = \sqrt{n}\,(\tilde{F}_n(\phi(\mu)) - \tilde{F}(\phi(\mu))) + o_P(1)$$

$$= \frac{1}{\sqrt{n}} \sum_{j=1}^{n} \left(\rho^\alpha(X_j, \mu) - \mathrm{E}[\rho^\alpha(X_1, \mu)] \right) + o_P(1). \quad (3.20)$$

By the CLT for the i.i.d. sequence $\{\rho^\alpha(X_j, \mu)\}$, $\sqrt{n}(V_n - V)$ converges in distribution to $N(0, \mathrm{Var}(\rho^\alpha(X_1, \mu))$. $\quad\square$

Remark 3.12 Although Proposition 3.3 does not require uniqueness of the Fréchet mean of Q for V_n to be a consistent estimator of V, Theorem 3.11 requires the Fréchet mean of Q to exist for the sample variation to be asymptotically Normal. It may be shown by examples (see Section 4.8) that it fails to give the correct distribution when there is no unique mean.

Using Theorem 3.11, we can construct the following confidence interval I for V:

$$I = \left\{ V \in [V_n - \frac{s}{\sqrt{n}} Z_{1-\frac{\theta}{2}}, V_n + \frac{s}{\sqrt{n}} Z_{1-\frac{\theta}{2}}] \right\}. \quad (3.21)$$

The interval I has an asymptotic confidence level of $(1 - \theta)$. Here s^2 is the sample variance of $\rho^\alpha(X_j, \mu_n)$, $j = 1, \ldots, n$, and $Z_{1-\theta/2}$ denotes the upper $(1 - \frac{\theta}{2})$-quantile of a standard Normal distribution. From the confidence interval I, we can also construct a pivotal bootstrap confidence interval for V, the details of which are left to the reader.

3.6 An example: the unit circle

Perhaps the simplest interesting example of a nonflat manifold is the unit circle $S^1 = \{(x, y) \in \mathbb{R}^2 : x^2 + y^2 = 1\}$. The goal in this section is to briefly illustrate the notions introduced in Sections 3.2–3.5 with $M = S^1$. A comprehensive account of circular statistics, with many fascinating data-based examples, may be found in Fisher (1993).

A convenient parametrization of S^1 is given by the map $\theta \mapsto (\cos\theta, \sin\theta)$, $-\pi \le \theta < \pi$. One may refer to θ as the angular coordinate of $(\cos\theta, \sin\theta)$. The geodesic, or intrinsic, distance ρ_I between two points on the circle is given by the arc length between them (the smaller of the lengths of the two arcs joining the points). This map is a local isometry (with respect to linear distance on $[-\pi, \pi)$ and arc length on S^1). As long as two points $\theta_1 < \theta_2$ in $[-\pi, \pi)$ are at a distance no more than π from each other, the arc length between the corresponding points on the circle is the same as the linear distance $\theta_2 - \theta_1$.

Consider the following cases of distributions Q. Unless stated otherwise, we let $f(u) = u^2$ in equation (3.1).

1. Q is uniform on S^1; that is, Q is the normalized length measure on S^1, assigning probability $l/(2\pi)$ to each arc of length l. It is easy to see, by symmetry, that the Fréchet (or intrinsic) mean set of Q is $C_Q = S^1$.

2. Q is uniform on $S^1 \setminus A$, where A is a nonempty open arc ($A \neq S^1$). A fairly simple direct calculation of the Fréchet function shows that the midpoint of the arc $S^1 \setminus A$ is the unique intrinsic mean.

3. Q has support contained in an arc $A = \{(\cos\theta, \sin\theta): \theta_1 \leq \theta \leq \theta_2\}$ ($\theta_1 < \theta_2$) of length no more than $\pi/3$. The intrinsic mean is unique. To see this, note that the map $\theta \mapsto (\cos\theta, \sin\theta)$ is an isometry of $[\theta_1, \theta_2]$ (as a subset of $\mathbb{R}^1$) onto A, since $\theta_2 - \theta_1 \leq \pi/3 < \pi$. If the Fréchet mean of (the image of) Q on $[\theta_1, \theta_2]$ is θ_0 (with ρ as the linear distance), that is, θ_0 is the usual mean of Q regarded as a distribution on $[\theta_1, \theta_2]$, then the Fréchet mean on A is $\mu_I = (\cos\theta_0, \sin\theta_0)$ ($= \mu$), and $F(\mu_I) < \pi^2/9$. Also, $(\cos\theta_0, \sin\theta_0)$ is the (local) minimizer of the Fréchet function F, restricted to the arc B of arc length π, corresponding to the linear interval $[\theta_1 - c, \theta_2 + c]$, with $c = \{\pi - (\theta_2 - \theta_1)\}/2 \geq \pi/3$. Here B is treated as the metric space M with the distribution Q on it. Because every point p of S^1 outside of B is at a distance larger than $\pi/3$ from A, $F(p) > (\pi/3)^2$. It follows that $(\cos\theta_0, \sin\theta_0)$ is indeed the intrinsic mean of Q on S^1.

4. Let Q be discrete with

$$Q(\{(1, 0)\}) = \alpha, \quad Q(\{(-1, 0)\}) = 1 - \alpha \quad (0 < \alpha < 1).$$

Then $C_Q = \{p_1, p_2\}$, where p_1 lies on the half-circle joining $(1, 0)$ to $(-1, 0)$ counterclockwise, while p_2 lies on the half-circle joining the points clockwise. This follows by restricting Q to each of these half-circles and finding the Fréchet mean on each half-circle viewed as the metric space M. The computation of p_1, p_2 is simple, using the isometry between a half-circle and its angular image on a line of length π.

5. If Q is absolutely continuous with a continuous nonconstant density g, then there are reasonably broad conditions under which μ_I is unique (see Chapter 5, Section 5.7). For example, if g is greater than $1/(2\pi)$ on an (open) arc, equals $1/(2\pi)$ at its end points, and is smaller than $1/(2\pi)$ in the complementary arc, then the intrinsic mean is unique.

Let us now turn briefly to the case $f(u) = u$. Then the Fréchet mean minimizes the mean expected distance ρ_I under Q. One sometimes refers to it as the intrinsic median of Q, if unique. It is easy to see that when Q is uniform on S^1, the intrinsic median set is S^1, the same as the intrinsic mean set (see case 1). Similarly, the intrinsic median is unique and equals the intrinsic mean, as in case 2. The intrinsic median suffers from the same issues of nonuniqueness in the case of discrete distributions on S^1 as it does on $\mathbb{R}^1$.

Consider next the embedding of S^1 into $\mathbb{R}^2$ by the inclusion map $i(m) = (x, y)$, $m = (x, y) \in S^1$. The Euclidean distance ρ_E inherited by S^1 from this embedding is referred to as the extrinsic distance on S^1: $\rho_E((x_1, y_1), (x_2, y_2)) = \{(x_1 - x_2)^2 + (y_1 - y_2)^2)\}^{1/2}$. Thus ρ_E is the length of the line segment joining the two points, and is sometimes called the chord distance (while ρ_I is the arc distance). It will be shown in the next chapter that the extrinsic mean μ_E exists as a unique minimizer of the expected squared extrinsic distance if and only if the mean μ, say, of Q, regarded as a distribution on $\mathbb{R}^2$, is not the origin $(0, 0)$, and in that case one has the extrinsic mean given by $\mu_E = \mu/\|\mu\|$. Unfortunately, such a simple and broad criterion does not exist for the intrinsic mean, thus making the use of this mean somewhat complex.

Coming to the asymptotic distribution of the sample means, let Q be absolutely continuous, with a continuous density g and unique intrinsic mean μ_I. Let $X_1, \ldots, X_n$ be i.i.d. observations with common distribution Q. Consider the open subset of S^1 given by $U = S^1 \setminus \{\mu_I\}$ mapped onto the line segment $(-\pi, \pi)$ using an angular coordinate ϕ around μ_I: $\phi^{-1}(\theta) = (\cos(\theta_0 + \theta), \sin(\theta_0 + \theta))$, where $\mu_I = (\cos(\theta_0), \sin(\theta_0))$. Then, with $\alpha = 2$ and $\rho = \rho_I$, the conditions of Theorem 3.10 are satisfied. The function h in the theorem is given by

$$h(u, \theta) = \begin{cases} (u - \theta)^2 & \text{for } -\pi + \theta < u < \pi, \\ (2\pi + u - \theta)^2 & \text{for } -\pi < u < -\pi + \theta \ (\theta \geq 0); \end{cases}$$

$$h(u, \theta) = \begin{cases} (u - \theta)^2 & \text{for } \pi + \theta < u < \pi, \\ (2\pi - u + \theta)^2 & \text{for } -\pi < u < \pi + \theta \ (\theta < 0). \end{cases}$$

Note that $[(\partial/\partial\theta)h(u, \theta)]_{\theta=0} = -2u$. Hence

$$\sigma^2 \equiv \text{Var}\{Dh(\phi^{-1}(X_1), 0)\} = 4\text{Var}\{\phi^{-1}(X_1)\}.$$

Also, $[(\partial^2/\partial\theta^2)h(u, \theta)]_{\theta=0} = 2$. Hence, by Theorem 3.10, we have

$$\sqrt{n}\, \{\phi^{-1}(\mu_{nI}) - \phi^{-1}(\mu_I)\} \xrightarrow{\mathcal{L}} N\left(0, \text{Var}\{\phi^{-1}(X_1)\}\right).$$

The asymptotic distribution of the extrinsic sample mean will be discussed in detail in the next chapter.

3.7 Data example on S^1

We return now to Example 2.1 on wind direction and ozone concentration observed at a weather station for 19 days. Table 2.1 displays the wind directions in degrees, and the data plot is given in Figure 3.1. The average squared arc distance of the 19 points on the circle is minimized when measured from the point at an angle of 5.68 degrees, or $*i = .099$ radian, measured counterclockwise from the point $(1, 0)$ on the circle. The 95% confidence region (arc interval) for the intrinsic mean may be constructed either directly or by using Theorem 3.10. This is plotted in Figure 3.1. For the extrinsic sample mean, one converts the data to rectangular coordinates

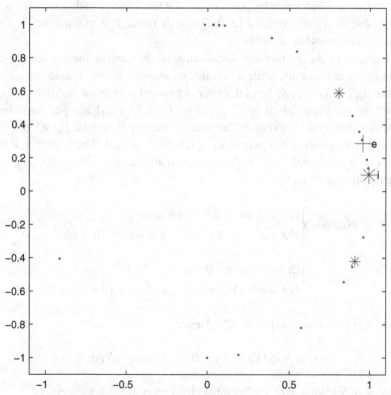

Figure 3.1 Wind directions from Table 2.1 (·), extrinsic mean direction (+e), intrinsic mean direction ($*i$), 95% confidence region end points ($*$).

$(\cos\theta, \sin\theta)$, whose mean occurs at a point inside the unit circle, and its projection on the circle is the extrinsic sample mean whose radian measure is $+e = .29$ (or 16.71 in degrees).

3.8 References

Fréchet (1948) defined the mean of a probability measure Q on a metric space as the minimizer, if unique, of the expected squared distance from a point. Ziezold (1977) has a general result for Fréchet mean sets under a squared distance (i.e., $f(u) = u^2$) on general separable metric spaces M. From this it follows that if M is compact and the mean set is a singleton $\{\mu\}$, then the sample Fréchet means converge to μ a.s. The somewhat stronger version implied by Corollary 3.4 (see Remark 3.5) was obtained in Bhattacharya and Patrangenaru (2003). Theorem 3.10 is due to Bhattacharya and Patrangenaru (2005). The results on the asymptotics of the sample dispersion are due to Bhattacharya (2008a).

4

Extrinsic analysis on manifolds

In this chapter we introduce the extrinsic distance on a manifold M obtained by embedding M into some Euclidean space, compute the Fréchet mean and variation, and lay the foundations of the corresponding statistical analysis, called extrinsic analysis on M. As we shall see in the following sections, extrinsic analysis is often simpler, both mathematically and computationally, than analysis based on other distances such as the intrinsic distance.

4.1 Extrinsic mean and variation

We assume that M is a *differentiable manifold* of dimension d. Consider an *embedding* of M into some Euclidean space E of dimension D (which can be identified with $\mathbb{R}^D$) via an injective differentiable map $j: M \to E$ whose derivative, or differential, is also injective. The dimension D is usually much higher than d. Appendix A gives a detailed description of differentiable manifolds and their submanifolds, tangent spaces, differentiable maps, and embeddings. The embedding j induces the distance

$$\rho(x, y) = \| j(x) - j(y) \| \tag{4.1}$$

on M, where $\|.\|$ denotes the usual Euclidean norm. The distance ρ is called the *extrinsic distance* on M. Given a probability distribution Q on M, we consider the Fréchet function

$$F(x) = \int_M \rho^2(x, y) Q(dy) \tag{4.2}$$

with ρ as in equation (4.1). This choice of Fréchet function makes the Fréchet mean and variation computable in a number of important examples using Proposition 4.2 below.

Definition 4.1 Let (M, ρ) and j be as above. Let Q be a probability distribution with finite Fréchet function F. The Fréchet mean set of Q is called the *extrinsic mean set* of Q and the Fréchet variation of Q is called the *extrinsic variation* of Q. If X_i, $i = 1, \ldots, n$, are i.i.d. observations from Q and $Q_n = \frac{1}{n} \sum_{i=1}^{n} \delta_{X_i}$ is the corresponding empirical distribution, then the Fréchet mean set of Q_n is called the *sample extrinsic mean set* and the Fréchet variation of Q is called the *sample extrinsic variation*.

We say that Q has an *extrinsic mean* μ_E if the extrinsic mean set of Q is a singleton. Proposition 4.2 gives a necessary and sufficient condition for Q to have an extrinsic mean. It also provides an analytic expression for the extrinsic mean set and extrinsic variation of Q. In the statement of the proposition, we assume that $j(M) = \tilde{M}$ is a closed subset of E. Then, for every $u \in E$, there exists a compact set of points in $\tilde{M}$ whose distance from u is the smallest among all points in $\tilde{M}$. We call this set the *projection set* of u and denote it by $P_{\tilde{M}}(u)$, or simply as $P(u)$ if the context makes $\tilde{M}$ clearly understood. It is given by

$$P_{\tilde{M}}(u) = \{x \in \tilde{M} : \|x - u\| \leq \|y - u\| \ \forall y \in \tilde{M}\}. \tag{4.3}$$

If this set is a singleton, u is said to be a *nonfocal point* of E (with respect to $\tilde{M}$); otherwise it is said to be a *focal point* of E.

We define the *sample extrinsic mean* μ_{nE} to be any measurable selection from the sample extrinsic mean set, in case μ_E is uniquely defined.

Proposition 4.2 *Let $\tilde{Q} = Q \circ j^{-1}$ be the image of Q in E. (a) If $\mu = \int_E u\tilde{Q}(du)$ is the mean of $\tilde{Q}$, then the extrinsic mean set of Q is given by $j^{-1}(P_{\tilde{M}}(\mu))$. (b) The extrinsic variation of Q equals*

$$V = \int_E \|x - \mu\|^2 \tilde{Q}(dx) + \|\mu - \tilde{\mu}\|^2,$$

where $\tilde{\mu} \in P_{\tilde{M}}(\mu)$. (c) If μ is a nonfocal point of E, then the extrinsic mean μ_E of Q exists, and the sample extrinsic mean μ_{nE} is a consistent estimator of μ_E.

Proof For $c \in \tilde{M}$, one has

$$F(c) = \int_{\tilde{M}} \|x - c\|^2 \tilde{Q}(dx) = \int_E \|x - \mu\|^2 \tilde{Q}(dx) + \|\mu - c\|^2,$$

which is minimized on $\tilde{M}$ by $c \in P_{\tilde{M}}(\mu)$. This proves the expression for V in part (b), and also part (a). Part (c) follows from general consistency (Corollary 3.4, Chapter 3). □

4.2 Asymptotic distribution of the sample extrinsic mean

From now on we assume that the extrinsic mean μ_E of Q is uniquely defined. It follows from Theorem 3.10 that, under suitable assumptions, the vector of coordinates of the sample extrinsic mean μ_{nE} has an asymptotic Gaussian distribution. However, apart from other assumptions, the theorem requires Q to have support in a single coordinate patch and the expression of the asymptotic covariance depends on what coordinates we choose. In this section, we derive the asymptotic Normality of μ_{nE} via Proposition 4.3 below. This proposition makes less restrictive assumptions on Q than Theorem 3.10 and the expression for the asymptotic covariance is easier to compute, as we shall see in the following sections and chapters. When the mean μ of $\tilde{Q}$ is a nonfocal point of E, the projection set in equation (4.3) is a singleton and we can define a *projection map* P on a neighborhood U of μ into $\tilde{M}$:

$$P:U \rightarrow \tilde{M}, \quad P(x) = \arg \min_{p \in \tilde{M}} \|x - p\|. \tag{4.4}$$

Also, in a neighborhood of a nonfocal point such as μ, P is smooth. Let $\bar{X} = \frac{1}{n} \sum_{i=1}^{n} \pi(X_i)$ be the mean of the embedded sample. Since $\bar{X}$ converges to μ a.s., for sample size large enough, $\bar{X}$ may be taken to be nonfocal. If we view P as a map on U into $E \approx \mathbb{R}^D$, and we let $J(\cdot)$ denote the $D \times D$ Jacobian matrix of P in a neighborhood of μ, then we may write

$$\sqrt{n} \{P(\bar{X}) - P(\mu)\} = \sqrt{n} J(\mu)(\bar{X} - \mu) + o_P(1). \tag{4.5}$$

Since $\sqrt{n}(\bar{X} - \mu)$ has an asymptotic Gaussian distribution, from equation (4.5) it follows that $\sqrt{n}\{P(\bar{X}) - P(\mu)\}$ has an asymptotic mean zero D-dimensional singular Gaussian distribution on the tangent space $T_{P(\mu)}E$, whose support is contained in the d-dimensional subspace $T_{P(\mu)}\tilde{M}$. Henceforth we view P as a map on U into the d-dimensional manifold $\tilde{M}$ and denote by $d_x P$ the differential of this map P (on $T_x E \approx \mathbb{R}^D$ into $T_{P(x)}\tilde{M} \approx \mathbb{R}^d$). To find a matrix representation of $d_x P$ one chooses an orthonormal basis (frame) $\{F_1, \ldots, F_d\}$ of $T_{P(x)}\tilde{M}$. In particular, expressing $d_\mu P(e_j) = \sum_{i=1}^{d} b_{ji} F_i$ (for $\{e_1, \ldots, e_d\}$ a standard orthonormal basis of $E \approx \mathbb{R}^D$), one has $d_\mu P(\bar{X} - \mu) = \sum_{j=1}^{D} \sum_{i=1}^{d} b_{ji} (\bar{X} - \mu)^{(j)} F_i = \sum_{i=1}^{d} (\sum_{j=1}^{D} b_{ji} (\bar{X} - \mu)^{(j)}) F_i$. Thus, in terms of the $D \times d$ matrix $B = ((b_{ji}))$, we may write $d_\mu P(\bar{X} - \mu) = \bar{T}$ $(\bar{T}^{(i)} = \sum_{j=1}^{D} b_{ji} (\bar{X} - \mu)$ for $i = 1, \ldots, d)$, which has an asymptotic d-dimensional Normal distribution. This is stated in Proposition 4.3, where $\text{Cov}(j(X_1))$ denotes the covariance matrix of $j(X_1)$ as a random element in $\mathbb{R}^D$.

Proposition 4.3 *Suppose μ is a nonfocal point of E and P is continuously differentiable in a neighborhood of μ. Then, if $Q \circ j^{-1}$ has finite second moments,*

$$\sqrt{n} \, d_\mu P(\bar{X} - \mu) = \sqrt{n} \, \bar{T} \xrightarrow{\mathcal{L}} N_d(0, \Sigma),$$

where $\Sigma = B'\mathrm{Cov}(j(X_1))B$, with $B = B(\mu)$ as the $D \times d$ matrix of $d_\mu P$ with respect to given orthonormal bases of $T_\mu E$ and $T_{P(\mu)}\tilde{M}$.

Using this proposition, an asymptotic confidence region for the population extrinsic mean μ_E is derived in Corollary 4.4.

Corollary 4.4 *In addition to the hypothesis of Proposition 4.3, assume that Σ is nonsingular. Then the probability of the confidence region,*

$$\Big\{ \mu_E \in M :$$

$$n\{(d_{\bar{X}}P(\bar{X} - \mu))'\hat{\Sigma}^{-1}(d_{\bar{X}}P(\bar{X} - \mu))\} \le \mathcal{X}_d^2(1 - \alpha)\Big\}, \quad (4.6)$$

converges to $(1 - \alpha)$ a.s. Here $\hat{\Sigma}$ is the empirical version of Σ.

The assumption that Σ is nonsingular holds in particular if $\mathrm{Cov}(j(X_1))$ is nonsingular, which in turn holds when the distribution $\tilde{Q}$ on $\tilde{M}$ does not have support in an affine subspace of E of dimension smaller than D.

For sample sizes not large enough, a pivotal bootstrap confidence region can be more effective; it is obtained by replacing $\mathcal{X}_d^2(1 - \alpha)$ by the upper $(1 - \alpha)$-quantile $c^*(1 - \alpha)$ of the bootstrapped values of the asymptotic chi-squared displayed in equation (4.6)

4.3 Asymptotic distribution of the sample extrinsic variation

Let V and V_n denote the extrinsic variations of Q and Q_n, respectively. We can deduce the asymptotic distribution of V_n from Theorem 3.11 in Chapter 3. However, for the hypothesis of that theorem to hold, we need to make a number of assumptions, including that Q has support in a single coordinate patch. Theorem 4.5 proves the asymptotic Normality of V_n under less restrictive assumptions. In the statement of the theorem, ρ denotes the extrinsic distance as defined in equation (4.1).

Theorem 4.5 *If Q has extrinsic mean μ_E and if $\mathrm{E}\rho^4(X_1, \mu_E) < \infty$, then*

$$\sqrt{n}\,(V_n - V) = \frac{1}{\sqrt{n}} \sum_{i=1}^{n} \{\rho^2(X_i, \mu_E) - V\} + o_P(1), \quad (4.7)$$

which implies that

$$\sqrt{n}\,(V_n - V) \xrightarrow{\mathcal{L}} N(0, \mathrm{Var}(\rho^2(X_1, \mu_E))).$$

Proof From the definitions of V_n and V, it follows that

$$
\begin{aligned}
V_n - V &= \tfrac{1}{n} \sum_{i=1}^n \rho^2(X_i, \mu_{nE}) - \int_M \rho^2(x, \mu_E) Q(dx) \\
&= \tfrac{1}{n} \sum_{i=1}^n \rho^2(X_i, \mu_{nE}) - \tfrac{1}{n} \sum_{i=1}^n \rho^2(X_i, \mu_E) \\
&\quad + \tfrac{1}{n} \sum_{i=1}^n \rho^2(X_i, \mu_E) - \mathrm{E}\rho^2(X_1, \mu_E),
\end{aligned}
\tag{4.8}
$$

where μ_{nE} is the sample extrinsic mean, that is, some measurable selection from the sample extrinsic mean set. Denote by X_i the embedded sample $j(X_i)$, $i = 1, \ldots, n$. By the definition of extrinsic distance,

$$
\begin{aligned}
\frac{1}{n} \sum_{i=1}^n \rho^2(X_i, \mu_{nE}) &= \frac{1}{n} \sum_{i=1}^n \| X_i - P(\bar{X}) \|^2 \\
&= \frac{1}{n} \sum_{i=1}^n \| X_i - P(\mu) \|^2 + \| P(\mu) - P(\bar{X}) \|^2 - 2\langle \bar{X} - P(\mu), P(\bar{X}) - P(\mu) \rangle,
\end{aligned}
\tag{4.9}
$$

where $\langle . \rangle$ denotes the Euclidean inner product. Substitute equation (4.9) into equation (4.8) to get

$$
\begin{aligned}
\sqrt{n}\,(V_n - V) &= \sqrt{n}\,\big(\| P(\bar{X}) - P(\mu) \|^2 - 2\langle \bar{X} - P(\mu), P(\bar{X}) - P(\mu) \rangle \big) \\
&\quad + \sqrt{n}\,\Big(\frac{1}{n} \sum_{i=1}^n \rho^2(X_i, \mu_E) - \mathrm{E}\rho^2(X_1, \mu_E) \Big).
\end{aligned}
\tag{4.10}
$$

Denote the two terms in equation (4.10) as T_1 and T_2, that is,

$$
T_1 = \sqrt{n}\, \| P(\bar{X}) - P(\mu) \|^2 - 2\sqrt{n}\, \langle \bar{X} - P(\mu), P(\bar{X}) - P(\mu) \rangle,
$$

$$
T_2 = \sqrt{n}\, \Big(\frac{1}{n} \sum_{i=1}^n \rho^2(X_i, \mu_E) - \mathrm{E}\rho^2(X_1, \mu_E) \Big).
$$

From the classical CLT, if $\mathrm{E}\rho^4(X_1, \mu_E) < \infty$, then

$$
T_2 \xrightarrow{\mathcal{L}} N(0, \mathrm{Var}(\rho^2(X_1, \mu_E))).
\tag{4.11}
$$

Compare the expression of T_1 with identity (4.5) to get

$$
T_1 = -2\langle d_\mu P(\bar{X} - \mu), \mu - P(\mu) \rangle + o_P(1).
\tag{4.12}
$$

From the definition of P, $P(\mu) = \arg\min_{p \in \tilde{M}} \| \mu - p \|^2$. Hence the Euclidean derivative of $\| \mu - p \|^2$ at $p = P(\mu)$ must be orthogonal to $T_{P(\mu)}\tilde{M}$, or $\mu - P(\mu) \in (T_{P(\mu)}\tilde{M})^\perp$. Because $d_\mu P(\bar{X} - \mu) \in T_{P(\mu)}\tilde{M}$, the first term in the expression for T_1 in (4.12) is 0, and hence $T_1 = o_P(1)$. From equations (4.10) and (4.11), we conclude that

$$\sqrt{n}\,(V_n - V) =$$

$$\frac{1}{\sqrt{n}} \sum_{i=1}^{n} \{\rho^2(X_i, \mu_E) - E\rho^2(X_1, \mu_E)\} + o_P(1) \xrightarrow{\mathcal{L}} N(0, \operatorname{Var}(\rho^2(X_1, \mu_E))).$$

This completes the proof. □

Remark 4.6 Although Proposition 3.8 does not require uniqueness of the extrinsic mean of Q for V_n to be a consistent estimator of V, Theorem 4.5 breaks down in the case of nonuniqueness. This is illustrated in Section 4.8.

Using Theorem 4.5, one can construct an asymptotic confidence interval

$$[V_n - \frac{s}{\sqrt{n}} Z_{1-\frac{\alpha}{2}}, V_n + \frac{s}{\sqrt{n}} Z_{1-\frac{\alpha}{2}}] \tag{4.13}$$

for V with an asymptotic confidence level of $(1 - \alpha)$. Here s^2 is the sample variance of $\rho^2(x, \mu_{nE})$ and $Z_{1-\alpha/2}$ denotes the upper $(1 - \frac{\alpha}{2})$-quantile of the $N(0, 1)$ distribution. From equation (4.13), we can also construct a bootstrap confidence interval for V, the details of which are left to the reader.

4.4 Asymptotic joint distribution of the sample extrinsic mean and variation

In many applications, especially on noncompact manifolds such as size-and-shape spaces, it is more effective to perform inference using the joint distribution of location and spread rather than the marginals.

Proposition 4.7 *Under the assumptions of Proposition 4.3 and Theorem 4.5,*

$$\sqrt{n}\,(d_\mu P(\bar{X} - \mu), V_n - V) \xrightarrow{\mathcal{L}} N_{d+1}(0, \Sigma)$$

with $\Sigma = \begin{pmatrix} \Sigma_{11} & \Sigma_{12} \\ \Sigma'_{12} & \sigma^2 \end{pmatrix}$, $\Sigma_{11} = \operatorname{Cov}(T)$, $T = (d_\mu P)(j(X_1) - \mu)$, $\Sigma_{12} = \operatorname{Cov}(T, \rho^2(X_1, \mu_E))$, *and* $\sigma^2 = \operatorname{Var}(\rho^2(X_1, \mu_E))$.

Proof It is shown in Theorem 4.5 that

$$\sqrt{n}\,(V_n - V) = \sqrt{n}\left(\frac{1}{n} \sum_{i=1}^{n} \rho^2(X_i, \mu_E) - E\rho^2(X_1, \mu_E)\right) + o_P(1).$$

Now the result is immediate. □

4.5 Two-sample extrinsic tests

In this section, we will use the asymptotic distribution of the sample extrinsic mean and variation to construct nonparametric tests to compare two probability distributions Q_1 and Q_2 on M.

4.5.1 Independent samples

Let $X_1, \ldots, X_{n_1}$ and $Y_1, \ldots, Y_{n_2}$ be two i.i.d. samples from Q_1 and Q_2, respectively, that are mutually independent. Let μ_{iE} and V_i denote the extrinsic means and variations of Q_i, $i = 1, 2$, respectively. Similarly denote by $\hat{\mu}_{iE}$ and $\hat{V}_i$ the sample extrinsic means and variations. We want to test the hypothesis $H_0 : Q_1 = Q_2$.

We start by comparing the sample extrinsic means. Let $\tilde{X}_i = j(X_i)$, $\tilde{Y}_i = j(Y_i)$ be the embeddings of the sample points into E. Let μ_i be the mean of $\tilde{Q}_i = Q_i \circ j^{-1}$ ($i = 1, 2$). Then, under H_0, $\mu_1 = \mu_2 = \mu$ (say). Let $\hat{\mu}_i$, $i = 1, 2$, be the sample means of $\{\tilde{X}_i\}$ and $\{\tilde{Y}_i\}$, respectively. Then, from Proposition 4.3, it follows that, if $n_i \to \infty$ such that $\frac{n_i}{n_1 + n_2} \to p_i$, $0 < p_i < 1$, $p_1 + p_2 = 1$, then

$$\sqrt{n}\, d_\mu P(\hat{\mu}_1 - \mu) - \sqrt{n}\, d_\mu P(\hat{\mu}_2 - \mu) \xrightarrow{\mathcal{L}} N\left(0, \frac{\Sigma^1}{p_1} + \frac{\Sigma^2}{p_2}\right). \tag{4.14}$$

Here $n = n_1 + n_2$ is the pooled sample size and Σ^i, $i = 1, 2$, are the covariance matrices of $d_\mu P(\tilde{X}_1 - \mu)$ and $d_\mu P(\tilde{Y}_1 - \mu)$. We estimate μ by the pooled sample mean $\hat{\mu} = \frac{1}{n}(n_1 \hat{\mu}_1 + n_2 \hat{\mu}_2)$ and Σ^i by $\hat{\Sigma}^i$, $i = 1, 2$. Then, if H_0 is true, the statistic

$$T_1 = (d_{\hat{\mu}} P(\hat{\mu}_1 - \hat{\mu}_2))' \left(\frac{1}{n_1}\hat{\Sigma}^1 + \frac{1}{n_2}\hat{\Sigma}^2\right)^{-1} d_{\hat{\mu}} P(\hat{\mu}_1 - \hat{\mu}_2) \tag{4.15}$$

converges in distribution to a $\mathcal{X}_d^2$ distribution, where d is the dimension of M. Hence we reject H_0 at an asymptotic level α if $T_1 > \mathcal{X}_d^2(1 - \alpha)$. Note that one could think of testing directly the equality of the means μ_i ($i = 1, 2$) of $\tilde{Q}_i = Q_i \circ j^{-1}$, to discriminate between Q_1 and Q_2. However, the dimension D of E is almost always much larger than d, and the estimates of the covariance matrices of $\tilde{Q}_i$ will be singular or nearly singular for moderate sample sizes. The test based on T_1 is, therefore, more effective.

Next we test the null hypothesis $H_0 : \mu_{1E} = \mu_{2E} = \mu_E$ against the alternative $H_a : \mu_{1E} \neq \mu_{2E}$, say. Since the samples are independent, under H_0, one has

$$\sqrt{n}\, d_{\mu_1} P(\hat{\mu}_1 - \mu_1) - \sqrt{n}\, d_{\mu_2} P(\hat{\mu}_2 - \mu_2) \xrightarrow{\mathcal{L}} N_d\left(0, \frac{\Sigma_1}{p_1} + \frac{\Sigma_2}{p_2}\right).$$

Here Σ_i, $i = 1, 2$, are the covariance matrices of $d_{\mu_1} P(\tilde{X}_1 - \mu_1)$ and $d_{\mu_2} P(\tilde{Y}_1 - \mu_2)$. Note that, in contrast to Σ_1, Σ^1 is the covariance matrix of $d_\mu P(\tilde{X}_1 - \mu)$ when $\mu_1 = \mu_1 = \mu$. Similarly, Σ_2 and Σ^2 are to be distinguished. Let $L = L_{\mu_E}$ denote the orthogonal linear projection of vectors in $T_{\mu_E} E \equiv E$ onto $T_{\mu_E} \tilde{M}$. Usually it is easier to compute $L(P(\hat{\mu}_i) - P(\mu_i))$ than $d_{\mu_i} P(\hat{\mu}_i - \mu_i)$. Note that

$$L\{P(\hat{\mu}_1) - P(\mu_1)\} = L\{d_{\mu_i} P(\hat{\mu}_i - \mu_i)\} + o(\|\hat{\mu}_i - \mu_i\|)$$
$$= d_{\mu_i} P(\hat{\mu}_i - \mu_i) + o(\|\hat{\mu}_i - \mu_i\|) \quad (i = 1, 2),$$
$$\sqrt{n}\, L\{P(\hat{\mu}_1) - P(\hat{\mu}_2)\} = \sqrt{n}\, L\{P(\hat{\mu}_1) - P(\mu_1)\} - \sqrt{n}\, L\{P(\hat{\mu}_2) - P(\mu_2)\}$$
$$= \sqrt{n}\, d_{\mu_1} P(\hat{\mu}_1 - \mu_1) - \sqrt{n}\, d_{\mu_2} P(\hat{\mu}_2 - \mu_2) + o_P(1).$$
$$(4.16)$$

Hence, if H_0 is true, then $P(\mu_1) = P(\mu_2)$ and

$$\sqrt{n}\, L\{P(\hat{\mu}_1) - P(\hat{\mu}_2)\} \xrightarrow{\mathcal{L}} N(0, (1/p_1)\Sigma_1 + (1/p_2)\Sigma_2). \qquad (4.17)$$

Using this one can construct the test statistic

$$T_2 = [\hat{L}(P(\hat{\mu}_1) - P(\hat{\mu}_2))]'((1/n_1)\hat{\Sigma}_1 + (1/n_2)\hat{\Sigma}_2)^{-1}\hat{L}(P(\hat{\mu}_1) - P(\hat{\mu}_2)) \quad (4.18)$$

to test if H_0 is true. In the statistic T_2, $\hat{L}$ is the linear projection from E onto $T_{\hat{\mu}_E} \tilde{M}$; $\hat{\mu}_E$ is the pooled sample estimate of μ_E; and $\hat{\Sigma}_i$, $i = 1, 2$, denote the sample covariance matrices of $d_{\hat{\mu}_1} P(\tilde{X}_j - \hat{\mu}_1)$ and $d_{\hat{\mu}_2} P(\tilde{Y}_j - \hat{\mu}_2)$, respectively. Under H_0, $T_2 \xrightarrow{\mathcal{L}} \mathcal{X}_d^2$. Hence we reject H_0 at an asymptotic level α if $T_2 > \mathcal{X}_d^2(1 - \alpha)$. In all our numerical examples, the two statistics (4.15) and (4.18) yield values that are quite close to each other.

When the sample sizes are not very large, Efron's bootstrap procedure generally provides better estimates of the coverage probability than the CLT-based methods. We describe now bootstrapping for the test T_1. For this, we first construct a confidence region for $\mu_2 - \mu_1$. Let δ belong to a neighborhood of $0 \in \mathbb{R}^{d+1}$, and consider $H_0 : \mu_2 = \mu_1 + \delta$. The test statistic $T_{1\delta}$, say, is analogous to T_1. Let $\tilde{X}_{i,\delta} = \tilde{X}_i + (n_2/n)\delta$, $\tilde{Y}_{i,-\delta} = \tilde{Y}_i - (n_1/n)\delta$. Then, under H_0, $E\tilde{X}_{1,\delta} = \mu_1 + (n_2/n)\delta = \mu_1 + \delta - (n_1/n)\delta = E\tilde{Y}_{1,-\delta}$. Let $T_{1\delta}$ be the test statistic obtained by replacing $\tilde{X}_i$ by $\tilde{X}_{i,\delta}$ ($i \le n_1$) and $\tilde{Y}_i$ by $\tilde{Y}_{i,-\delta}$ ($i \le n_2$). Note that the pooled estimate of the common mean for the new variables is $(n_1/n)(\hat{\mu}_1 + (n_2/n)\delta) + (n_2/n)(\hat{\mu}_2 - (n_1/n)\delta) = (n_1/n)\hat{\mu}_1 + (n_2/n)\hat{\mu}_2 = \hat{\mu}$, the same as that for the original data. The set of δ such that $T_{1\delta}$ accepts the new $H_0 : \mu_2 = \mu_1 + \delta$ is

$$\{\delta : T_{1,\delta} < c\},$$

where $c = \mathcal{X}_d^2(1 - \alpha)$ for the chi-squared-based procedure. For bootstrapping, use bootstrapped data $\tilde{X}_i^*$ and $\tilde{Y}_i^*$ and let $\tilde{X}_{i,\delta}^* = \tilde{X}_i^* + (n_2/n)\delta$,

$\tilde{Y}^*_{i,-\delta} = \tilde{Y}^*_i - (n_1/n)\delta$, with $\delta = \hat{\mu}_2 - \hat{\mu}_1$, and use the bootstrapped estimate of the probability $P^*(T^*_{1,\delta} \leq c)$, with c as the observed value of T_1. The bootstrap-estimated p-value of the test is $1 - P^*(T^*_{1,\delta} \leq c)$. Another general method of bootstrapping in the two-sample problem is described in Section 4.6.

Next we test whether Q_1 and Q_2 have the same extrinsic variations, that is, $H_0 : V_1 = V_2$. From Theorem 4.5 and using the fact that the samples are independent, we get, under H_0,

$$\sqrt{n}\,(\hat{V}_1 - \hat{V}_2) \xrightarrow{\mathcal{L}} N\left(0, \frac{\sigma_1^2}{p_1} + \frac{\sigma_2^2}{p_2}\right)$$

$$\Rightarrow \frac{\hat{V}_1 - \hat{V}_2}{\sqrt{\frac{s_1^2}{n_1} + \frac{s_2^2}{n_2}}} \xrightarrow{\mathcal{L}} N(0, 1), \qquad (4.19)$$

where $\sigma_1^2 = \text{Var}(\rho^2(X_1, \mu_{1E}))$, $\sigma_2^2 = \text{Var}(\rho^2(Y_1, \mu_{2E}))$ and s_1^2, s_2^2 are their sample estimates. Hence, to test if H_0 is true, we can use the test statistic

$$T_3 = \frac{\hat{V}_1 - \hat{V}_2}{\sqrt{\frac{s_1^2}{n_1} + \frac{s_2^2}{n_2}}}. \qquad (4.20)$$

For a test of asymptotic size α, we reject H_0 if $|T_3| > Z_{1-\frac{\alpha}{2}}$, where $Z_{1-\frac{\alpha}{2}}$ is the upper $(1 - \frac{\alpha}{2})$-quantile of an $N(0, 1)$ distribution. We can also construct a bootstrap confidence interval for $V_1 - V_2$ and use that to test if $V_1 - V_2 = 0$. The details of that are left to the reader.

4.5.2 Matched pair samples

Next consider the case when $(X_1, Y_1), \ldots, (X_n, Y_n)$ is an i.i.d. sample from some distribution Q on $\bar{M} = M \times M$. Such samples arise when, for example, we have two different observations from each subject (see Section 9.6).

Let the X_j have distribution Q_1 while the Y_j come from some distribution Q_2 on M. Our objective is to distinguish Q_1 from Q_2 by comparing the sample extrinsic means and variations. Because the X and Y samples are not independent, we cannot apply the methods of the earlier section. Instead we do our analyses on $\bar{M}$. Note that $\bar{M}$ is a differentiable manifold that can be embedded into $E^N \times E^N$ via the map

$$\bar{j} : \bar{M} \to E^N \times E^N, \quad \bar{j}(x, y) = (j(x), j(y)).$$

Let $\tilde{Q} = Q \circ \bar{j}^{-1}$. Then, if $\tilde{Q}_i$ has mean μ_i, $i = 1, 2$, $\tilde{Q}$ has mean $\bar{\mu} = (\mu_1, \mu_2)$. The projection of $\bar{\mu}$ on $\bar{M} \equiv \bar{j}(\bar{M})$ is given by $\bar{P}(\mu) = (P(\mu_1), P(\mu_2))$. Hence, if Q_i has extrinsic mean μ_{iE}, $i = 1, 2$, then $\tilde{Q}$ has extrinsic mean

$\bar{\mu}_E = (\mu_{1E}, \mu_{2E})$. Denote the paired sample as $Z_i \equiv (X_i, Y_i)$, $i = 1, \ldots, n$, and let $\hat{\bar{\mu}} = (\hat{\mu}_1, \hat{\mu}_2)$, $\hat{\bar{\mu}}_E = (\hat{\mu}_{1E}, \hat{\mu}_{2E})$ be the sample estimates of $\bar{\mu}$ and $\bar{\mu}_E$, respectively. Hence, if $H_0 : \mu_1 = \mu_2 = \mu$, then under H_0, writing L for $L_{P(\mu)}$ and B for an orthonormal basis of $T_{P(\mu)}\tilde{M}$,

$$\sqrt{n}\left(\begin{array}{c} L\{P(\hat{\mu}_1) - P(\mu)\} \\ L\{P(\hat{\mu}_2) - P(\mu)\} \end{array} \right) = \sqrt{n}\left(\begin{array}{c} d_\mu P(\hat{\mu}_1 - \mu) \\ d_\mu P(\hat{\mu}_2 - \mu) \end{array} \right) + o_P(1)$$

$$\xrightarrow{\mathcal{L}} N\left(0, \Sigma = \left(\begin{array}{cc} \Sigma^1 & \Sigma^{12} \\ \Sigma^{21} & \Sigma^2 \end{array} \right)\right). \qquad (4.21)$$

In equation (4.21), $\Sigma^1 = \text{Cov } d_\mu P(X_1 - \mu)$, $\Sigma^2 = \text{Cov } d_\mu P(Y_1 - \mu)$, and $\Sigma^{12} = (\Sigma^{21})'$ is the covariance between $d_\mu P(\tilde{X}_1 - \mu)$ and $d_\mu P(\tilde{Y}_1 - \mu)$. From equation (4.21), it follows that

$$\sqrt{n}\, d_\mu P(\hat{\mu}_1 - \hat{\mu}_2) \xrightarrow{\mathcal{L}} N(0, \Sigma^1 + \Sigma^2 - \Sigma^{12} - \Sigma^{21}).$$

This gives rise to the test statistic

$$T_{1p} = n(d_{\hat{\mu}} P(\hat{\mu}_1 - \hat{\mu}_2))'(\hat{\Sigma}^1 + \hat{\Sigma}^2 - \hat{\Sigma}^{12} - \hat{\Sigma}^{21})^{-1} d_{\hat{\mu}} P(\hat{\mu}_1 - \hat{\mu}_2), \qquad (4.22)$$

where, with $\hat{\mu}$ as in case of T_1, $\hat{\Sigma}^1$, $\hat{\Sigma}^2$, and $\hat{\Sigma}^{12} = (\hat{\Sigma}^{21})'$ are the sample versions of Σ^1, Σ^2, and Σ^{12}, respectively. If H_0 is true, T_{1p} converges in distribution to the $\mathcal{X}_d^2$ distribution. Hence we reject H_0 at the asymptotic level α if $T_{1p} > \mathcal{X}_d^2(1 - \alpha)$.

If we are testing $H_0 : \mu_{1E} = \mu_{2E} = \mu_E$ (say), then from equation (4.21) it follows that, under H_0, writing L for $L_{\pi(\mu_E)}$, one has

$$\sqrt{n}\, L\{P(\hat{\mu}_1) - P(\hat{\mu}_2)\}$$

$$= \sqrt{n}\, L\{d_{\mu_1} P(\hat{\mu}_1 - \mu_1)\} - \sqrt{n}\, L\{d_{\mu_2} P(\hat{\mu}_2 - \mu_2)\} + o_P(1)$$

$$\xrightarrow{\mathcal{L}} N(0, \Sigma = \Sigma_1 + \Sigma_2 - \Sigma_{12} - \Sigma_{21}). \qquad (4.23)$$

In equation (4.23), $\Sigma_1 = \text{Cov } d_{\mu_1} P(\tilde{X}_1 - \mu_1)$, $\Sigma_2 = \text{Cov } d_{\mu_2} P(\tilde{Y}_1 - \mu_2)$, and $\Sigma_{12} = \Sigma_{21}'$ denotes the covariance between $d_{\mu_1} P(\tilde{X}_1 - \mu_1)$ and $d_{\mu_2} P(\tilde{Y}_1 - \mu_2)$. Hence, to test whether H_0 is true, one can use the test statistic

$$T_{2p} = n[\hat{L}\{P(\hat{\mu}_1) - P(\hat{\mu}_2)\}]'\hat{\Sigma}^{-1}\hat{L}\{P(\hat{\mu}_1) - P(\hat{\mu}_2)\}, \quad \text{where} \qquad (4.24)$$

$$\hat{\Sigma} = \hat{\Sigma}_1 + \hat{\Sigma}_2 - \hat{\Sigma}_{12} - \hat{\Sigma}_{21}. \qquad (4.25)$$

In the statistic T_{2p}, $\hat{L}$ and $\hat{\Sigma}_i$, $i = 1, 2$, are as in equation (4.18) and $\hat{\Sigma}_{12} = \hat{\Sigma}_{21}'$ denotes the sample covariance estimate of Σ_{12}. Under H_0, $T_{2p} \xrightarrow{\mathcal{L}} \mathcal{X}_d^2$. Hence we reject H_0 at an asymptotic level α if $T_{2p} > \mathcal{X}_d^2(1 - \alpha)$. In the application considered in Section 9.6, the values for the two statistics T_{1p} and T_{2p} are very close to each other.

Let V_1 and V_2 denote the extrinsic variations of Q_1 and Q_2 and let $\hat{V}_1$, $\hat{V}_2$ be their sample analog. Suppose we want to test the hypothesis $H_0 : V_1 = V_2$. From (4.7) we get that

$$\begin{pmatrix} \sqrt{n}(\hat{V}_1 - V_1) \\ \sqrt{n}(\hat{V}_2 - V_2) \end{pmatrix} = \frac{1}{\sqrt{n}} \begin{pmatrix} \sum_{j=1}^{n} [\rho^2(X_j, \mu_{1E}) - \mathrm{E}\rho^2(X_1, \mu_{1E})] \\ \sum_{j=1}^{n} [\rho^2(Y_j, \mu_{2E}) - \mathrm{E}\rho^2(Y_1, \mu_{2E})] \end{pmatrix} + o_P(1)$$

$$\xrightarrow{\mathcal{L}} N\left(0, \begin{pmatrix} \sigma_1^2 & \sigma_{12} \\ \sigma_{12} & \sigma_2^2 \end{pmatrix}\right), \qquad (4.26)$$

where $\sigma_{12} = \mathrm{Cov}(\rho^2(X_1, \mu_{1E}), \rho^2(Y_1, \mu_{2E}))$ and σ_1^2 and σ_2^2 are as in equation (4.19). Hence, if H_0 is true, then

$$\sqrt{n}\,(\hat{V}_1 - \hat{V}_2) \xrightarrow{\mathcal{L}} N(0, \sigma_1^2 + \sigma_2^2 - 2\sigma_{12}).$$

This gives rise to the test statistic

$$T_{3p} = \frac{\sqrt{n}\,(\hat{V}_1 - \hat{V}_2)}{\sqrt{s_1^2 + s_2^2 - 2s_{12}}}, \qquad (4.27)$$

where s_1^2, s_2^2, s_{12} are sample estimates of $\sigma_1^2, \sigma_2^2, \sigma_{12}$, respectively. We reject H_0 at an asymptotic level α if $|T_{3p}| > Z_{1-\frac{\alpha}{2}}$. We can also get a $(1 - \alpha)$-level confidence interval for $V_1 - V_2$ using bootstrap simulations and use that to test whether H_0 is true.

4.6 Hypothesis testing using extrinsic mean and variation

Suppose we have s samples $x = \{X_{ij}\}$, $i = 1, \ldots, n_j$, with n_j being the sample j size, $j = 1, \ldots, s$, on M. Observations $\{x_{.j}\}$ in sample j are assumed to be drawn independently from common distribution Q_j, $j = 1, \ldots, s$, and those distributions are unknown. In this section, we construct nonparametric tests to compare those distributions or a subcollection of them by using the asymptotic distribution of the sample extrinsic mean and variation.

4.6.1 Independent samples

Consider the case when the s samples are jointly independent. We want to test the null hypothesis $H_0 : Q_j$ are all same against the alternative H_1, which is its complement. We start by comparing the sample extrinsic means. Denote by $X = \{X_{ij}\}$ the embedded sample and by $\bar{X}_j$ the jth embedded sample mean. Under the null, $\{X_{ij}\}$ are i.i.d. as Q (say). Let μ be the mean of $Q \circ \pi^{-1}$ and $\bar{X} = (1/n) \sum_{j=1}^{s} n_j \bar{X}_j$ be its pooled sample estimate. From Proposition 4.3, it follows that, as $n_j \to \infty$,

$$\sqrt{n_j} \Sigma^{-1/2} T_j \xrightarrow{\mathcal{L}} N_d(0, I_d) \text{ independently} , \quad \forall j \leq s.$$

Here $T_j = d_\mu P(\bar{X}_j - \mu)$, Σ is the covariance of $d_\mu P(X_{11} - \mu)$ under the null, and I_d denotes the $d \times d$ identity matrix. This implies that if $n_j/n \to p_j$, $0 < p_j < 1$, then

$$\sum_{j=1}^{s} (n_j d_{\bar{X}} P(\bar{X}_j - \bar{X}))' \hat{\Sigma}^{-1} d_{\bar{X}} P(\bar{X}_j - \bar{X}) \xrightarrow{\mathcal{L}} \mathcal{X}^2_{(s-1)d}, \quad (4.28)$$

with $\hat{\Sigma}$ being the sample covariance of $d_{\bar{X}} P(X)$. Hence we can reject H_0 with type 1 error at-most α if the asymptotic p-value $\Pr(T > T_{\text{obs}})$ turns out to be smaller than α where $T \sim \mathcal{X}^2_{(s-1)d}$ and T_{obs} is the observed value of the statistic in equation (4.28). Similarly we can construct asymptotic chi-squared tests for comparing the extrinsic means of a subcollection of samples.

When the sample sizes are not very large, it is more efficient to perform hypothesis testing by bootstrap methods. Suppose we want to test the point null $H_0 : \theta = 0$ by bootstrap means. To do so, we find a statistic $T(X, \theta)$ whose asymptotic distribution is θ free and construct the asymptotic level $(1 - \alpha)$ confidence region

$$\{\theta : T(X, \theta) < c_{1-\alpha}\}$$

for θ. Denoting by $\hat{\theta}$ a consistent sample estimate of θ, the corresponding bootstrap confidence region will be

$$\{\theta : T(X, \theta) < c^*_{1-\alpha}\},$$

where, denoting by X^* the bootstrap sample from X,

$$\Pr(T(X^*, \hat{\theta}) < c^*_{1-\alpha} | X) = 1 - \alpha.$$

We reject H_0 at an asymptotic level α if 0 is not in the above confidence region. The greatest level at which we can reject H_0 is then

$$\Pr(T(X^*, \hat{\theta}) > T(X, 0) | X), \quad (4.29)$$

which is known as the *bootstrap p-value*.

When we have two samples and are testing equality of extrinsic means, that is, $H_0 : P(\mu_1) = P(\mu_2)$, then $\theta = P(\mu_1) - P(\mu_2)$. We take

$$T(X, \theta) = L\{P(\bar{X}_1) - P(\bar{X}_2) - \theta\}' \hat{\Sigma}^{-1} L\{P(\bar{X}_1) - P(\bar{X}_2) - \theta\}$$

$$= [\{(L \circ P)(\bar{X}_1) - (L \circ P)(\mu_1)\} - \{(L \circ P)(\bar{X}_2) - (L \circ P)(\mu_2)\}]' \hat{\Sigma}^{-1}$$

$$[\{(L \circ P)(\bar{X}_1) - (L \circ P)(\mu_1)\} - \{(L \circ P)(\bar{X}_2) - (L \circ P)(\mu_2)\}], \quad (4.30)$$

with L denoting the linear projection from E ($\approx \mathbb{R}^D$) into $T_m \tilde{M}$ ($\approx \mathbb{R}^d$) for some $m \in \tilde{M}$, $\hat{\Sigma} = \sum_{i=1}^{2} \hat{\Sigma}_i / n_i$, with $\hat{\Sigma}_i$ being the sample estimate of

$J(\mu_i)'\text{Cov}(X_{1i})J(\mu_i)$, where $J(p)$ is the $D \times d$ Jacobian matrix of $L \circ P$ at p. Then it follows that this $T(X, \theta)$ has an asymptotic $\mathcal{X}_d^2$ distribution irrespective of the choice of m or θ. Therefore, the bootstrap p-value can be expressed as $\Pr(T(X^*, \hat{\theta}) > T(x, 0)|X)$, where $T(X, 0) = L\{P(\bar{X}_1) - P(\bar{X}_2)\}'\hat{\Sigma}^{-1}L\{P(\bar{X}_1) - P(\bar{X}_2)\}$ and $\hat{\theta} = P(\bar{X}_1) - P(\bar{X}_2)$.

Using the asymptotic distribution of the sample extrinsic variation derived in Theorem 4.5, we can construct multi-sample tests to compare the spreads. Suppose we are in the general set-up where we have s independent random samples on M. Under the null hypothesis that all the samples come from some common distribution Q (say), it follows that

$$\sigma^{-1}\sqrt{n_j}(\hat{V}_j - V) \xrightarrow{\mathcal{L}} N(0, 1) \text{ independently}, \quad \forall j \leq s.$$

Here $\hat{V}_j$ is the sample j extrinsic variation, V is the extrinsic variation of Q, and $\sigma^2 = \text{Var}\{\rho^2(X_{11}, \mu_E)\}$, with μ_E being the extrinsic mean of Q. This implies that if $n_j/n \to p_j$, $0 < p_j < 1$, then the null distribution of the test statistic

$$\hat{\sigma}^{-2} \sum_{j=1}^{s} n_j(\hat{V}_j - \hat{V})^2 \tag{4.31}$$

is asymptotically $\mathcal{X}_{s-1}^2$. Here $\hat{V} = (1/n) \sum_j n_j \hat{V}_j$ and $\hat{\sigma}^2$ is the pooled sample variance of $\{\rho^2(X, \hat{\mu}_E)\}$, with $\hat{\mu}_E$ being the pooled sample extrinsic mean.

Proposition 4.7 can be used to compare the group means and variations jointly via an asymptotic $\mathcal{X}_{(s-1)(d+1)}^2$ statistic. The details are left to the reader.

4.7 Equivariant embedding

Among the possible embeddings, we seek out *equivariant embeddings* that preserve many of the geometric features of M. Recall that a *Lie group G* is a group that is also a differentiable manifold such that the group operations $(g_1, g_2) \to g_1 g_2$, $g \to g^{-1}$ are infinitely differentiable. We consider Lie groups G of diffeomorphisms $g: M \to M$ with multiplication being defined as composition of maps. Such *groups* are said to *act on M*.

Definition 4.8 For a Lie group H acting on a manifold M, an embedding $j: M \to \mathbb{R}^D$ is *H-equivariant* if there exists a group homomorphism $\phi : H \to GL(D, \mathbb{R})$ such that

$$j(hp) = \phi(h)j(p), \quad \forall p \in M, \quad \forall h \in H.$$

Here $GL(D, \mathbb{R})$ is the *general linear group* of all $D \times D$ nonsingular matrices.

One may think of an H-equivariant embedding $j: M \to \mathbb{R}^D$ as one that replicates the group action H on M by a corresponding action on $\tilde{M} = j(M)$, thus reproducing the geometry represented by the group action. For example, on a Riemannian manifold M, H may be a group of isometries and $\phi(H)$ may be a subgroup of the special orthogonal group $SO(\mathbb{R}^D)$.

4.8 Extrinsic analysis on the unit sphere S^d

An important and perhaps the simplest non-Euclidean manifold is the space of all directions in $\mathbb{R}^{d+1}$, which can be identified with the unit sphere S^d. Directional data analysis finds lots of applications in paleomagnetism and spatial statistics, and we shall see some such applications in this chapter.

The sphere can be embedded in $\mathbb{R}^{d+1}$ via the inclusion map

$$i: S^d \to \mathbb{R}^{d+1}, \quad i(x) = x.$$

The mapping $j = i$ is equivariant under the group $H = SO(\mathbb{R}^{d+1})$ acting on S^d, with $\phi(g) = g$ acting on $\mathbb{R}^{d+1}$. The extrinsic mean set of a probability measure Q on S^d is then the projection set of $\mu = \int_{\mathbb{R}^{d+1}} x \tilde{Q}(dx)$ on S^d, where $\tilde{Q}$ is Q regarded as a probability measure on $\mathbb{R}^{d+1}$. Note that μ is nonfocal if and only if $\mu \neq 0$. Then $P(\mu) = \frac{\mu}{\|\mu\|}$ and if $\mu = 0$, then its projection set is the entire sphere. The extrinsic variation of Q is

$$V = \int_{\mathbb{R}^{d+1}} \|x - \mu\|^2 \tilde{Q}(dx) + (\|\mu\| - 1)^2 = 2(1 - \|\mu\|).$$

If V_n denotes the sample extrinsic variation, it is easy to check that $\sqrt{n}(V_n - V)$ is asymptotically Normal if and only if $\mu \neq 0$.

The projection map $P: \mathbb{R}^{d+1} \to S^d$ is smooth on $\mathbb{R}^{d+1} \setminus \{0\}$. The Jacobian of its derivative at $\mu \in \mathbb{R}^{d+1}$ can be derived to be $\|\mu\|^{-1}(I_{d+1} - \|\mu\|^{-2}\mu\mu^T)$, where I_R is the identity matrix of order R. We will use just I for identity when its order is obvious. The tangent space at $m \in S^d$ is

$$T_m S^d = \{v \in \mathbb{R}^{d+1} : v'm = 0\},$$

that is, all vectors orthogonal to m. Then the derivative, or the differential, of P at μ can be expressed as

$$d_\mu P: \mathbb{R}^{d+1} \to T_{P(\mu)} S^d, \quad d_\mu P(v) = \|\mu\|^{-1}(I_{d+1} - \|\mu\|^{-2}\mu\mu')v.$$

Let $B(\mu)'$ denote the $d \times (d + 1)$ matrix of $d_\mu(P)$ with respect to the standard basis at $T_\mu \mathbb{R}^{d+1} \approx \mathbb{R}^{d+1}$. Note that $\mu' B(\mu) = 0$.

Suppose we have s samples X on S^d and want to perform hypothesis testing as in Sections 4.5 and 4.6. The asymptotic chi-squared statistic derived in equation (4.28) simplifies to

$$T_1 = \sum_{j=1}^{s} n_j \bar{X}'_j B (B'SB)^{-1} B' \bar{X}_j, \qquad (4.32)$$

where $\bar{X}_j = (1/n_j) \sum_i X_{ij}$, $\bar{X} = (1/n) \sum_j n_j \bar{X}_j$, $B \equiv B(\bar{X})$ is an (consistent) orthonormal basis for $T_{P(\bar{X})} S^d$, and $S = (1/n) \sum_{ij} (X_{ij} - \bar{X})(X_{ij} - \bar{X})'$.

When we have two groups and are only interested in testing the hypothesis that the group extrinsic means are the same, the asymptotic X_d^2 test statistic derived in equation (4.18) simplifies to

$$T_2 = (\bar{X}_1/\|\bar{X}_1\| - \bar{X}_2/\|\bar{X}_2\|)' B \hat{\Sigma}^{-1} B' (\bar{X}_1/\|\bar{X}_1\| - \bar{X}_2/\|\bar{X}_2\|), \qquad (4.33)$$

where B is as before,

$$\hat{\Sigma} = B' \left\{ \sum_j n_j^{-1} \|\bar{X}_j\|^{-2} (I - \|\bar{X}_j\|^{-2} \bar{X}_j \bar{X}'_j) S_j (I - \|\bar{X}_j\|^{-2} \bar{X}_j \bar{X}'_j) \right\} B,$$

and $S_j = (1/n_j) \sum_i (X_{ij} - \bar{X}_j)'(X_{ij} - \bar{X}_j)$, $j = 1, 2$. The bootstrap p-value in equation (4.29) can be expressed as

$$\Pr(T_2(X^*) > T_2(X)|X), \qquad (4.34)$$

where $\bar{X}_j^*$ denotes the mean of a bootstrap sample $X_{\cdot j}^*$ drawn with replacement from the jth sample X_{ij} ($i = 1, \ldots, n_j, j = 1, 2$), $\hat{\Sigma}^*$ is similar to $\hat{\Sigma}$ but with X replaced by resample X^*, and T_{2o} is the observed value of T_2 in equation (4.33). The other asymptotic X_d^2 statistic for comparing the two distributions derived in equation (4.15) simplifies to

$$T_1 = (\bar{X}_1 - \bar{X}_2)' B \hat{\Sigma}^{-1} B' (\bar{X}_1 - \bar{X}_2), \qquad (4.35)$$

where $\hat{\Sigma}$ is now $B'(\sum_j n_j^{-1} S_j) B$. Note that T_2 becomes $\tilde{T}_1$ when we replace $\|\bar{X}_j\|$ by $\|\bar{X}\|$ everywhere.

The asymptotic X_{s-1}^2 test statistic for comparing the extrinsic variations in equation (4.31) can be expressed as

$$T_3 = 4 \hat{\sigma}^{-2} \sum_{j=1}^{s} n_j \left(\|\bar{X}_j\| - (1/n) \sum n_j \|\bar{X}_j\| \right)^2, \qquad (4.36)$$

where $\hat{\sigma}^2$ is the pooled sample variance of $\left\| X - \bar{X}/\|\bar{X}\| \right\|^2$.

Since the extrinsic mean is the direction of the Euclidean mean while the variation is a bijection of its magnitude, comparing the mean and variation jointly is equivalent to comparing the Euclidean means via a $X_{(s-1)(d+1)}^2$ test.

In the next section, we present some real-life data on the two-dimensional sphere, where we apply these estimation and inference techniques.

An extension of the sphere is the manifold of all k mutually orthogonal directions in $\mathbb{R}^m$, which is known as the Stiefel manifold $V_{k,m}$ or $St_k(m)$ and is

$$V_{k,m} = St_k(m) = \{A \in M(m,k): A'A = I_k\}, \quad k \leq m,$$

where $M(m,k)$ is the space of all $m \times k$ real matrices. Hence $S^d = St_1(d+1)$. We will study this manifold in detail and perform statistical analysis on it in Chapter 10.

4.9 Applications on the sphere

4.9.1 Magnetization direction data

In this example from Embleton and McDonnell (1980), measurements of remanent magnetization in red silts and claystones are made at four locations. This results in independent samples from four group of directions on the sphere S^2, and the sample sizes are 36, 39, 16, and 16. Figure 4.1 shows the three-dimensional plot of the sample clouds.

The goal is to compare the magnetization direction distributions across the groups and test for any significant difference. We use the test statistic T_1 derived in equation (4.32) to compare the extrinsic mean directions and obtain the asymptotic X_6^2 p-value. The p-value is 0.06, suggesting that not enough evidence exists of significant differences in magnetization directions across the four locations at a 5% level of significance. The test statistic value and p-value are listed in Table 4.1. In example 7.7 of Fisher et al. (1987), a coordinate-based parametric test is conducted to compare mean directions in these data using a X_6^2 statistic whose p-value turns out to be

Table 4.1 *Test results from Section 4.9.1*

Group: (1,2,3,4)		
T_1=12.03	T_3=10.09	T_5=24.96
p-value=0.06	p-value=0.02	p-value=0.003
Group: (1,2)		
$\tilde{T}_1=T_2$=1.94	T_4=1.24	
X^2 p-value=0.38	Z p-value=0.22	
bootstrap p-value=0.37	bootstrap p-value=0.37	

0.089. Hence they reach the same conclusion. However, according to that book, because the sizes (16, 16) of two of the samples are small, in Example 7.10 a two-sample parametric test is performed to compare groups 1 and 2, obtaining a p-value around 0.05 (the exact value is not presented). The authors find the evidence inconclusive about what decision to take as they comment, "the hypothesis of a common mean direction is just tenable." When we compare the extrinsic means using the test statistics T_2 and T_1 derived in equations (4.33) and (4.35), respectively, we obtain an asymptotic X_2^2 p-value of 0.38, which is very high. The two statistics yield the same value up to four decimal places, with T_2 being slightly smaller. Hence we reach the conclusion that there is no difference in mean directions. Just to make sure, we also find the bootstrap p-value derived in equation (4.34). Based on 10^5 simulations, its value is 0.3684. We repeat the simulations several times, and each time the p-value exceeds 0.35. It is natural to expect no differences in distributions of a subcollection after having reached this conclusion from comparing all the groups simultaneously. This is strongly reflected by our various p-values. However, the analysis in Fisher et al. (1987) could not reach the same conclusion, suggesting that nonparametric methods perform better than the parametric analogs.

Figure 4.1, however, suggests some differences in spreads. To test that statistically, we next compare the extrinsic means and variations for the four groups jointly, which is equivalent to comparing the group Euclidean means. We get a very low p-value using an asymptotic X_9^2 test, suggesting strong evidence for differences. In Table 4.1, the statistic is called T_5. To

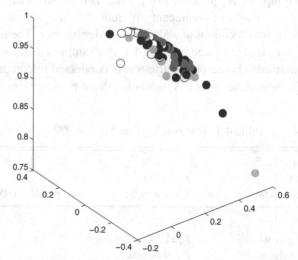

Figure 4.1 Three-dimensional coordinates of 4 groups in Section 4.9.1: 1(*black*), 2(*dark gray*), 3(*light gray*), 4(*open circle*).

confirm that this difference is due to spread and not mean, we use the χ_3^2 statistic T_3 obtained in equation (4.36) to compare the four group extrinsic variations. The asymptotic p-value is low, but not as low as that of the former test.

Our final conclusion is that there are significant differences in magnetization directional distributions across the four locations caused primarily by differences in spreads and not means. This example is interesting in that it is the only data set we have looked at where we find differences in distributions caused by variations and not means. We will return to this example in Chapter 14, where we use full likelihood-based nonparametric Bayes methods for discrimination.

When comparing the extrinsic variations for the first two groups, the asymptotically Normal p-value and various bootstrap p-values from 10^5 simulations are pretty high, suggesting no difference.

4.9.2 Volcano Location Data

The NOAA National Geophysical Data Center Volcano Location Database (1994) contains information on locations and characteristics of volcanoes across the globe. The locations are plotted in Figure 4.2 using latitude–longitude coordinates. We are interested in testing whether there is any association between the location and the type of volcano. We consider

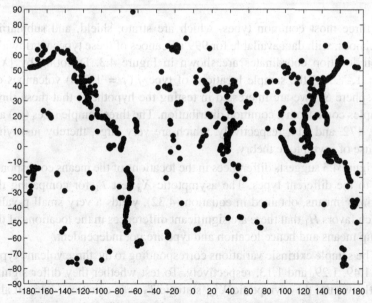

Figure 4.2 Latitude–longitude coordinates of volcano locations in Section 4.9.2.

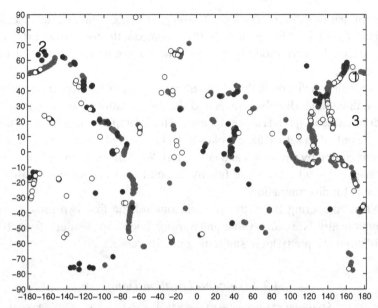

Figure 4.3 Coordinates of volcano locations by three major
types: strato circles (*gray*), shield circles (*black*), and submarine
(*open circles*). Sample extrinsic mean locations: 1, 2, 3. Full
sample extrinsic mean: open circle enclosing 1.

the three most common types, which are strato, shield, and submarine
volcanoes, with data available for 999 volcanoes of these types worldwide.
Their location coordinates are shown in Figure 4.3. Denoting by $\{X_{ij}\}$
$(i = 1, \ldots, n_j)$ the sample locations of type j $(j = 1, 2, 3)$ volcanoes on
the sphere S^2, we are interested in testing the hypothesis that these three
samples come from a common distribution. The three sample sizes (n_j) are
713, 172, and 114, respectively, which are very large, thereby justifying
the use of asymptotic theory.

Figure 4.3 suggests differences in the locations of the means corresponding to the different types. The asymptotic $\mathcal{X}_4^2$ test T_1 for comparing the
extrinsic means, obtained in equation (4.32), yields a very small p-value
which favors H_1 that there are significant differences in the locations of the
group means and hence location and type are not independent.

The sample extrinsic variations corresponding to the three volcano types
are 1.49, 1.29, and 1.13, respectively. To test whether they differ significantly, we use the asymptotic $\mathcal{X}_{s-1}^2$ statistic T_3 as in equation (4.36). It also
yields a very low p-value, suggesting significant differences.

Table 4.2 *Test results from Section 4.9.2*

Test statistic	$T_1 = 35.54$	$T_3 = 12.04$	$T_6 = 17.07$
p-value	$3.6e{-}7$	$2.4e{-}3$	$1.9e{-}3$

For comparison, we perform a coordinate-based test by comparing the means of the latitude–longitude coordinates of the three samples using a $\mathcal{X}_4^2$ statistic; we call it T_6. The asymptotic p-value is larger by orders of magnitude than its coordinate-free counterpart, but still significant. Coordinate-based methods, however, can be very misleading because of the discontinuity at the boundaries. They heavily distort the geometry of the sphere, which is evident from the figures. The values of the statistics and p-values are listed together in Table 4.2.

We will return to this example in Chapter 14, where we compare the distributions via nonparametric Bayes methods.

4.10 References

Nonparametric inference for location, in terms of what is called the extrinsic mean in this book, was carried out by Hendriks and Landsman (1998) on regular submanifolds of the Euclidean space $\mathbb{R}^k$ with applications to spheres and Stiefel manifolds. Independently, Patrangenaru (1998) derived nonparametric extrinsic inference for general manifolds, with further developments provided in Bhattacharya and Patrangenaru (2003, 2005). The latter articles also developed applications to the shape spaces of Kendall (1984). Some of the precise forms of the statistics and their bootstrap versions derived in this chapter are new, while others are slight modifications of those obtained in Bhattacharya (2008a) and Bhattacharya and Bhattacharya (2008a, 2009). Also see Bandulasiri et al. (2009).

The authors thank David Dunson for providing them with the volcano data.

There is an extensive prior literature on directional statistics and on shape spaces, using parametric models for the most part. Comprehensive accounts of these may be found in Mardia and Jupp (2000) and Dryden and Mardia (1998). Among many significant contributions, one may mention Watson (1965, 1983), Prentice (1984), Kent (1992, 1994), Dryden and Mardia (1992), Dryden et al. (1997), Goodall (1991), Prentice and Mardia (1995); and Bookstein (1978, 1989, 1991) – made pioneering contributions to statistical shape analysis with applications to morphometrics. Bookstein used transformations to bring some of the landmarks to a standard position and used the positions of the other landmarks so transformed, to study the differences in shapes. Beran (1968) obtained general nonparametric tests for uniformity on compact homogeneous spaces, and Beran and Fisher (1998) developed nonparametric tests for comparing distributions on the axial space $\mathbb{R}P^d$.

5

Intrinsic analysis on manifolds

Let (M, g) be a complete connected Riemannian manifold of dimension d with metric tensor g. Then a natural choice for the distance metric ρ discussed in Chapter 3 is the geodesic distance d_g on M. Statistical analysis on M using this distance is called *intrinsic analysis*. Unless otherwise stated, we consider $f(u) = u^2$ in the definition of the Fréchet function in equation (3.1). However, we will consider Fréchet functions determined by $f(u) = u^\alpha$ as well for suitable $\alpha \geq 1$ (see Section 8.5). Asymptotic distributions of the intrinsic sample mean and variation are derived and deployed for estimation and two-sample tests.

5.1 Intrinsic mean and variation

Let Q be a probability distribution on M with the finite Fréchet function

$$F(p) = \int_M d_g^2(p, m) Q(dm). \tag{5.1}$$

Let $X_1, \ldots, X_n$ be an i.i.d. sample from Q.

Definition 5.1 The Fréchet mean set of Q for the Fréchet function (5.1) is called the *intrinsic mean set* of Q, and the Fréchet variation of Q is called the *intrinsic variation* of Q. The Fréchet mean set of the empirical distribution Q_n is called the *sample intrinsic mean set*, and the sample Fréchet variation is called the *sample intrinsic variation*.

Before proceeding further, let us define a few technical terms related to Riemannian manifolds that we will use extensively in this chapter. For an introduction to Riemannian manifolds, see Appendix B, and for details, see Do Carmo (1992), Gallot et al. (1990), or Lee (1997). For an introductory account, see Millman and Parker (1977, chapters 4 and 7).

1. *Geodesics*: These are curves γ on the manifold with zero acceleration. They are locally length-minimizing curves. For example, great circles are the geodesics on the sphere and straight lines are the geodesics in $\mathbb{R}^d$.

2. *Exponential map*: For $p \in M$, $v \in T_pM$, we define $\exp_p(v) = \gamma(1)$, where γ is a geodesic with $\gamma(0) = p$ and $\dot{\gamma}(0) = v$.

3. *Cut-locus*: For a point $p \in M$, we define the cut-locus $C(p)$ of p as the set of points of the form $\gamma(t_0)$, where γ is a unit speed geodesic starting at p and t_0 is the supremum of all $t > 0$ such that γ is length minimizing from p to $\gamma(t)$. For example, $C(p) = \{-p\}$ on the sphere.

4. *Sectional curvature*: For a somewhat informal definition, recall the notion of the Gaussian curvature of a two-dimensional surface. On a Riemannian manifold M, choose a pair of linearly independent vectors $u, v \in T_pM$. A two-dimensional submanifold of M, or a section π, is swept out by the set of all geodesics starting at p and with initial velocities lying in the two-dimensional subspace of T_pM spanned by u, v. The Gaussian curvature at p of this submanifold is called the sectional curvature at p of the section π.

5. *Injectivity radius*: The injectivity radius of M is defined as

$$\mathrm{inj}(M) = \inf\{d_g(p, C(p)) : p \in M\}.$$

For example, the sphere of radius 1 has injectivity radius equal to π.

6. *Convex set*: A subset S of M is said to be convex if, for any $x, y \in S$, there exists a unique shortest geodesic in M joining x and y that lies entirely in S.

Also let $r_* = \min\{\mathrm{inj}(M), \frac{\pi}{\sqrt{\overline{C}}}\}$, where $\overline{C}$ is the least upper bound of sectional curvatures of M if this upper bound is positive, and $\overline{C} = 0$ otherwise. The exponential map at p is injective on $\{v \in T_p(M) : |v| < r_*\}$. By $B(p, r)$ we will denote an open ball with center $p \in M$ and geodesic radius r, and $\overline{B}(p, r)$ will denote its closure. It is known that $B(p, r)$ is convex whenever $r \leq \frac{r_*}{2}$ (see Kendall, 1990).

When Q has a unique intrinsic mean μ_I, it follows from Corollary 3.4 and Remark 3.5 that the *sample intrinsic mean* μ_{nI}, that is, a measurable selection from the sample intrinsic mean set, is a consistent estimator of μ_I. Broad conditions for the existence of a unique intrinsic mean are not known. From results due to Karcher (1977) and Le (2001), it follows that if the support of Q is in a geodesic ball of radius $\frac{r_*}{4}$, that is, $\mathrm{supp}(Q) \subseteq B(p, \frac{r_*}{4})$, then Q has a unique intrinsic mean. This result has been substantially extended by Kendall (1990), who showed that if $\mathrm{supp}(Q) \subseteq B(p, \frac{r_*}{2})$, then there is a unique local minimum of the Fréchet function F in that ball. We define the *local intrinsic mean* of Q in a neighborhood $V \subset M$ as the minimizer, if unique, of the Fréchet function (5.1) restricted to V, assuming $Q(V) = 1$.

Proposition 5.2   *Let Q have support in $B(p, \frac{r_*}{2})$ for some $p \in M$. Then Q has a unique local intrinsic mean μ_I in $B(p, \frac{r_*}{2})$ and the local sample intrinsic mean μ_{nI} in $B(p, \frac{r_*}{2})$ is a strongly consistent estimator of μ_I.*

Proof   Follows from Kendall (1990) and Corollary 3.4.   □

5.2 Asymptotic distribution of the sample intrinsic mean

The asymptotic distribution of the sample intrinsic mean follows from Theorem 3.10 once we verify assumptions A1–A6. Theorem 5.3 gives sufficient conditions for those assumptions to hold. In the statement of the theorem, the usual partial order $A \geq B$ between the $d \times d$ symmetric matrices A, B, means that $A - B$ is nonnegative definite. For Jacobi fields, used in the proof of Theorem 5.3, see Do Carmo (1992, chapter 5).

Theorem 5.3   *Suppose* $\mathrm{supp}(Q) \subseteq B(p, \frac{r_*}{2})$ *for some* $p \in M$. *Let* $\phi = \exp_{\mu_I}^{-1} \colon B(p, \frac{r_*}{2}) \longrightarrow T_{\mu_I}M (\approx \mathbb{R}^d)$. *Then the map* $y \mapsto h(x, y) = d_g^2(\phi^{-1}x, \phi^{-1}y)$ *is twice continuously differentiable in a neighborhood of* 0. *In terms of normal coordinates with respect to a chosen orthonormal basis for* $T_{\mu_I}M$, *one has*

$$D_r h(x, 0) = -2x^r, \quad 1 \leq r \leq d, \tag{5.2}$$

$$[D_r D_s h(x, 0)] \geq 2\left[\left\{\left(\frac{1 - f(|x|)}{|x|^2}\right) x^r x^s + f(|x|)\delta_{rs}\right\}\right]_{1 \leq r, s \leq d}. \tag{5.3}$$

Here $x = (x^1, \ldots, x^d)'$, $|x| = \sqrt{(x^1)^2 + (x^2)^2 + \cdots + (x^d)^2}$ *and*

$$f(y) = \begin{cases} 1 & \text{if } \overline{C} = 0, \\[2mm] \sqrt{\overline{C}}\, y \dfrac{\cos(\sqrt{\overline{C}}y)}{\sin(\sqrt{\overline{C}}y)} & \text{if } \overline{C} > 0, \\[2mm] \sqrt{-\overline{C}}\, y \dfrac{\cosh(\sqrt{-\overline{C}}y)}{\sinh(\sqrt{-\overline{C}}y)} & \text{if } \overline{C} < 0. \end{cases} \tag{5.4}$$

There is equality in equation (5.3) when M has a constant sectional curvature $\overline{C}$, and in this case Λ in Theorem 3.10 has the expression

$$\Lambda_{rs} = 2\mathrm{E}\left\{\left(\frac{1 - f(|\tilde{X}_1|)}{|\tilde{X}_1|^2}\right)\tilde{X}_1^r \tilde{X}_1^s + f(|\tilde{X}_1|)\delta_{rs}\right\}, \quad 1 \leq r, s \leq d, \tag{5.5}$$

It is positive definite if $\mathrm{supp}(Q) \subseteq B(\mu_I, \frac{r_*}{2})$.

Proof   Let $\gamma(s)$ be a geodesic, $\gamma(0) = \mu_I$. Define $c(s, t) = \exp_m(t \exp_m^{-1} \gamma(s))$, $s \in [0, \epsilon]$, $t \in [0, 1]$, as a smooth variation of γ through geodesics lying entirely in $B(p, \frac{r_x}{2})$. Let $T = \frac{\partial}{\partial t}c(s,t)$, $S = \frac{\partial}{\partial s}c(s,t)$. Since $c(s,0) = m$, $S(s,0) = 0$; and since $c(s,1) = \gamma(s)$, $S(s,1) = \dot{\gamma}(s)$. Also, $\langle T, T \rangle = d_g^2(\gamma(s),m)$ is independent of t, and the covariant derivative $D_t T$ vanishes because $t \mapsto c(s,t)$ is a geodesic (for each s). Then

$$d_g^2(\gamma(s),m) = \langle T(s,t), T(s,t) \rangle = \int_0^1 \langle T(s,t), T(s,t) \rangle dt.$$

Hence $d_g^2(\gamma(s),m)$ is C^∞ smooth, and using the symmetry of the connection on a parametrized surface (see lemma 3.4, p. 68, in Do Carmo, 1992), we get

$$\frac{d}{ds}d_g^2(\gamma(s),m) = 2\int_0^1 \langle D_s T, T \rangle dt = 2\int_0^1 \frac{d}{dt}\langle T, S \rangle dt$$

$$= 2\langle T(s,1), S(s,1) \rangle = -2\langle \exp_{\gamma(s)}^{-1} m, \dot{\gamma}(s) \rangle. \qquad (5.6)$$

Substituting $s = 0$ in equation (5.6), we get expressions for $D_r h(x,0)$ as in equation (5.2). Also,

$$\frac{d^2}{ds^2}d_g^2(\gamma(s),m) = 2\langle D_s T(s,1), S(s,1) \rangle$$

$$= 2\langle D_t S(s,1), S(s,1) \rangle = 2\langle D_t J_s(1), J_s(1) \rangle, \qquad (5.7)$$

where $J_s(t) = S(s,t)$. Note that J_s is a Jacobi field along $c(s,\cdot)$ with $J_s(0) = 0$, $J_s(1) = \dot{\gamma}(s)$. Let $J_s^\perp$ and J_s^- be the normal and tangential components of J_s. Let η be a unit speed geodesic in M and J a normal Jacobi field along η, $J(0) = 0$. Define

$$u(t) = \begin{cases} t & \text{if } \bar{C} = 0, \\ \frac{\sin(\sqrt{\bar{C}}t)}{\sqrt{\bar{C}}} & \text{if } \bar{C} > 0, \\ \frac{\sinh(\sqrt{-\bar{C}}t)}{\sqrt{-\bar{C}}} & \text{if } \bar{C} < 0. \end{cases}$$

Then $u''(t) = -\bar{C}u(t)$ and

$$(|J|'u - |J|u')'(t) = (|J|'' + \bar{C}|J|)u(t).$$

By exact differentiation and the Schwarz inequality, it is easy to show that $|J|'' + \bar{C}|J| \geq 0$. Hence $(|J|'u - |J|u')'(t) \geq 0$ whenever $u(t) \geq 0$. This implies that $|J|'u - |J|u' \geq 0$ if $t \leq t_0$, where u is positive on $(0,t_0)$. Also, $|J|' = \frac{\langle J', J \rangle}{|J|}$. Therefore, $\langle J(t), D_t J(t) \rangle \geq \frac{u'(t)}{u(t)}|J(t)|^2$, $\forall t < t_0$. If we drop the unit speed assumption on η, we get

$$\langle J(1), D_t J(1) \rangle \geq |\dot{\eta}|\frac{u'(|\dot{\eta}|)}{u(|\dot{\eta}|)}|J(1)|^2 \quad \text{if } |\dot{\eta}| < t_0. \qquad (5.8)$$

Here $t_0 = \infty$ if $\bar{C} \le 0$ and equals $\frac{\pi}{\sqrt{\bar{C}}}$ if $\bar{C} > 0$. When M has a constant sectional curvature $\bar{C}$, $J(t) = u(t)E(t)$, where E is a parallel normal vector field along η. Hence

$$\langle J(t), D_t J(t) \rangle = u(t)u'(t)|E(t)|^2 = \frac{u'(t)}{u(t)}|J(t)|^2.$$

If we drop the unit speed assumption, we get

$$\langle J(t), D_t J(t) \rangle = |\dot{\eta}| \frac{u'(|\dot{\eta}|t)}{u(|\dot{\eta}|t)}|J(t)|^2. \tag{5.9}$$

Since $J_s^\perp$ is a normal Jacobi field along the geodesic $c(s, \cdot)$, from equations (5.8) and (5.9) it follows that

$$\langle J_s^\perp(1), D_t J_s^\perp(1) \rangle \ge f(d(\gamma(s), m))|J_s^\perp(1)|^2 \tag{5.10}$$

with equality in equation (5.10) when M has a constant sectional curvature $\bar{C}$, with f being defined in equation (5.4).

Next suppose J is a Jacobi field along a geodesic η, $J(0) = 0$ and let $J^-(t)$ be its tangential component. Then $J^-(t) = \lambda t \dot{\eta}(t)$, where $\lambda t = \frac{\langle J(t), \dot{\eta}(t) \rangle}{|\dot{\eta}|^2}$, with λ being independent of t. Hence

$$
\begin{aligned}
(D_t J)^-(t) &= \frac{\langle D_t J(t), \dot{\eta}(t) \rangle}{|\dot{\eta}|^2} \dot{\eta}(t) \\
&= \frac{d}{dt}\left(\frac{\langle J(t), \dot{\eta}(t) \rangle}{|\dot{\eta}|^2}\right)\dot{\eta}(t) = \lambda \dot{\eta}(t) = D_t(J^-)(t)
\end{aligned}
\tag{5.11}
$$

and

$$
\begin{aligned}
D_t|J^-|^2(1) &= 2\lambda^2|\dot{\eta}|^2 = 2\frac{\langle J(1), \dot{\eta}(1) \rangle^2}{|\dot{\eta}(1)|^2} \\
&= D_t\langle J, J^- \rangle(1) = \langle D_t J(1), J^-(1) \rangle + |J^-(1)|^2,
\end{aligned}
$$

which implies that

$$\langle D_t J(1), J^-(1) \rangle = 2\frac{\langle J(1), \dot{\eta}(1) \rangle^2}{|\dot{\eta}(1)|^2} - |J^-(1)|^2 = \frac{\langle J(1), \dot{\eta}(1) \rangle^2}{|\dot{\eta}(1)|^2}. \tag{5.12}$$

Apply equations (5.11) and (5.12) to the Jacobi field J_s to get

$$D_t(J_s^-)(1) = (D_t J_s)^-(1) = J_s^-(1) = \frac{\langle J_s(1), T(s, 1) \rangle}{|T(s, 1)|^2}T(s, 1), \tag{5.13}$$

$$\langle D_t J_s(1), J_s^-(1) \rangle = \frac{\langle J_s(1), T(s, 1) \rangle^2}{|T(s, 1)|^2}. \tag{5.14}$$

Using equations (5.10), (5.13), and (5.14), equation (5.7) becomes

$$\frac{d^2}{ds^2}d_g^2(\gamma(s),m) = 2\langle D_t J_s(1), J_s(1)\rangle$$

$$= 2\langle D_t J_s(1), J_s^-(1)\rangle + 2\langle D_t J_s(1), J_s^\perp(1)\rangle$$

$$= 2\langle D_t J_s(1), J_s^-(1)\rangle + 2\langle D_t(J_s^\perp)(1), J_s^\perp(1)\rangle$$

$$\geq 2\frac{< J_s(1), T(s,1) >^2}{|T(s,1)|^2} + 2f(|T(s,1)|)|J_s^\perp(1)|^2 \qquad (5.15)$$

$$= 2\frac{\langle J_s(1), T(s,1)\rangle^2}{|T(s,1)|^2} + 2f(|T(s,1)|)|J_s(1)|^2$$

$$\quad - 2f(|T(s,1)|)\frac{\langle J_s(1), T(s,1)\rangle^2}{|T(s,1)|^2}$$

$$= 2f(d_g(\gamma(s),m))|\dot\gamma(s)|^2$$

$$\quad + 2(1 - f(d_g(\gamma(s),m)))\frac{\langle\dot\gamma(s), \exp_{\gamma(s)}^{-1} m\rangle^2}{d_g^2(\gamma(s),m)} \qquad (5.16)$$

with equality in (5.15) when M has a constant sectional curvature $\bar C$. Substituting $s = 0$ in equation (5.16), we get a lower bound for $[D_r D_s h(x, 0)]$ as in equation (5.3) and an exact expression for $D_r D_s h(x, 0)$ when M has a constant sectional curvature. To see this, let $\dot\gamma(0) = v$. Then, writing $m = \phi^{-1}(x)$ and $\gamma(s) = \phi^{-1}(sv)$, one has

$$\frac{d^2}{ds^2}d_g^2(\gamma(s),m)\Big|_{s=0} = \frac{d^2}{ds^2}d_g^2(\phi^{-1}(x), \phi^{-1}(sv))\Big|_{s=0}$$

$$= \frac{d^2}{ds^2}h(x, sv)\Big|_{s=0} = \sum_{r,s=1}^d v_r v_s D_r D_s h(x, 0).$$

Because $d^2(\gamma(s), m)$ is twice continuously differentiable and Q has compact support, using the Lebesgue dominated convergence theorem (DCT), we get

$$\frac{d^2}{ds^2}F(\gamma(s))\Big|_{s=0} = \int \frac{d^2}{ds^2}d^2(\gamma(s),m)\Big|_{s=0}Q(dm). \qquad (5.17)$$

Then equation (5.5) follows from equation (5.16). If $\text{supp}(Q) \subseteq B(\mu_I, \frac{r_*}{2})$, then the expression in equation (5.16) is strictly positive at $s = 0$ for all $m \in \text{supp}(Q)$, and hence Λ is positive definite. This completes the proof. $\quad\square$

Corollary 5.4 *Suppose* $\text{supp}(Q) \subseteq B(\mu_I, \frac{r_*}{2})$, *with* μ_I *being the local intrinsic mean of* Q. *Let* $X_1, \ldots, X_n$ *be an i.i.d. sample from* Q *and let* $\tilde X_j = \phi(X_j)$, $j = 1, \ldots, n$, *be the normal coordinates of the sample with* ϕ *as*

in Theorem 5.3. Let μ_{nI} be the sample intrinsic mean in $B(\mu_I, \frac{r_*}{2})$. Then (a)
$E(\tilde{X}_1) = 0$ *and (b)*

$$\sqrt{n}\,\phi(\mu_{nI}) \xrightarrow{\mathcal{L}} N(0, \Lambda^{-1}\Sigma\Lambda^{-1}),$$

where $\Sigma = 4E(\tilde{X}_1\tilde{X}_1')$ and Λ is as derived in Theorem 5.3.

Proof Follows from Theorem 5.3 and Theorem 3.10.   □

Remark 5.5 As the proof shows, result (a) of Corollary 5.4 is true even without the support restriction on Q as long as $Q(C(\mu_I)) = 0$, where μ_I is a local minimum of the Fréchet function (5.1) and $C(\mu_I)$ denotes its cut-locus. This holds, for example, on compact Riemannian manifolds for all absolutely continuous distributions Q. Also see Oller and Corcuear (1995).

From Corollary 5.4, it follows that the sample intrinsic mean μ_{nI} satisfies $\frac{1}{n}\sum_{i=1}^{n} \exp_{\mu_{nI}}^{-1}(X_i) = 0$ and hence is a fixed point of $f: M \rightarrow M$, $f(p) = \exp_p\{\frac{1}{n}\sum_{i=1}^{n} \exp_p^{-1}(X_i)\}$. Using this fact, we can build a fixed point algorithm to compute μ_{nI}. This is derived in Le (2001). There it is also shown using the Banach Fixed Point Theorem that this algorithm will converge if the data lie in a geodesic ball of radius $\frac{r_*}{8}$.

As in Sections 3.4, 4.5, and 4.6, if Σ is nonsingular, we can construct asymptotic chi-squared statistics and pivotal bootstrapped confidence regions for μ_I. Note that Σ is nonsingular if $Q \circ \phi^{-1}$ has support in no smaller dimensional affine subspace of $\mathbb{R}^d$. That condition holds if, for example, Q has a density with respect to the volume measure on M.

Alternatively, one may consider the statistic

$$T_n = d_g^2(\mu_{nI}, \mu_I).$$

Then $T_n = \|\phi(\mu_{nI})\|^2$, and hence, from Corollary 5.4, it follows that

$$nT_n \xrightarrow{\mathcal{L}} \sum_{i=1}^{d} \lambda_i Z_i^2,$$

where $\lambda_1 \leq \lambda_2 \leq \cdots \leq \lambda_d$ are the eigenvalues of $\Lambda^{-1}\Sigma\Lambda^{-1}$ and $Z_1, \ldots, Z_d$ are i.i.d. $N(0, 1)$. Using this statistic, an asymptotic level $(1 - \alpha)$ confidence set for μ_I is given by

$$\{\mu_I : nT_n \leq \hat{c}_{1-\alpha}\}, \tag{5.18}$$

where $\hat{c}_{1-\alpha}$ is the estimated upper $(1 - \alpha)$-quantile of the distribution of $\sum_{i=1}^{d} \hat{\lambda}_i Z_i^2$, with $\hat{\lambda}_i$ being the sample estimate of λ_i, $i = 1, 2, \ldots, d$ and $(Z_1, Z_2, \ldots)$ a sequence of i.i.d. $N(0, 1)$ random variables independent of the original sample $X_1, \ldots, X_n$. A corresponding bootstrapped confidence region can be obtained by replacing $\hat{c}_{1-\alpha}$ by the upper $(1-\alpha)$-quantile of the bootstrapped values of nT_n. The advantage of using this confidence region over the region in equation (3.15) is that it is easier to compute and visualize, and it does not require Σ to be nonsingular. However, the test based on the CLT is more efficient and is to be preferred under the hypothesis of Corollary 5.4.

5.3 Intrinsic analysis on S^d

Consider the space of all directions in $\mathbb{R}^{d+1}$. Because any direction has a unique point of intersection with the unit sphere S^d in $\mathbb{R}^{d+1}$, this space can be identified with S^d as

$$S^d = \{p \in \mathbb{R}^{d+1}: \|p\| = 1\}.$$

At each $p \in S^d$, we endow the tangent space

$$T_p S^d = \{v \in \mathbb{R}^{d+1}: v'p = 0\}$$

with the metric tensor $g_p: T_p S^d \times T_p S^d \to \mathbb{R}$ as the restriction of the scalar product at p of the tangent space of $\mathbb{R}^d$: $g_p(v_1, v_2) = v_1' v_2$. Then g is a smooth metric tensor on the tangent bundle

$$T S^d = \{(p, v): p \in S^d, v \in \mathbb{R}^{d+1}, v'p = 0\}.$$

The geodesics are the great circles,

$$\gamma_{p,v}(t) = \cos(t)p + \sin(t)v, \quad -\pi < t \le \pi.$$

Here $\gamma_{p,v}(.)$ is the great circle starting at p at $t = 0$ in the direction of the unit vector v. The exponential map, $\exp: T_p S^d \to S^d$, is given by

$$\exp_p(0) = p,$$

$$\exp_p(v) = \cos(\|v\|)p + \sin(\|v\|)\frac{v}{\|v\|}, \quad v \ne 0.$$

The inverse of the exponential map on $S^d \backslash \{-p\}$ into $T_p S^d$ has the expression

$$\exp_p^{-1}(q) = \frac{\arccos(p'q)}{\sqrt{1 - (p'q)^2}}[q - (p'q)p] \quad (q \ne p, -p),$$

$$\exp_p^{-1}(p) = 0.$$

The geodesic distance between p and q is given by

$$d_g(p, q) = \arccos(p'q),$$

which lies in $[0, \pi]$. Hence S^d has an injectivity radius of π. It also has a constant sectional curvature of 1, and therefore $r_* = \pi$.

Let Q be a probability distribution on S^d. It follows from Proposition 5.2 that if supp(Q) lies in an open geodesic ball of radius $\frac{\pi}{2}$, then it has a unique intrinsic mean μ_I in that ball. If $X_1, \ldots, X_n$ is an i.i.d. random sample from Q, then the sample intrinsic mean μ_{nI} in that ball is a strongly consistent estimator of μ_I. From Corollary 5.4 it follows that

$$\sqrt{n}\ \phi(\mu_{nI}) \overset{\mathcal{L}}{\longrightarrow} N(0, \Lambda^{-1}\Sigma\Lambda^{-1}),$$

where $\Sigma = 4E[\phi(X_1)\phi(X_1)']$. To get an expression for ϕ, pick an orthonormal basis $\{v_1, \ldots, v_d\}$ for $T_{\mu_I}S^d$. For $x \in S^d$, $|x'\mu_I| < 1$, we have

$$\exp_{\mu_I}^{-1}(x) = \frac{\arccos(x'\mu_I)}{\sqrt{1 - (x'\mu_I)^2}}[x - (x'\mu_I)\mu_I].$$

Then

$$\phi(x) = y \equiv (y^1, \ldots, y^d)',$$

where $\exp_{\mu_I}^{-1}(x) = \sum_{r=1}^d y^r v_r$, so that

$$y^r = \frac{\arccos(x'\mu_I)}{\sqrt{1 - (x'\mu_I)^2}}(x'v_r), \quad r = 1, 2, \ldots, d.$$

From Theorem 5.3 we get the expression for Λ as

$$\Lambda_{rs} = 2E\left[\frac{1}{[1 - (X_1'\mu_I)^2]}\left(1 - \frac{\arccos(X_1'\mu_I)}{\sqrt{1 - (X_1'\mu_I)^2}}(X_1'\mu_I)\right)(X_1'v_r)(X_1'v_s) \right.$$
$$\left. + \frac{\arccos(X_1'\mu_I)}{\sqrt{1 - (X_1'\mu_I)^2}}(X_1'\mu_I)\delta_{rs}\right], \quad 1 \le r \le s \le d.$$

Note that Λ is nonsingular if supp(Q) $\subseteq B(\mu_I, \frac{\pi}{2})$.

5.4 Two-sample intrinsic tests

In this section we will construct nonparametric tests to compare the intrinsic means and variations of two probability distributions Q_1 and Q_2 on M. This can be used to distinguish between the two distributions.

5.4.1 Independent samples

Let $X_1, \ldots, X_{n_1}$ and $Y_1, \ldots, Y_{n_2}$ be two i.i.d. samples from Q_1 and Q_2, respectively, that are mutually independent. Let μ_i and V_i denote the intrinsic mean and variation of Q_i, $i = 1, 2$, respectively. Similarly denote by $\hat{\mu}_i$ and $\hat{V}_i$ the sample intrinsic mean and variation.

First we test the hypothesis $H_0 : \mu_1 = \mu_2 = \mu$, say, against $H_1 : \mu_1 \neq \mu_2$. We assume that under H_0 both Q_1 and Q_2 have support in $B(\mu, \frac{r}{2})$, so that the normal coordinates of the sample intrinsic means have an asymptotic Normal distribution. Let $\phi(\hat{\mu}_i)$, $i = 1, 2$, where $\phi = \exp_\mu^{-1}$ are the normal coordinates of the sample means in $T_\mu M$ ($\approx \mathbb{R}^d$). It follows from Corollary 5.4 that

$$\sqrt{n_i}\, \phi(\hat{\mu}_i) \xrightarrow{\mathcal{L}} N(0, \Lambda_i^{-1}\Sigma_i\Lambda_i^{-1}), \quad i = 1, 2 \tag{5.19}$$

as $n_i \to \infty$. Let $n = n_1 + n_2$ be the pooled sample size. Then, if $\frac{n_1}{n} \to \theta$, $0 < \theta < 1$, it follows from equation (5.19) assuming H_0 to be true that

$$\sqrt{n}\, (\phi(\hat{\mu}_1) - \phi(\hat{\mu}_2)) \xrightarrow{\mathcal{L}} N\left(0, \frac{1}{\theta}\Lambda_1^{-1}\Sigma_1\Lambda_1^{-1} + \frac{1}{1-\theta}\Lambda_2^{-1}\Sigma_2\Lambda_2^{-1}\right). \tag{5.20}$$

Estimate μ by the pooled sample intrinsic mean $\hat{\mu}$, coordinates ϕ by $\hat{\phi} \equiv \exp_{\hat{\mu}}^{-1}$, and Λ_i and Σ_i by their sample analogs $\hat{\Lambda}_i$ and $\hat{\Sigma}_i$, respectively. Denote by μ_{ni} the coordinates $\hat{\phi}(\hat{\mu}_i)$, $i = 1, 2$, of the two sample intrinsic means. Because under H_0, $\hat{\mu}$ is a consistent estimator of μ, it follows from equation (5.20) that the statistic

$$T_{n1} = n(\mu_{n1} - \mu_{n2})'\hat{\Sigma}^{-1}(\mu_{n1} - \mu_{n2}), \tag{5.21}$$

where

$$\hat{\Sigma} = n\left(\frac{1}{n_1}\hat{\Lambda}_1^{-1}\hat{\Sigma}_1\hat{\Lambda}_1^{-1} + \frac{1}{n_2}\hat{\Lambda}_2^{-1}\hat{\Sigma}_2\hat{\Lambda}_2^{-1}\right),$$

converges in distribution to the chi-squared distribution with d degrees of freedom, with d being the dimension of M, that is,

$$T_{n1} \xrightarrow{\mathcal{L}} \mathcal{X}_d^2.$$

Hence we reject H_0 at the asymptotic level α if $T_{n1} > \mathcal{X}_d^2(1 - \alpha)$.

Next we test the hypothesis $H_0 : V_1 = V_2 = V$, say, against $H_1 : V_1 \neq V_2$. We assume that the hypothesis of Theorem 3.11 holds so that the sample intrinsic variations have asymptotic Normal distributions. Then, under H_0, as $n_i \to \infty$,

$$\sqrt{n_i}\, (\hat{V}_i - V) \xrightarrow{\mathcal{L}} N(0, \sigma_i^2), \tag{5.22}$$

where $\sigma_i^2 = \int_M (d_g^2(x, \mu_i) - V)^2 Q_i(dx)$, $i = 1, 2$. Suppose $\frac{n_1}{n} \to \theta$, $0 < \theta < 1$. Then it follows from equation (5.22) assuming H_0 to be true that

$$\sqrt{n}\, (\hat{V}_1 - \hat{V}_2) \xrightarrow{\mathcal{L}} N\left(0, \frac{\sigma_1^2}{\theta} + \frac{\sigma_2^2}{1-\theta}\right),$$

so that

$$T_{n2} = \frac{\hat{V}_1 - \hat{V}_2}{\sqrt{\frac{s_1^2}{n_1} + \frac{s_2^2}{n_2}}} \xrightarrow{\mathcal{L}} N(0, 1)$$

as $n \to \infty$. Here $s_1^2 = \frac{1}{n_1} \sum_{j=1}^{n_1} (d_g^2(X_j, \mu_1) - \hat{V}_1)^2$ and $s_2^2 = \frac{1}{n_2} \sum_{j=1}^{n_2} (d_g^2(Y_j, \mu_2) - \hat{V}_2)^2$ are the sample estimates of σ_1^2 and σ_2^2, respectively. For a test of asymptotic size α, we reject H_0 if $|T_{n2}| > Z_{1-\frac{\alpha}{2}}$, where $Z_{1-\frac{\alpha}{2}}$ is the upper $(1 - \frac{\alpha}{2})$-quantile of the standard Normal distribution.

5.4.2 Matched pair samples

Next consider the case when $(X_1, Y_1), \ldots, (X_n, Y_n)$ is an i.i.d. sample from some distribution Q on $\bar{M} = M \times M$. Such a sample is called a matched pair sample and arises when, for example, two different treatments are applied to each subject in the sample. An example of a matched pair sample of shapes is considered in Chapter 9.

Let the X_j come from some distribution Q_1 while the Y_j come from some distribution Q_2 on M. Our objective is to distinguish between Q_1 and Q_2 by comparing the sample intrinsic means and variations. Because the X and Y samples are not independent, we cannot apply the methods of Section 5.4.1. Instead we do our analyses on the Riemannian manifold $\bar{M}$. As in Section 5.4.1, we will denote by μ_i and V_i the intrinsic mean and variation of Q_i $(i = 1, 2)$, respectively, and by $\hat{\mu}_i$ and $\hat{V}_i$ the sample intrinsic mean and variation.

First we test the hypothesis $H_0 : \mu_1 = \mu_2 = \mu$, say, against $H_1 : \mu_1 \neq \mu_2$. We assume that under H_0 both Q_1 and Q_2 have support in $B(\mu, \frac{r}{2})$. Consider the coordinate map Φ on $\bar{M}$ given by

$$\Phi(m_1, m_2) = (\phi(m_1), \phi(m_2)), \quad m_1, m_2 \in M,$$

where $\phi = \exp_\mu^{-1}$. It follows from Corollary 5.4 that under H_0

$$\sqrt{n} \begin{pmatrix} \phi(\hat{\mu}_1) \\ \phi(\hat{\mu}_2) \end{pmatrix} \xrightarrow{\mathcal{L}} N(0, \Gamma), \tag{5.23}$$

where $\Gamma = \Lambda^{-1} \Sigma \Lambda^{-1}$ and Σ, Λ are obtained from Theorem 5.3 as follows. For $x = (x_1, x_2)', y = (y_1, y_2)', x_1, \dot{x}_2, y_1, y_2 \in \mathbb{R}^d$, define

$$\begin{aligned} H(x, y) &= d_g^2(\Phi^{-1}(x), \Phi^{-1}(y)) \\ &= d_g^2(\phi^{-1}(x_1), \phi^{-1}(y_1)) + d_g^2(\phi^{-1}(x_2), \phi^{-1}(y_2)) \\ &= h(x_1, y_1) + h(x_2, y_2). \end{aligned}$$

Then

$$\Lambda = E[(D_r D_s H(\Phi(X_1, Y_1), 0))] = \begin{pmatrix} \Lambda_1 & 0 \\ 0 & \Lambda_2 \end{pmatrix}$$

and

$$\Sigma = \text{Cov}[(D_r H(\Phi(X_1, Y_1), 0))] = \begin{pmatrix} \Sigma_1 & \Sigma_{12} \\ \Sigma_{21} & \Sigma_2 \end{pmatrix}.$$

Note that $\Lambda_1, \Lambda_2, \Sigma_1, \Sigma_2$ are as in Section 5.4.1 and

$$\Sigma_{12} = \Sigma'_{21} = \text{Cov}[(D_r h(\phi(X_1), 0)), D_r h(\phi(Y_1), 0))].$$

Therefore

$$\Gamma = \Lambda^{-1}\Sigma\Lambda^{-1} = \begin{pmatrix} \Lambda_1^{-1} & 0 \\ 0 & \Lambda_2^{-1} \end{pmatrix} \begin{pmatrix} \Sigma_1 & \Sigma_{12} \\ \Sigma_{21} & \Sigma_2 \end{pmatrix} \begin{pmatrix} \Lambda_1^{-1} & 0 \\ 0 & \Lambda_2^{-1} \end{pmatrix}$$

$$= \begin{pmatrix} \Lambda_1^{-1}\Sigma_1\Lambda_1^{-1} & \Lambda_1^{-1}\Sigma_{12}\Lambda_2^{-1} \\ \Lambda_2^{-1}\Sigma_{21}\Lambda_1^{-1} & \Lambda_2^{-1}\Sigma_2\Lambda_2^{-1} \end{pmatrix}.$$

It follows from equation (5.23) that if H_0 is true, then

$$\sqrt{n}\,(\phi(\hat{\mu}_1) - \phi(\hat{\mu}_1)) \xrightarrow{\mathcal{L}} N(0, \tilde{\Sigma}),$$

where

$$\tilde{\Sigma} = \Lambda_1^{-1}\Sigma_1\Lambda_1^{-1} + \Lambda_2^{-1}\Sigma_2\Lambda_2^{-1} - (\Lambda_1^{-1}\Sigma_{12}\Lambda_2^{-1} + \Lambda_2^{-1}\Sigma_{21}\Lambda_1^{-1}).$$

Estimate $\phi(\hat{\mu}_i)$ by μ_{ni}, $i = 1, 2$, as in Section 5.4.1 and $\tilde{\Sigma}$ by its sample analog $\hat{\tilde{\Sigma}}$. Then, under H_0, the test statistic

$$T_{n3} = n(\mu_{n1} - \mu_{n2})'\hat{\tilde{\Sigma}}^{-1}(\mu_{n1} - \mu_{n2})$$

converges in distribution to a chi-squared distribution with d degrees of freedom, that is, $T_{n3} \xrightarrow{\mathcal{L}} \mathcal{X}_d^2$. Therefore one rejects H_0 at an asymptotic level α if $T_{n3} > \mathcal{X}_d^2(1 - \alpha)$.

Next we test the null hypothesis $H_0 : V_1 = V_2$ against the alternative $H_1 : V_1 \neq V_2$. From equation (3.20) it follows that

$$\begin{pmatrix} \sqrt{n}(\hat{V}_1 - V_1) \\ \sqrt{n}(\hat{V}_2 - V_2) \end{pmatrix} = \frac{1}{\sqrt{n}} \begin{pmatrix} \sum_{j=1}^{n}[d_g^2(X_j, \mu_1) - V_1] \\ \sum_{j=1}^{n}[d_g^2(Y_j, \mu_2) - V_2] \end{pmatrix} + o_P(1)$$

$$\xrightarrow{\mathcal{L}} N\left(0, \begin{pmatrix} \sigma_1^2 & \sigma_{12} \\ \sigma_{12} & \sigma_2^2 \end{pmatrix}\right),$$

where $\sigma_1^2 = \text{Var}(d_g^2(X_1, \mu_1))$, $\sigma_2^2 = \text{Var}(d_g^2(Y_1, \mu_2))$ and $\sigma_{12} = \text{Cov}(d_g^2(X_1, \mu_1), d_g^2(Y_1, \mu_2))$. Hence, if H_0 is true, then

$$\sqrt{n}\,(\hat{V}_1 - \hat{V}_2) \xrightarrow{\mathcal{L}} N(0, \sigma_1^2 + \sigma_2^2 - 2\sigma_{12}),$$

which implies that the statistic

$$T_{n4} = \frac{\sqrt{n}(\hat{V}_1 - \hat{V}_2)}{\sqrt{s_1^2 + s_2^2 - 2s_{12}}}$$

has an asymptotic standard Normal distribution. Here s_1^2, s_2^2, and s_{12} are sample estimates of σ_1^2, σ_2^2, and σ_{12}, respectively. Therefore we reject H_0 at the asymptotic level α if $|T_{n4}| > Z_{1-\frac{\alpha}{2}}$.

5.5 Data example on S^2

From the lava flow of 1947–1948, nine specimens on the directions of flow were collected. The data can be viewed as an i.i.d. sample on the manifold S^2 and can be found in Fisher (1953). Figure 5.1 shows the data plots. The sample extrinsic and intrinsic means are very close, namely, at a geodesic

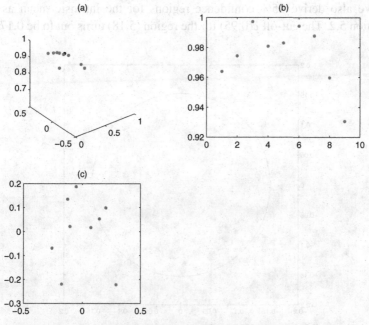

Figure 5.1 Lava flow directions on S^2 in §5.5. (a) Three-dimensional sample cloud along with extrinsic (*black*), intrinsic (*gray*) sample means. (b) Sample projected along $\hat{\mu}_E$. (c) Projections onto $T_{\hat{\mu}_E} S^2$.

distance of 0.0007 from each other. They are

$$\hat{\mu}_E = (0.2984, 0.1346, 0.9449)' \quad \text{and} \quad \hat{\mu}_I = (0.2990, 0.1349, 0.9447)',$$

respectively. They are indistinguishable in Figure 5.1.

In Fisher (1953), a von Mises–Fisher distribution (see Appendix D) is fitted to the data and a 95% confidence region based on the MLEs is obtained for the mean direction of flow (extrinsic or intrinsic). It turns out to be

$$\{p \in S^2 : d_g(\hat{\mu}_E, p) \le 0.1536\}. \tag{5.24}$$

Our asymptotic confidence region for the population extrinsic mean derived in Chapter 4 turns out to be

$$\{p \in S^2 : p'\bar{x} > 0, \ n|\bar{x}|^2 p' B(B'S B)^{-1} B' p \le X_2^2(0.95) = 5.9915\}. \tag{5.25}$$

The linear projection of this region onto $T_{\hat{\mu}_E} S^2$ is an ellipse centered around the origin while that of equation (5.24) is a disc. Figure 5.2 plots those projections. As it shows, the latter nearly contains the former and is considerably larger.

We also derive 95% confidence regions for the intrinsic mean as in Section 5.2. The cut-off $\hat{c}(0.95)$ for the region (5.18) turns out to be 0.1778

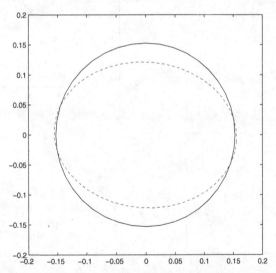

Figure 5.2 Ninety-five percent confidence region for extrinsic mean lava flow direction in Section 5.5. Region (5.24) (*solid*) and region (5.25) (*dashed*).

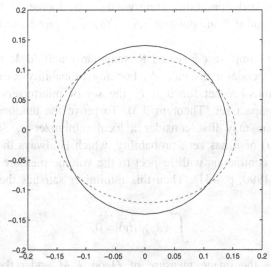

Figure 5.3 95% confidence region for intrinsic mean lava flow direction in Section 5.5. Region (5.26) (*solid*) and region (5.27) (*dashed*).

and hence the region is

$$\{\mu_I : d_g(\mu_I, \mu_{nI}) \le 0.1405\}, \tag{5.26}$$

which is smaller than that in equation (5.24). The ellipsoidal region in equation (3.15) becomes

$$\{\mu_I : n\phi(\mu_I)'\hat{\Lambda}\hat{\Sigma}^{-1}\hat{\Lambda}\phi(\mu_I) \le 5.992\}, \tag{5.27}$$

where ϕ gives the normal coordinates into $T_{\mu_{nI}}S^2$ (identified with $\mathbb{R}^2$),

$$\hat{\Lambda} = \begin{bmatrix} 1.987 & -0.001 \\ -0.001 & 1.979 \end{bmatrix} \text{ and } \hat{\Sigma} = \begin{bmatrix} 0.143 & -0.005 \\ -0.005 & 0.087 \end{bmatrix}.$$

As Figure 5.3 suggests, the two regions are close to each other in area.

5.6 Some remarks on the uniqueness of the intrinsic mean and the nonsingularity of the asymptotic distribution of the sample mean

An outstanding problem in intrinsic inference is to find broad conditions for the uniqueness of the minimizer of the Fréchet function (5.1) with respect to the geodesic distance d_g. Applications of intrinsic analysis are hindered by the lack of such conditions. Here we describe sources of the difficulties that arise and indicate possible routes that may prove fruitful for their resolution.

Consider a complete connected Riemannian manifold M with metric tensor g and geodesic distance d_g. For any probability measure Q on M with a finite Fréchet function F, the set of minimizers of F is a nonempty compact set (Theorem 3.3). To prove the uniqueness of the minimizer, one may first consider a local minimizer μ. Suppose the cut-locus $C(\mu)$ of μ has zero probability, which is always the case if Q is absolutely continuous with respect to the volume measure on M (see Gallot et al., 1990, p. 141). Then this minimizer satisfies the first-order condition

$$\int_M v\lambda_{\mu,Q}(dv) = 0, \tag{5.28}$$

where $\lambda_{\mu,Q}$ is the image measure of Q on $T_\mu M$ under the exponential map $\exp_\mu$ (see Remark 5.5, or Bhattacharya and Patrangenaru, 2003, theorem 2.1). It is important to find conditions that guarantee that such a minimizer is also the global minimizer, or that at least the minimizer is the minimum of F over a reasonably large geodesic ball. The smoothness of F is also important in deriving the asymptotic distribution of the sample Fréchet mean. The most general result for uniqueness of such a local minimizer is due to Karcher (1977), with an improvement due to Kendall (1990) that requires rather restrictive conditions on the support of Q. These restrictions arise mainly from the fact that d_g^2 is smooth and convex (along geodesics) only in a relatively small ball.

In general, $p \to d_g^2(p, p')$ is neither twice continuously differentiable nor convex (along geodesics), for all $p' \in M \backslash C(p)$. Therefore, the corresponding properties of the Fréchet function F are also in question. Concerning smoothness, note that the squared geodesic distance from p generally loses its smoothness as it reaches the critical point at its cut-point along any given geodesic. The simplest example, namely, that of a circle, already illustrates this. Let $p = (\cos\theta_0, \sin\theta_0)$ be a point on the unit circle; then its squared geodesic distance, say $f(x)$, to the point $p = (\cos(\theta_0 + \pi + x), \sin(\theta_0 + \pi + x))$ is given, for x close to zero, by $f(x) = (\pi + x)^2$ if $x < 0$ and $(\pi - x)^2$ if $x > 0$. Thus $f'(x)$ is discontinuous at $x = 0$. In general, discontinuities occur at the cut-locus. Suppose then that one considers an open geodesic ball $B = B(p_0, r_I)$ of center p_0 and radius equal to the injectivity radius r_I, thus excluding the cut-locus of p_0. If the dimension d of the Riemannian manifold is larger than 1, then the squared distance between the points in this ball is still not smooth, because there are many points in B whose cut-points are still in *B*. (Consider, e.g., the sphere S^2; *B* is then the sphere minus the antipodal point p_0. There are many great circles entirely contained in *B*, together with their antipodal points.) In an open geodesic ball of radius $\frac{r_i}{2}$, this cannot happen, in view of the triangle inequality, and therefore the squared distance is smooth in it. For strict convexity of the squared distance along geodesics, one needs a further restriction that the radius of the geodesic ball be no more than $\frac{\pi}{4\sqrt{C}}$, where *C* is the least upper bound of sectional curvatures on *M*, if nonnegative, or zero if all sectional curvatures of *M* are negative. From this, the classic result of Karcher (1977) follows for the existence of a unique (local) minimizer μ in the closed ball of radius $\frac{r^*}{2}$, where $r^* = \min\{\frac{r_i}{2}, \frac{\pi}{2\sqrt{C}}\}$. In an improvement of this, Kendall (1990) has shown, by a Jensen's inequality–type argument, that if the support is contained in a geodesic ball of radius r^*, then there is a unique minimizer μ of *F* in this ball. Hence, by the triangle inequality, it follows that the local minimizer under Karcher's condition is also the global minimizer of *F*. The last support condition is still restrictive, but it cannot be relaxed in general, even in the case of a circle. Kendall's result, however, does not imply convexity of *F* in the larger ball, and without this the Hessian Λ (at μ) of the averaged squared geodesic distance in normal coordinates at μ may not be of full rank, and, therefore, the asymptotic distribution of the sample Fréchet mean may be singular. It is shown in Theorem 5.3 that the nonsingularity condition holds if the support of *Q* is contained in a geodesic ball with center μ and radius r^*, and this existence of a unique (local) minimizer in a geodesic ball containing the support of *Q* is adequate for statistical purposes.

For most statistical applications, one may assume the existence of a density *q* of *Q* (with respect to the volume measure). It may be noted that the only significant result in that direction has been of Le, who showed that, for the special case of the planar shape space Σ_2^k, a sufficient condition for a point μ to be the unique minimizer of *F* is that *q* is a decreasing function of the geodesic distance from μ (see Kendall et al., 1999, p. 211). Le's proof is elegant and makes use of the symmetries of Σ_2^k, but the very restrictive hypothesis of a radial *q* makes it unusable in real statistical problems. One plausible approach to circumventing this is to pursue a different general path whose underlying ideas are the following:

(1) *Smoothness of F*. If *q* is the density of *Q* with respect to the volume measure dv on *M*, then although d_g^2 is not smooth on all of *M*, the Fréchet function

$$F(p) = \int_M d_g^2(p, p') q(p') dv(p') \tag{5.29}$$

is expected to be smooth (i.e., twice continuously differentiable) on all of M if q is smooth. For example, we check below that for the circle S^1, F is smooth if q is continuous. One may think of this situation as somewhat analogous to that of convolution of a nonsmooth function with a smooth one. Because the Kendall shape spaces are quotient spaces of the form S^d/G, where G is the special orthogonal group $SO(m)$ ($m = 2$ or 3), one may arrive at results for these spaces by using densities q on S^d that are invariant under the action of the group $SO(m)$.

We now turn to the special case of the circle S^1. Here, writing Log_{p_0} for the inverse exponential map $\exp_{p_0}^{-1}$, one has for $p_0 := (\cos\theta_0, \sin\theta_0)$, $\text{Log}_{p_0}(\cos(\theta_0 + \theta), \sin(\theta_0 + \theta)) = \theta v_0$, $\theta \in (-\pi, \pi)$, where $v_0 := (-\sin\theta_0, \cos\theta_0)$ is a unit tangent vector at p_0. We identify θv_0 with θ. The Fréchet function for the case of a continuous density q on the tangent space at p_0 is then given by

$$F(\theta) = \int_{-\pi+\theta}^{\pi} (u - \theta)^2 q(u)\, du + \int_{-\pi}^{-\pi+\theta} (2\pi + u - \theta)^2 q(u)\, du;$$

$$F'(\theta) = -2\int_{-\pi}^{\pi} (u - \theta)q(u)\, du - 4\pi\int_{-\pi}^{-\pi+\theta} q(u)\, du, \quad 0 \le \theta \le \pi;$$

$$F(\theta) = \int_{\pi+\theta}^{\pi} (u - \theta)^2 q(u)\, du + \int_{-\pi}^{\pi+\theta} (2\pi + u - \theta)^2 q(u)\, du;$$

$$F'(\theta) = -2\int_{-\pi}^{\pi} (u - \theta)q(u)\, du + 4\pi\int_{\pi+\theta}^{\pi} q(u)\, du, \quad -\pi < \theta < 0. \quad (5.30)$$

It is simple to check that F' is continuous on $(-\pi, \pi]$. Also,

$$F''(\theta) = \begin{cases} 2 - 4\pi q(-\pi + \theta) & \text{for } 0 \le \theta \le \pi, \\ 2 - 4\pi q(\pi + \theta) & \text{for } -\pi < \theta < 0. \end{cases} \quad (5.31)$$

Thus the second derivative of F is also continuous on $(-\pi, \pi]$. Also, F is convex at p if and only if $q(-p) < \frac{1}{2\pi}$, and concave if $q(-p) > \frac{1}{2\pi}$.

It follows from equation (5.30) that a necessary condition for F to have a local minimum at p_0 is that

$$\int_{-\pi}^{\pi} uq(u)\, du = 0. \quad (5.32)$$

For p_0 to be the unique global minimizer one must check that $F(0) < F(\theta)$ for all $\theta > 0$ and for all $\theta < 0$. Taylor's theorem in calculus then yields the conditions (5.33) below. Thus, irrespective of the region of convexity of F, a necessary and sufficient condition for a local minimizer p_0 to be the *global minimizer* is that, in addition to equation (5.32), the following must hold under the log map Log_{p_0}:

$$\int_0^1 (1 - v)q(-\pi + v\theta)dv < \frac{1}{4\pi}, \quad \theta > 0,$$

$$\int_0^1 (1 - v)q(\pi + v\theta)dv < \frac{1}{4\pi}, \quad \theta < 0, \quad (5.33)$$

that is, if and only if the *average value* of q with respect to $(2/\theta)(1 - [u + \pi]/\theta)du$ on $(-\pi, \theta - \pi)$, for $\theta > 0$ (and similarly on $(\theta + \pi, \pi)$, for $\theta < 0$) is less than the uniform density $\frac{1}{2\pi}$.

It is perhaps reasonable to expect that, for a general Riemannian manifold, the existence of a Fréchet mean (i.e., of a unique global minimizer) may be related to similar average values of q.

(2) *Convexity of F.* In general, F is not convex on all of M. However, all that is needed for nonsingularity of the asymptotic distribution of the sample mean shape are (i) nondegeneracy of Q in normal coordinates at the (local) minimizer, which is automatic when Q has a density, and (ii) nonsingularity of the average Hessian of the squared geodesic distance measured from this minimizer.

In the case of the circle, with the arc θ measured from a fixed point p_0, $F''(\theta) = 2 - 4\pi q(\pi + \theta)$, $\theta \le 0$, and $F''(\theta) = 2 - 4\pi q(-\pi + \theta)$, $\theta > 0$, from which one can easily determine the region of convexity. It follows that F'' is positive at the (local) minimizer if and only if at the cut-lotus of this minimizer the density is less than $\frac{1}{2\pi}$.

The existing approach (see, e.g., Karcher, 1977) of establishing strict convexity pointwise (i.e., of the squared distance along each geodesic) requires the imposition of a severe restriction on the support of Q, and it is clearly unsuitable for absolutely continuous Q with a continuous density.

5.7 References

Karcher (1977) was the first author to derive uniqueness of the (local) intrinsic mean of distribution Q supported in a geodesic ball $B(p, \frac{r_*}{4})$ of radius $\frac{r_*}{4}$. Kendall (1990) substantially extended this result to a ball of radius $\frac{r_*}{2}$, from which global uniqueness of the intrinsic mean of Q supported in $B(p, \frac{r_*}{4})$ follows. Also see Le (2001) in this regard. The CLT for the sample intrinsic mean is derived in Bhattacharya and Patrangenaru (2005). Theorem 5.3 is obtained in Bhattacharya and Bhattacharya

(2008b). Do Carmo (1992) contains a good account of Jacobi fields. The first-order condition $E(\tilde{X}_1) = 0$ (in the statement of part (a) of Corollary 5.4) may be found in Oller and Corcuear (1995) and Bhattacharya and Patrangenaru (2003). The intrinsic two-sample and matched pair tests may be found in Bhattacharya (2008a) and Bhattacharya and Bhattacharya (2008a). Section 5.6 is taken from Bhattacharya (2007).

6

Landmark-based shape spaces

Manifolds of greatest interest in this book are spaces of shapes of k-ads in $\mathbb{R}^m$, with a k-ad being a set of k labeled points, or landmarks, on an object in $\mathbb{R}^m$. This chapter introduces these shape spaces.

6.1 Introduction

The statistical analysis of shape distributions based on random samples is important in many areas such as morphometrics, medical diagnostics, machine vision, and robotics. In this chapter and the chapters that follow, we will be interested mainly in the analysis of shapes of landmark-based data, in which each observation consists of $k > m$ points in m dimensions, representing k landmarks on an object, called a k-ad. The choice of landmarks is generally made with expert help in the particular field of application. Depending on the way the data are collected and recorded, the appropriate shape of a k-ad is the maximal invariant specified by its orbit under a group of transformations.

For example, one may look at k-ads modulo size and Euclidean rigid body motions of translation and rotation. The analysis of this invariance class of shapes was pioneered by Kendall (1977, 1984) and Bookstein (1978). Bookstein's approach is primarily registration-based, requiring two or three landmarks to be brought into a standard position by translating, rotating and scaling the k-ad. We would prefer Kendall's more invariant view of a shape identified with the orbit under rotation (in m dimensions) of the k-ad centered at the origin and scaled to have a unit size. The resulting shape spaces are called *similarity shape spaces*. A fairly comprehensive account of parametric inference on these spaces especially for $m = 2$, with many references to the literature, may be found in Dryden and Mardia (1998). We also explore other shape spaces such as the *reflection similarity shape spaces*, whose orbits are generated by the class of all orthogonal transformations – rotations and reflection – of centered and scaled k-ads.

77

Recently there has been much emphasis on the statistical analysis of other notions of shapes of k-ads, namely, *affine shapes* invariant under affine transformations and *projective shapes* invariant under projective transformations. Reconstructing a scene from two (or more) aerial photographs taken from a plane is one of the research problems in affine shape analysis. Potential applications of projective shape analysis include face recognition and robotics.

In this chapter, we will briefly describe the geometry of the above shape spaces and return to them one by one in the subsequent chapters.

6.2 Geometry of shape manifolds

Many differentiable manifolds M naturally occur as submanifolds, or surfaces or hypersurfaces, of a Euclidean space. One example is the sphere $S^d = \{p \in \mathbb{R}^{d+1} : \|p\| = 1\}$. The shape spaces of interest here are not of this type. They are quotients of a Riemannian manifold N under the action of a transformation group G, that is, $M = N/G$. A number of them are quotient spaces of $N = S^d$ under the action of a compact group G; that is, the elements of the space are orbits in S^d traced out by the application of G. Among important examples of this kind are Kendall's shape spaces and reflection shape spaces.

When the *action of the group is free*, which means that $gp = p$ for some p only holds for the identity element $g = e$, the elements of the orbit $O_p = \{gp : g \in G\}$ are in one-to-one correspondence with the elements of G. Then one can identify the orbit with the group and the orbit inherits the differential structure of G. The tangent space T_pN at a point p may then be decomposed into a *vertical subspace* V_p of dimension equal to that of the group G along the orbit space to which p belongs, and a *horizontal subspace* H_p that is orthogonal to it. The vertical subspace is isomorphic to the tangent space of G and the horizontal subspace can be identified with the tangent space of M at the orbit O_p. With this identification, M is a differentiable manifold of dimension equal to that of N minus the dimension of G.

To carry out *extrinsic analysis* on M, we use a smooth map j from M into some Euclidean space E that is an embedding of M into E. Then the image $j(M)$ of M is a differentiable submanifold of E. The tangent space at $j(O_p)$ is $dj(H_p)$, where dj is the differential of the map $j: M \to E$. Among all possible embeddings, we choose j to be equivariant under the action of a large group H on M. In most cases, H is compact.

For *intrinsic analysis* on $M = N/G$, one considers a Riemannian structure on N, providing a metric tensor smoothly on its tangent spaces. The Lie group G has its own natural Riemannian structure. If G acts as isometries of N, then the projection σ, $\sigma(p) = O_p$, is a *Riemannian submersion* of N onto the quotient space M. In other words, $\langle d\sigma(v), d\sigma(w) \rangle_{\sigma(p)} = \langle v, w \rangle_p$ for horizontal vectors $v, w \in T_p N$, where $d\sigma \colon T_p N \to T_{\sigma(p)} M$ denotes the differential of the projection σ. With this metric tensor, M has the natural structure of a Riemannian manifold. This provides the framework for carrying out intrinsic analysis.

6.2.1 Similarity shape spaces Σ_m^k

Consider a k-ad in two or three dimensions with not all landmarks identical. Its similarity shape is what remains after removing the effects of translation, one-dimensional scaling, and rotation. The space of all similarity shapes forms the *similarity shape space* Σ_m^k, with m being the dimension of the Euclidean space where the landmarks lie, which is usually 2 or 3. Similarity shape analysis finds many applications in morphometrics – classification of biological species based on their shapes; medical diagnostics – disease detection based on change in shape of an organ due to disease or deformation; evolution studies – studying the change in shape of an organ or organism with time, age, and so on; and many more. Some such applications will be considered in subsequent chapters.

For $m = 2$, the action of similarity transformations is free on the space of planar k-ads (excluding those with all k elements of the k-ad identical). The resulting quotient space Σ_2^k is then a compact differentiable manifold.

6.2.2 Reflection similarity shape spaces $R\Sigma_m^k$

When the k-ads lie in $\mathbb{R}^m$ for some $m > 2$, the action of the group of similarity transformations is not free. In other words, in different parts of Σ_m^k, the orbits have different dimensions, and hence Σ_m^k is not a manifold. In this case one considers the *reflection similarity shape* of a k-ad, that is, features invariant under translation, scaling, and *all* orthogonal transformations. After excluding a singular set, it is possible to embed the resulting shape space in some higher dimensional Euclidean space and carry out extrinsic analysis. Such an embedding, which is equivariant under a large group action, makes it possible to extend the results of nonparametric inference on shapes from 2 to m (in particular 3) dimensions.

6.2.3 Affine shape spaces $A\Sigma_m^k$

An application in bioinformatics consists in matching two marked electrophoresis gels. Proteins are subjected to stretches in two directions.

Due to their molecular mass and electrical charge, the amount of stretching depends on the strength and duration of the electrical fields applied. For this reason, the same tissue analyzed by different laboratories may yield different constellations of protein spots. The two configurations differ by a change of coordinates that can be given approximately by an *affine transformation* which may not be a similarity transformation as considered in Sections 6.2.1 and 6.2.2.

Another application of *affine shape* analysis is in scene recognition: to reconstruct a larger image from partial views in a number of aerial images of that scene. For a remote scene, the image acquisition process will involve a parallel projection, which in general is not orthogonal. For example, a rectangle may appear as a parallelogram with the adjacent sides not perpendicular to each other. Two common parts of the same scene seen in different images will essentially differ by an affine transformation but not a similarity.

6.2.4 Projective shape spaces $P\Sigma_m^k$

In machine vision, if images are taken from a great distance, affine shape analysis is appropriate. Otherwise, *projective shape* analysis is a more appropriate choice. If images are obtained through a central projection, a ray is received as a point on the image plane. Since axes in three dimensions comprise the *projective space* $\mathbb{R}P^2$, k-ads in this view are valued in $\mathbb{R}P^2$. To have invariance with regard to camera angles, one may first look at the original three-dimensional k-ad and achieve affine invariance by its affine shape and finally take the corresponding equivalence class of axes in $\mathbb{R}P^2$, to define the projective shape of the k-ad invariant with respect to all projective transformations on $\mathbb{R}P^2$. Potential applications of projective shape analysis arise in robotics, particularly in machine vision for robots to visually recognize a scene, avoid an obstacle, and so on.

For a remote view, the rays falling on the image plane are more or less parallel, and then a projective transformation can be approximated by an affine transformation. Further, if it is assumed that the rays fall perpendicular to the image plane, then similarity or reflection similarity shape space analysis becomes appropriate.

6.3 References

In addition to the references already mentioned in this chapter, we note the recent work by Huckemann et al. (2010), which introduced *generalized principal components analysis* (GPCA) on $M = N/G$, where N is a Riemannian manifold and G is a Lie group of isometries of N. These authors also develop the notion of generalized geodesics, which coincide with the usual geodesics in the absence of singularities (i.e., when M is a manifold, with the group action being free) but are defined on singular regions of M as occur, for example, in Σ_m^k for $m > 2$. The first generalized principal component (GPC) is the generalized geodesic that has the smallest expected squared distance from an M-valued random variable X with the given distribution Q. The second GPC is the generalized geodesic with the smallest expected squared distance from X among all those that intersect the first GPC orthogonally, and so on.

For references to reflection similarity shape spaces, and affine and projective shape spaces, see the chapters devoted to them.

7

Kendall's similarity shape spaces Σ_m^k

Landmark-based similarity shape spaces Σ_m^k were introduced briefly in Chapter 6. It was also mentioned there that while for $m = 2$ the planar shape space Σ_2^k is a compact differentiable manifold, Σ_m^k for $m > 2$ fails to be a manifold due to the presence of singularities. The present chapter describes the geometry of the manifold Σ_{0m}^k ($m > 2$) obtained after the removal of the singularities.

7.1 Introduction

Kendall's shape spaces are quotient spaces S^d/G, under the action of the special orthogonal group $G = SO(m)$ of $m \times m$ orthogonal matrices with determinant $+1$. Important cases include $m = 2, 3$.

For the case $m = 2$, consider the space of all planar k-ads $(z_1, z_2, \ldots, z_k)$ ($z_j = (x_j, y_j)$), $k > 2$, excluding those with k identical points. The set of all centered and normed k-ads, say $u = (u_1, u_2, \ldots, u_k)$, comprises a unit sphere in a $(2k - 2)$-dimensional vector space and is, therefore, a $(2k - 3)$-dimensional sphere S^{2k-3}, called the *preshape sphere*. The group $G = SO(2)$ acts on the sphere by rotating each landmark by the same angle. The orbit under G of a point u in the preshape sphere can thus be seen to be a circle S^1, so that Kendall's *planar shape space* Σ_2^k can be viewed as the quotient space $S^{2k-3}/G \sim S^{2k-3}/S^1$, a $(2k - 4)$-dimensional compact manifold. An algebraically simpler representation of Σ_2^k is given by the complex projective space $\mathbb{C}P^{k-2}$. We will return to this shape space in the next chapter.

When $m > 2$, we will denote a k-ad by the $m \times k$ matrix, $x = (x_1, \ldots, x_k)$, where x_i, $i = 1, \ldots, k$, are the k landmarks from the object of interest. Assume $k > m$. The *similarity shape* of the k-ad is what remains after we remove the effects of translation, rotation, and scaling. To remove translation, we subtract the mean $\bar{x} = \frac{1}{k} \sum_{i=1}^{k} x_i$ from each landmark to get the centered k-ad $u = (x_1 - \bar{x}, \ldots, x_k - \bar{x})$. We remove the effect of scaling by dividing u by its Euclidean norm to get

$$z = \left(\frac{x_1 - \bar{x}}{\|u\|}, \ldots, \frac{x_k - \bar{x}}{\|u\|}\right) = (z_1, z_2, \ldots, z_k). \tag{7.1}$$

This z is called the *preshape* of the k-ad x and it lies in the unit sphere S_m^k in the hyperplane $H_m^k = \{z \in \mathbb{R}^{m \times k} : z\mathbf{1}_k = 0\}$. Hence

$$S_m^k = \{z \in \mathbb{R}^{m \times k} : \text{Trace}(zz') = 1, \ z\mathbf{1}_k = 0\}. \tag{7.2}$$

Here $\mathbf{1}_k$ denotes the $k \times 1$ vector of all ones. Thus the *preshape sphere* S_m^k may be identified with the sphere S^{km-m-1}. Then the shape of the k-ad x is the orbit of z under left multiplication by $m \times m$ rotation matrices. In other words, $\Sigma_m^k = S_m^k/SO(m)$. One can also remove the effect of translation from the original k-ad x by postmultiplying the centered k-ad u by a *Helmert* matrix H, which is a $k \times (k-1)$ matrix satisfying $H'H = I_{k-1}$ and $\mathbf{1}_k'H = 0$. The resulting k-ad $\tilde{u} = uH$ lies in $\mathbb{R}^{m \times (k-1)}$ and is called the *Helmertized k-ad*. Then the preshape of x or $\tilde{u}$ is $\tilde{z} = \tilde{u}/\|\tilde{u}\|$ and the preshape sphere is

$$S_m^k = \{z \in \mathbb{R}^{m \times (k-1)} : \text{Trace}(zz') = 1\}. \tag{7.3}$$

The advantage of using this representation of S_m^k is that there is no linear constraint on the coordinates of z and hence analysis becomes simpler. However, now the choice of the preshape depends on the choice of H, which can vary. In most cases, including applications, we will represent the preshape of x as in equation (7.1) and the preshape sphere as in equation (7.2).

7.2 Geometry of similarity shape spaces

In this section, we study the topological and geometrical properties of Σ_m^k represented as $S_m^k/SO(m)$. We are interested in the case when $m > 2$. The case $m = 2$ is studied in Chapter 8.

For $m > 2$, the direct similarity shape space Σ_m^k fails to be a manifold. That is because the action of $SO(m)$ is not in general free. Indeed, the orbits of preshapes under $SO(m)$ have different dimensions in different regions. To see this consider the case $m = 3$. If the k landmarks of a (translated and scaled) k-ad lie on a line, then the subgroup (of $SO(3)$) of rotations G that rotate the k-ad (a point on the preshape sphere) around this line in $\mathbb{R}^3$ keep the k-ad fixed: $gp = p$. That is, the action of $SO(3)$ is not free on the singular set of collinear k-ads. To avoid that, one may consider the shape of only those k-ads whose preshapes have rank at least $m - 1$. Define

$$\bar{N}S_m^k = \{z \in S_m^k : \text{rank}(z) \geq m - 1\}$$

as the *nonsingular part* of S_m^k and let $\Sigma_{0m}^k = \bar{N}S_m^k/SO(m)$. Then, because the action of $SO(m)$ on $\bar{N}S_m^k$ is free, Σ_{0m}^k is a differentiable manifold of dimension $km - m - 1 - \frac{m(m-1)}{2}$. Also, because $SO(m)$ acts as isometries of the sphere, Σ_{0m}^k inherits the Riemannian metric tensor of the sphere and hence is a Riemannian manifold. However, it is not complete because of the "holes" created by removing the singular part.

Consider the projection map

$$\pi: \bar{N}S_m^k \to \Sigma_{0m}^k, \quad \pi(z) = \{Az : A \in SO(m)\}.$$

This map is a *Riemannian submersion* (see Appendix B and Do Carmo, 1992, pp. 185–187). This means that if we write $T_z S_m^k$ as the direct sum of the *horizontal subspace* H_z and the *vertical subspace* V_z, then $d\pi$ is an isometry from H_z into $T_{\pi(z)}\Sigma_{0m}^k$. The tangent space $T_z S_m^k$ is

$$T_z S_m^k = \{v \in H_m^k : \text{Trace}(vz') = 0\}, \tag{7.4}$$

recalling that S_m^k is the unit sphere in the hyperplane H_m^k. The vertical subspace V_z consists of initial velocity vectors of curves in S_m^k starting at z and remaining in the orbit $\pi(z)$. Such a curve will have the form $\gamma(t) = \tilde{\gamma}(t)z$, where $\tilde{\gamma}(t)$ is a curve in $SO(m)$ starting at the identity matrix I_m. Geodesics in $SO(m)$ starting at I_m have the form $\tilde{\gamma}(t) = \exp(tA)$, where

$$\exp(A) = I + A + \frac{A^2}{2} + \frac{A^3}{3!} + \cdots,$$

and A is skew-symmetric ($A + A' = 0$). The skew-symmetry follows from the requirement $\exp(tA)(\exp(tA))' = I_m$, that is, $\exp(t(A + A')) = I_m$ for all t. That $\tilde{\gamma}(t)$ is a geodesic follows from the fact that its "acceleration," or the covariant derivative of $d\tilde{\gamma}(t)/dt$ along $\tilde{\gamma}(t)$, vanishes for all t (see Appendix B). For such a curve, $\dot{\tilde{\gamma}}(0) = A$ and therefore $\dot{\gamma}(0) = Az$, which implies that

$$V_z = \{Az : A + A' = 0\}.$$

The horizontal subspace is its orthocomplement, which is

$$H_z = \{v \in H_m^k : \text{Trace}(vz') = 0, \ vz' = zv'\}. \tag{7.5}$$

To verify equation (7.5), fix $v \in H_z$. The first requirement $\text{Trace}(vz') = 0$ follows from equation (7.4). To check the second condition, let $y \in V_z$ be of the form $y = A_{ij}z$, where $A_{ij} = E_{ij} - E_{ji}$, with $E_{r,s}$ being the $m \times m$ matrix with 1 in the (r, s) position and zeros elsewhere. We must have $\text{Trace}(v'y) = 0$. An easy calculation yields $\text{Trace}(v'A_{ij}z) = ((vz'))_{ij} - ((zv'))_{ij}$.

From this, equation (7.4) follows. Because π is a Riemannian submersion, $T_{\pi(z)}\Sigma_{0m}^k$ is isometric to H_z.

The geodesic distance between two shapes $\pi(x)$ and $\pi(y)$, where $x, y \in S_m^k$, is given by

$$d_g(\pi(x), \pi(y)) = \min_{T \in SO(m)} d_{gs}(x, Ty).$$

Here $d_{gs}(\cdot, \cdot)$ is the geodesic distance on S_m^k, which is

$$d_{gs}(x, y) = \arccos(\text{Trace}(yx')).$$

Therefore,

$$d_g(\pi(x), \pi(y)) = \arccos(\max_{T \in SO(m)} \text{Trace}(Tyx')). \qquad (7.6)$$

Consider the *pseudo-singular-value decomposition* of yx', which is

$$yx' = U\Lambda V, \quad U, V \in SO(m),$$
$$\Lambda = \text{diag}(\lambda_1, \ldots, \lambda_m), \quad \lambda_1 \geq \lambda_2 \geq \cdots \geq \lambda_{m-1} \geq |\lambda_m|,$$
$$\text{sign}(\lambda_m) = \text{sign}(\det(yx'))$$

(see, e.g., Kendall et al., 1999, pp. 114, 115).

Then the value of T for which $\text{Trace}(Tyx')$ in equation (7.6) is maximized is $T = V'U'$ and then

$$d_g(\pi(x), \pi(y)) = \arccos(\text{Trace}(\Lambda)) = \arccos\Big(\sum_{j=1}^m \lambda_j\Big),$$

which lies between 0 and $\frac{\pi}{2}$.

Define the singular part D_{m-2} of S_m^k as the set of all preshapes with rank less than $m - 1$. Then it is shown in Kendall et al. (1999) that for $x \in S_m^k \setminus D_{m-2} \equiv \bar{N}S_m^k$, the cut-locus of $\pi(x)$ in Σ_{0m}^k is given by

$$C(\pi(x)) = \pi(D_{m-2}) \cup C_0(\pi(X)),$$

where $C_0(\pi(X))$ is defined to be the set of all shapes $\pi(y) \in \Sigma_{0m}^k$ such that there exists more than one length minimizing geodesic joining $\pi(x)$ and $\pi(y)$. It is also shown that the least upper bound of all sectional curvatures of Σ_{0m}^k is $+\infty$. Also, Σ_{0m}^k is not complete in the geodesic distance. Hence we cannot directly apply the results of Chapter 5 to carry out intrinsic analysis on this space. However, such analysis may be carried out when the supports of the probability measures under consideration are contained in a sufficiently small compact convex geodesic ball in Σ_{0m}^k.

Once we remove the effects of reflections along with rotations from the preshapes, we can embed the shape space into a higher dimensional

Euclidean space and carry out extrinsic analysis of shapes. This is done in Chapter 9.

7.3 References

As the main reference for this chapter we cite Kendall et al. (1999). Also see Small (1996).

8

The planar shape space Σ_2^k

This chapter develops in detail the geometry of both the planar shape space Σ_2^k and nonparametric inference on this space.

8.1 Introduction

We return to the set of all planar k-ads, $k > 2$, considered in Section 7.1. We will now denote a k-ad by k complex numbers $(z_j = x_j + iy_j : 1 \leq j \leq k)$, that is, we will represent k-ads on the complex plane. As before, the similarity shape of a k-ad $z = (z_1, z_2, \ldots, z_k)$ represents the equivalence class, or orbit of z under translation, one-dimensional scaling, and rotation. To remove translation, one subtracts

$$\langle z \rangle = (\bar{z}, \bar{z}, \ldots, \bar{z}), \qquad \bar{z} = \frac{1}{k} \sum_{j=1}^{k} z_j,$$

from z to get $z - \langle z \rangle$. Rotation of the k-ad by an angle θ and scaling by a factor $r > 0$ are achieved by multiplying $z - \langle z \rangle$ by the complex number $\lambda = re^{i\theta}$. Hence one may represent the shape of the k-ad as the complex line passing through $z - \langle z \rangle$, namely, $\{\lambda(z - \langle z \rangle)) : \lambda \in \mathbb{C}\backslash\{0\}\}$. Thus the space of similarity shapes of k-ads is the set of all complex lines on the (complex $(k - 1)$-dimensional) hyperplane, $H^{k-1} = \{w \in \mathbb{C}^k\backslash\{0\} : \sum_1^k w_j = 0\}$. Therefore, the similarity shape space Σ_2^k of planar k-ads has the structure of the *complex projective space* $\mathbb{C}P^{k-2}$, that is, the space of all complex lines through the origin in $\mathbb{C}^{k-1}$ (see Do Carmo, 1992, p. 188).

Readers interested in only extrinsic analysis may proceed directly to Section 8.6.

8.2 Geometry of the planar shape space

When identified with $\mathbb{C}P^{k-2}$, Σ_2^k is a compact connected Riemannian manifold of (real) dimension $2k - 4$. As in the case of $\mathbb{C}P^{k-2}$, it is convenient to represent the shape $\sigma(z)$ of a k-ad z by the curve

$$\sigma(z) = \pi(u) = \{e^{i\theta}u : -\pi < \theta \leq \pi\}, \quad u = \frac{z - \langle z \rangle}{\|z - \langle z \rangle\|}$$

on the unit sphere $\mathbb{C}S^{k-1}$ in H^{k-1}. The quantity u is called the *preshape* of the original k-ad z and it lies on the complex hypersphere

$$\mathbb{C}S^{k-1} = \{u \in \mathbb{C}^k : \sum_{j=1}^{k} u_j = 0, \ \|u\| = 1\}.$$

The map $\pi: \mathbb{C}S^{k-1} \to \Sigma_2^k$ is a Riemannian submersion. Hence, its derivative $d\pi$ is an isometry from H_u into $T_{\pi(u)}\Sigma_2^k$, where H_u is the horizontal subspace of the tangent space $T_u\mathbb{C}S^{k-1}$ of $\mathbb{C}S^{k-1}$ at u, namely,

$$H_u = \{v \in \mathbb{C}^k : uv^* = 0, \ v\mathbf{1}_k = 0\},$$

where, for $v = (v_1, \ldots, v_k)$, $v^* = (\bar{v}_1, \ldots, \bar{v}_k)'$ denotes its complex conjugate transpose and $\mathbf{1}_k$ is the column vector in $\mathbb{R}^k$ of all ones. The preshape sphere $\mathbb{C}S^{k-1}$ can be identified with the real sphere of dimension $2k - 3$, namely, S^{2k-3}. Hence, if exp denotes the exponential map of $\mathbb{C}S^{k-1}$ as derived in Chapter 5, then the exponential map of Σ_2^k is given by

$$\exp_{\pi(u)}: T_{\pi(u)}\Sigma_2^k \to \Sigma_2^k, \quad \exp_{\pi(u)} = \pi \circ \exp_u \circ d\pi_u^{-1}.$$

The geodesic distance between two shapes $\sigma(x)$ and $\sigma(y)$ is given by

$$d_g(\sigma(x), \sigma(y)) = d_g(\pi(z), \pi(w)) = \inf_{\theta \in (-\pi, \pi]} d_{gs}(z, e^{i\theta}w),$$

where x and y are two k-ads in $\mathbb{C}^k$, z and w are their preshapes in $\mathbb{C}S^{k-1}$, and $d_{gs}(\cdot, \cdot)$ denotes the geodesic distance on $\mathbb{C}S^{k-1}$, which is given by $d_{gs}(z, w) = \arccos(\text{Re}(zw^*))$ as mentioned in Section 5.3. Hence, geodesic distance on Σ_2^k has the following expression:

$$d_g(\pi(z), \pi(w)) = \inf_{\theta \in (-\pi, \pi]} \arccos(\text{Re}(e^{-i\theta}zw^*))$$

$$= \arccos \sup_{\theta \in (-\pi, \pi]} \text{Re}(e^{-i\theta}zw^*) = \arccos(|zw^*|).$$

Therefore, the geodesic distance between any pair of planar shapes lies between 0 and $\frac{\pi}{2}$, which means that Σ_2^k has an injectivity radius of $\frac{\pi}{2}$. The cut-locus $C(\pi(z))$ of $z \in \mathbb{C}S^{k-1}$ is given by

$$C(\pi(z)) = \{\pi(w) : w \in \mathbb{C}S^{k-1}, \; d_g(\pi(z), \pi(w)) = \frac{\pi}{2}\} = \{\pi(w) : zw^* = 0\}.$$

The exponential map $\exp_{\pi(z)}$ is invertible outside the cut-locus of z and its inverse is given by

$$\exp_{\pi(z)}^{-1} : \Sigma_2^k \backslash C(\pi(z)) \to T_{\pi(z)}\Sigma_2^k, \; \pi(w) \mapsto d\pi_{\pi(z)}\left\{\frac{r}{\sin(r)}(-\cos(r)z + e^{i\theta}w)\right\},$$

(8.1)

where $r = d_g(\pi(z), \pi(w))$ and $e^{i\theta} = \frac{zw^*}{|zw^*|}$. It has been shown in Kendall (1984) that Σ_2^k has constant holomorphic sectional curvature of 4.

Given two preshapes u and v, the *Procrustes coordinates* of v with respect to u are defined as

$$v^P = e^{i\theta}v,$$

where $\theta \in (-\pi, \pi]$ is chosen so as to minimize the Euclidean distance between u and $e^{i\theta}v$, namely, $d_P(\theta) = \|u - e^{i\theta}v\|$. In other words, one tries to rotate the preshape v so as to bring it closest to u. Then

$$d_P^2(\theta) = 2 - 2\operatorname{Re}(e^{i\theta}vu^*),$$

which is minimized when $e^{i\theta} = \frac{uv^*}{|uv^*|}$, and then the minimum value of the Euclidean distance turns out to be

$$d_P = \min_{\theta \in (-\pi, \pi]} d_P(\theta) = \sqrt{2(1 - |uv^*|)}.$$

This d_P is a distance metric on Σ_2^k, called the *Procrustes distance* (see Dryden and Mardia, 1998, for details). The Procrustes coordinates can be particularly useful for plotting shapes, as we shall see in the next section.

8.3 Examples

In this section, we discuss two applications of planar shape analysis. We will return to these examples in Section 8.11.

8.3.1 Gorilla skulls

For the skulls of 29 male and 30 female gorillas, consider eight landmarks chosen on the midline plane of each of the skulls. The data can be found in Dryden and Mardia (1998). It is of interest to study the shapes of the skulls and use them to detect differences between the sexes. This technique finds applications in morphometrics and other biological sciences. For the analysis, we consider the planar shapes of the k-ads of observations, which lie in Σ_2^k, $k = 8$. Figure 8.1(a) shows the Procrustes coordinates of the shapes of the female gorilla skulls. The coordinates are obtained with respect to a preshape of the sample extrinsic mean, which is defined in Section 8.7. Figure 8.1(b) shows the Procrustes coordinates of the shapes of the male gorilla skulls with respect to a preshape of the male sample extrinsic mean.

8.3.2 Schizophrenic patients

In this example from Bookstein (1991), 13 landmarks are recorded on a midsagittal two-dimensional slice from a magnetic resonance brain scan of each of 14 schizophrenic patients and 14 normal patients. It is of interest to study differences in the shapes of the brains between the two groups, which can be used to detect schizophrenia. This application is an example of disease detection. The shapes of the sample k-ads lie in Σ_2^k, $k = 13$. Figure 8.2(a) shows the Procrustes coordinates of the shapes for the schizophrenic patients while Figure 8.2(b) shows the coordinates for the normal patients. As in Section 8.3.1, the coordinates are obtained with respect to the preshapes of the respective sample extrinsic means.

8.4 Intrinsic analysis on the planar shape space

Let Q be a probability distribution on Σ_2^k. From Proposition 5.2, it follows that if the support of Q is contained in a geodesic ball of radius $\frac{\pi}{4}$, then it has a unique intrinsic mean in that ball. In this section we assume that $\text{supp}(Q) \subseteq B(p, \frac{\pi}{4})$ for some $p \in \Sigma_2^k$. Let $\mu_I = \pi(\mu)$ be the (local) intrinsic mean of Q in $B(p, \frac{\pi}{4})$, with μ being one of its preshapes. Let $X_1, \ldots, X_n$ be an i.i.d. sample from Q on Σ_2^k, and let μ_{nI} be the (local) sample intrinsic mean in $B(p, \frac{\pi}{4})$. From Proposition 5.2, it follows that μ_{nI} is a consistent estimator of μ_I. Furthermore, if we assume that $\text{supp}(Q) \subseteq B(\mu_I, \frac{\pi}{4})$, then Theorem 5.3 implies that the coordinates of μ_{nI} have an asymptotic Normal distribution. However, this theorem does not give an expression for the asymptotic parameter Λ because Σ_2^k does not have constant sectional curvature. Theorem 8.1 shows how to get the analytic expression for Λ and relaxes the support condition for its positive-definiteness.

Theorem 8.1 *Let $\phi: B(p, \frac{\pi}{4}) \to \mathbb{C}^{k-2} (\approx \mathbb{R}^{2k-4})$ be the coordinates of $d\pi_\mu^{-1} \circ \text{Exp}_{\mu_I}^{-1}: B(p, \frac{\pi}{4}) \to H_\mu$ with respect to some orthonormal basis $\{v_1, \ldots, v_{k-2}, iv_1, \ldots, iv_{k-2}\}$ for H_μ (over $\mathbb{R}$). Define $h(x, y) = d_g^2(\phi^{-1}(x), \phi^{-1}(y))$. Let $(D_r h)_{r=1}^{2k-4}$ and $(D_r D_s h)_{r,s=1}^{2k-4}$ be the matrix of first- and second-order derivatives of $y \mapsto h(x, y)$. Let $\tilde{X}_j = \phi(X_j) = (\tilde{X}_j^1, \ldots, \tilde{X}_j^{k-2})$; $j = 1, \ldots, n$ be the*

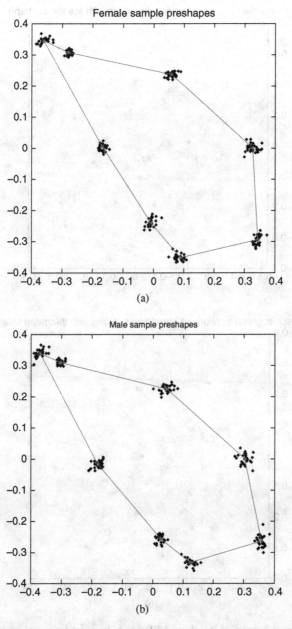

Figure 8.1 (a) and (b) show eight landmarks from skulls of 30 female and 29 male gorillas, respectively, along with the respective sample mean shapes. Gray diamonds indicate the mean shapes' landmarks.

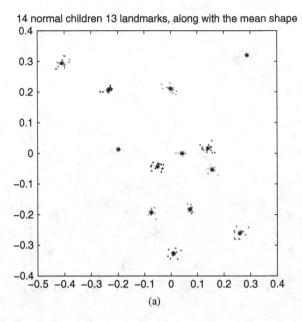

14 normal children 13 landmarks, along with the mean shape

(a)

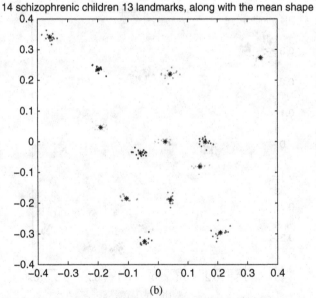

14 schizophrenic children 13 landmarks, along with the mean shape

(b)

Figure 8.2 (a) and (b) show 13 landmarks for 14 normal and 14 schizophrenic patients, respectively, along with the respective mean shapes. Asterisks ($*$) correspond to the mean shapes' landmarks.

coordinates of the sample observations. Define $\Lambda = E((D_r D_s h(\tilde{X}_1, 0)))_{r,s=1}^{2k-4}$.
Then Λ *is positive definite if the support of* Q *is contained in* $B(\mu_I, R)$,
where R *is the unique solution of* $\tan(x) = 2x$, $x \in (0, \frac{\pi}{2})$.

Proof For a geodesic γ starting at μ_I, write $\gamma = \pi \circ \tilde{\gamma}$, where $\tilde{\gamma}$ is a geodesic in $\mathbb{C}S^{k-1}$ starting at μ. From the proof of Theorem 5.3 we have, for $m = \pi(z) \in B(p, \frac{\pi}{4})$,

$$\frac{d}{ds} d_g^2(\gamma(s), m) = 2\langle T(s, 1), \dot{\gamma}(s) \rangle = 2\langle \tilde{T}(s, 1), \dot{\tilde{\gamma}}(s) \rangle \tag{8.2}$$

$$\frac{d^2}{ds^2} d_g^2(\gamma(s), m) = 2\langle D_s T(s, 1), \dot{\gamma}(s) \rangle = 2\langle D_s \tilde{T}(s, 1), \dot{\tilde{\gamma}}(s) \rangle, \tag{8.3}$$

where $\tilde{T}(s, 1) = d\pi_{\gamma(s)}^{-1}(T(s, 1))$. From equation (8.1), this has the expression

$$\tilde{T}(s, 1) = -\frac{\rho(s)}{\sin(\rho(s))} \left[-\cos(\rho(s))\tilde{\gamma}(s) + e^{i\theta(s)} z \right], \tag{8.4}$$

where $e^{i\theta(s)} = \frac{\tilde{\gamma}(s)z^*}{\cos(\rho(s))}$ and $\rho(s) = d_g(\gamma(s), m)$.

The inner product in equations (8.2) and (8.3) is the Riemannian metric on $T\mathbb{C}S^{k-1}$, which is $\langle v, w \rangle = \text{Re}(vw^*)$. Observe that $D_s \tilde{T}(s, 1)$ is $\frac{d}{ds}\tilde{T}(s, 1)$ projected onto $H_{\tilde{\gamma}(s)}$. Because $\langle \mu, \dot{\tilde{\gamma}}(0) \rangle = 0$, we get

$$\frac{d^2}{ds^2} d_g^2(\gamma(s), m)\Big|_{s=0} = 2\langle \frac{d}{ds} \tilde{T}(s, 1)\Big|_{s=0}, \dot{\tilde{\gamma}}(0) \rangle.$$

From equation (8.4) we have

$$\frac{d}{ds}\tilde{T}(s, 1)|_{s=0} = \left(\frac{d}{ds}(\frac{\rho(s)\cos(\rho(s))}{\sin(\rho(s))})\Big|_{s=0} \right) \mu + \left(\frac{\rho(s)\cos(\rho(s))}{\sin \rho(s)}\Big|_{s=0} \right) \dot{\tilde{\gamma}}(0)$$

$$- \left(\frac{d}{ds}(\frac{\rho(s)}{\sin(\rho(s))\cos(\rho(s))})\Big|_{s=0} \right) (\mu z^*) z$$

$$- \left(\frac{\rho(s)}{\sin(\rho(s))\cos(\rho(s))}\Big|_{s=0} \right) (\dot{\tilde{\gamma}}(0)z^*) z,$$

and along with equation (8.2) we get

$$\frac{d}{ds}\rho(s)\Big|_{s=0} = \frac{-1}{\sin(r)} \langle \dot{\tilde{\gamma}}(0), \frac{\mu z^*}{\cos(r)} z \rangle,$$

where $r = d_g(m, \mu_I)$. Hence

$$\left\langle \frac{d}{ds}\tilde{T}(s,1)\Big|_{s=0}, \dot{\bar\gamma}(0)\right\rangle = r\frac{\cos(r)}{\sin(r)}\|\dot{\bar\gamma}(0)\|^2 - \left(\frac{1}{\sin^2 r} - r\frac{\cos(r)}{\sin^3(r)}\right)(\mathrm{Re}(x))^2$$
$$+ \frac{r}{\sin(r)\cos(r)}(\mathrm{Im}(x))^2, \tag{8.5}$$

where

$$x = e^{i\theta}z\dot{\bar\gamma}(0)^*, \quad e^{i\theta} = \frac{\mu z^*}{\cos(r)}. \tag{8.6}$$

The value of x in equation (8.6) and hence the expression in equation (8.5) depend on z only through $m = \pi(z)$. Also, if $\gamma = \pi(\gamma_1) = \pi(\gamma_2)$, γ_1 and γ_2 being two geodesics on $\mathbb{C}S^{k-1}$ starting at μ_1 and μ_2, respectively, with $\pi(\mu_1) = \pi(\mu_2) = \pi(\mu)$, then $\gamma_1(t) = \lambda\gamma_2(t)$, where $\mu_2 = \lambda\mu_1$, $\lambda \in \mathbb{C}$. Now it is easy to check that the expression in equation (8.5) depends on μ only through $\pi(\mu) = \mu_I$.

Note that $|x|^2 < 1 - \cos^2(r)$. So when $|\dot{\bar\gamma}(0)| = 1$, equation (8.5) becomes

$$r\frac{\cos(r)}{\sin(r)} - \left(\frac{1}{\sin^2(r)} - r\frac{\cos(r)}{\sin^3(r)}\right)(\mathrm{Re}(x))^2 + \frac{r}{\sin(r)\cos(r)}(\mathrm{Im}(x))^2$$
$$> r\frac{\cos(r)}{\sin(r)} - \left(\frac{1}{\sin^2(r)} - r\frac{\cos(r)}{\sin^3(r)}\right)\sin^2(r)$$
$$= \frac{2r - \tan(r)}{\tan(r)}, \tag{8.7}$$

which is strictly positive if $r \leq R$, where

$$\tan(R) = 2R, \quad R \in (0, \frac{\pi}{2}).$$

Therefore, if $\mathrm{supp}(Q) \subseteq B(\mu_I, R)$, then $\frac{d^2}{ds^2}d^2(\gamma(s), m)|_{s=0} > 0$ and hence Λ is positive definite. $\qquad\square$

It can be shown that $R \in (\frac{\pi}{3}, \frac{2\pi}{5})$. It is approximately 0.37101π.

From Theorems 3.10 and 8.1, we conclude that if $\mathrm{supp}(Q) \subseteq B(p, \frac{\pi}{4}) \cap B(\mu_I, R)$ and if Σ is nonsingular (e.g., if Q is absolutely continuous), then the coordinates of the sample mean shape from an i.i.d. sample have an asymptotically Normal distribution with nonsingular covariance. Note that the coordinate map ϕ in Theorem 8.1 has the form

$$\phi(m) = (\tilde{m}^1, \ldots, \tilde{m}^{k-2}), \quad \tilde{m}^j = \frac{r}{\sin(r)}e^{i\theta}z\bar{v}_j,$$

where $m = \pi(z)$, $\mu_I = \pi(\mu)$, $r = \arccos(|\mu z^*|)$, and $e^{i\theta} = \frac{\mu z^*}{|\mu z^*|}$. Corollary 8.2 below derives expressions for Λ and Σ in terms of ϕ.

Corollary 8.2 *In Theorem 8.1, if Q has support in a geodesic ball of radius $\frac{\pi}{4}$, then Λ has the following expression:*

$$\Lambda = \begin{bmatrix} \Lambda_{11} & \Lambda_{12} \\ \Lambda'_{12} & \Lambda_{22} \end{bmatrix}, \tag{8.8}$$

where, for $1 \le r, s \le k - 2$,

$$(\Lambda_{11})_{rs} = 2\mathrm{E}\Big[d_1 \cot(d_1)\delta_{rs} - \frac{(1 - d_1 \cot(d_1))}{d_1^2}(\mathrm{Re}(\tilde{X}_1^r))(\mathrm{Re}(\tilde{X}_1^s))$$
$$+ \frac{\tan(d_1)}{d_1}(\mathrm{Im}(\tilde{X}_1^r))(\mathrm{Im}(\tilde{X}_1^s)) \Big],$$

$$(\Lambda_{22})_{rs} = 2\mathrm{E}\Big[d_1 \cot(d_1)\delta_{rs} - \frac{(1 - d_1 \cot(d_1))}{d_1^2}(\mathrm{Im}\tilde{X}_1^r)(\mathrm{Im}\tilde{X}_1^s)$$
$$+ \frac{\tan(d_1)}{d_1}(\mathrm{Re}\tilde{X}_1^r)(\mathrm{Re}\tilde{X}_1^s) \Big],$$

$$(\Lambda_{12})_{rs} = -2\mathrm{E}\Big[\frac{(1 - d_1 \cot(d_1))}{d_1^2}(\mathrm{Re}(\tilde{X}_1^r))(\mathrm{Im}(\tilde{X}_1^s))$$
$$+ \frac{\tan(d_1)}{d_1}(\mathrm{Im}(\tilde{X}_1^r))(\mathrm{Re}(\tilde{X}_1^s)) \Big],$$

with $d_1 = d_g(X_1, \mu_I)$. If we define $\Sigma = \mathrm{Cov}((D_r h(\tilde{X}_1, 0))_{r=1}^{2k-4}$, then it can be expressed as

$$\Sigma = \begin{bmatrix} \Sigma_{11} & \Sigma_{12} \\ \Sigma'_{12} & \Sigma_{22} \end{bmatrix}, \tag{8.9}$$

where, for $1 \le r, s \le k - 2$,

$$(\Sigma_{11})_{rs} = 4\mathrm{E}(\mathrm{Re}(\tilde{X}_1^r)\mathrm{Re}(\tilde{X}_1^s)),$$
$$(\Sigma_{12})_{rs} = 4\mathrm{E}(\mathrm{Re}(\tilde{X}_1^r)\mathrm{Im}(\tilde{X}_1^s)),$$
$$(\Sigma_{22})_{rs} = 4\mathrm{E}(\mathrm{Im}(\tilde{X}_1^r)\mathrm{Im}(\tilde{X}_1^s)).$$

Proof With respect to the orthonormal basis $\{v_1, \dots, v_{k-2}, iv_1, \dots, iv_{k-2}\}$ for H_μ, $\tilde{X}_j$ has coordinates

$$(\mathrm{Re}(\tilde{X}_j^1), \dots, \mathrm{Re}(\tilde{X}_j^{k-2}), \mathrm{Im}(\tilde{X}_j^1), \dots, \mathrm{Im}(\tilde{X}_j^{k-2}))$$

in $\mathbb{R}^{2k-4}$. Now the expression for Σ follows from Corollary 5.4. If one writes Λ as in equation (8.8) and if $\dot{\tilde{\gamma}}(0) = \sum_{j=1}^{k-2} x^j v_j + \sum_{j=1}^{k-2} y^j(iv_j)$, then

$$\mathrm{E}\left(\frac{d^2}{ds^2}d_g^2(\gamma(s), X_1)\right)\Big|_{s=0} = x\Lambda_{11}x' + y\Lambda_{22}y' + 2x\Lambda_{12}y',$$

where $x = (x^1, \dots, x^{k-2})$ and $y = (y^1, \dots, y^{k-2})$. Now expressions for Λ_{11}, Λ_{12}, and Λ_{22} follow from the proof of Theorem 8.1. $\square$

Using the expressions for Λ and Σ from Corollary 8.2, one can construct confidence regions for the population intrinsic mean as in Sections 3.4 and 5.2. One may also carry out two-sample tests as in Section 5.4 to distinguish between two probability distributions on Σ_2^k by comparing the sample intrinsic means.

8.5 Other Fréchet functions

Consider the general definition of a Fréchet function as in equation (3.1), with ρ being the geodesic distance on Σ_2^k, that is,

$$F(p) = \int_{\Sigma_2^k} d_g^\alpha(p, m) Q(dm).$$

In this section we investigate conditions for the existence of a unique Fréchet mean.

Suppose the support of Q is contained in a convex geodesic ball $B(p, \frac{\pi}{4})$. Let $m \in B(p, \frac{\pi}{4})$. Let $\gamma(s)$ be a geodesic in its closure $\bar{B}(p, \frac{\pi}{4})$. Then it is easy to show that

$$\frac{d}{ds} d_g^\alpha(\gamma(s), m) = \frac{\alpha}{2} d_g^{\alpha-2}(\gamma(s), m) \frac{d}{ds} d_g^2(\gamma(s), m),$$

$$\frac{d^2}{ds^2} d_g^\alpha(\gamma(s), m) = \frac{\alpha}{2}(\frac{\alpha}{2} - 1) d_g^{\alpha-4}(\gamma(s), m) \frac{d}{ds} d_g^2(\gamma(s), m)$$

$$+ \frac{\alpha}{2} d_g^{\alpha-2}(\gamma(s), m) \frac{d^2}{ds^2} d_g^2(\gamma(s), m).$$

We can get expressions for $\frac{d}{ds} d_g^2(\gamma(s), m)$ and $\frac{d^2}{ds^2} d_g^2(\gamma(s), m)$ from equations (8.2) and (8.3). For example, when $\alpha = 3$,

$$\frac{d}{ds} d_g^3(\gamma(s), m) = -3 d_g(\gamma(s), m) \langle \exp_{\gamma(s)}^{-1} m, \dot{\gamma}(s) \rangle$$

$$\frac{d^2}{ds^2} d_g^3(\gamma(s), m) = 3 d^2 \frac{\cos(d)}{\sin(d)} |\dot{\gamma}(s)|^2 + 3 d^2 \frac{\cos(d)}{\sin^3(d)} (\mathrm{Re}(z))^2$$

$$+ \frac{3 d^2}{\sin(d) \cos(d)} (\mathrm{Im}(z))^2,$$

where $d = d_g(\gamma(s), m)$, $z = e^{i\theta} \tilde{m} \dot{\tilde{\gamma}}(s)^*$, $e^{i\theta} = \frac{\tilde{\gamma}(s) \tilde{m}^*}{\cos(d)}$, $m = \pi(\tilde{m})$, and $\gamma(s) = \pi(\tilde{\gamma}(s))$. The expression for $\frac{d^2}{ds^2} d_g^3(\gamma(s), m)$ is strictly positive if $m \neq \gamma(s)$. Hence the Fréchet function of Q is strictly convex in $B(p, \frac{\pi}{4})$ and hence has a unique minimizer that is called the (local) Fréchet mean of Q and denoted by μ_F. Replace Q by the empirical distribution Q_n to get the (local) sample Fréchet mean μ_{nF}. This proves the following theorem.

Theorem 8.3 *Suppose* supp$(Q) \subseteq B(p, \frac{\pi}{4})$. *Consider the Fréchet function of Q:*

$$F(q) = \int_{\Sigma_2^k} d_g^3(q, m) Q(dm).$$

Then (a) Q has a unique (local) Fréchet mean μ_F in $B(p, \frac{\pi}{4})$ and if μ_{nF} denotes the (local) sample Fréchet mean from an i.i.d. random sample from Q, then (b) $\sqrt{n}\, \phi(\mu_{nF})$ has a asymptotic mean zero Normal distribution, with ϕ defined as in Theorem 8.1.

8.6 Extrinsic analysis on the planar shape space

For extrinsic analysis on the planar shape space, we embed it into the space $S(k, \mathbb{C})$ of all $k \times k$ complex Hermitian matrices. Here $S(k, \mathbb{C})$ is viewed as a (real) vector space with respect to the scalar field $\mathbb{R}$. The embedding is called the *Veronese–Whitney embedding* and is given by

$$j: \Sigma_2^k \to S(k, \mathbb{C}),$$
$$j(\sigma(z)) = j(\pi(u)) = u'\bar{u} \quad (u = (u_1, \ldots, u_k) \in \mathbb{C}S^{k-1})$$
$$= ((u_i \bar{u}_j))_{1 \le i, j \le k},$$

where $u = \frac{z - \langle z \rangle}{\|z - \langle z \rangle\|}$ is the preshape of the planar k-ad z. Define the extrinsic distance ρ on Σ_2^k by that induced from this embedding, namely,

$$\rho^2(\sigma(z), \sigma(w)) = \|u'\bar{u} - v'\bar{v}\|^2, \quad u \doteq \frac{z - \langle z \rangle}{\|z - \langle z \rangle\|}, \quad v \doteq \frac{w - \langle w \rangle}{\|w - \langle w \rangle\|},$$

where, for arbitrary $k \times k$ complex matrices A and B,

$$\|A - B\|^2 = \sum_{j,j'} \|a_{jj'} - b_{jj'}\|^2 = \text{Trace}((A - B)(A - B)^*)$$

is just the squared Euclidean distance between A and B regarded as elements of $\mathbb{C}^{k^2}$ (or $\mathbb{R}^{2k^2}$). Hence we get

$$\rho^2(\sigma(z), \sigma(w)) = 2(1 - |\bar{u}v'|^2).$$

The image of Σ_2^k under the Veronese–Whitney embedding is given by

$$j(\Sigma_2^k) = \{A \in S^+(k, \mathbb{C}) : \text{rank}(A) = 1, \ \text{Trace}(A) = 1, A\mathbf{1}_k = 0\}.$$

Here $S^+(k, \mathbb{C})$ is the space of all complex positive semi-definite matrices, "rank" denotes the complex rank, and $\mathbf{1}_k$ is the k-dimensional column vector of all ones. Thus the image is a compact submanifold of $S(k, \mathbb{C})$ of (real) dimension $2k - 4$.

We may also represent the preshape u using $(k-1)$ Helmertized coordinates and embed the resulting shape space into $S(k-1, \mathbb{C})$. Then

$$j(\Sigma_2^k) = \{A \in S^+(k-1, \mathbb{C}) : \text{rank}(A) = \text{Trace}(A) = 1\}.$$

Kendall (1984) shows that then the embedding j is equivariant under the action of the *special unitary group*:

$$SU(k-1) = \{A \in GL(k-1, \mathbb{C}) : AA^* = I, \det(A) = 1\}$$

which acts on the right: $A\pi(u) = \pi(uA)$. Indeed,

$$j(A\pi(u)) = A'u'\bar{u}\bar{A} = \phi(A)j(\pi(u)),$$

where

$$\phi(A) : S(k-1, \mathbb{C}) \to S(k-1, \mathbb{C}), \quad \phi(A)B = A'B\bar{A}$$

is an isometry.

8.7 Extrinsic mean and variation

Let Q be a probability measure on the shape space Σ_2^k, let $X_1, X_2, \ldots, X_n$ be an i.i.d. sample from Q, and let $\tilde{\mu}$ denote the mean vector of $\tilde{Q} \doteq Q \circ j^{-1}$, regarded as a probability measure on $\mathbb{C}^{k^2}$ (or $\mathbb{R}^{2k^2}$). Note that $\tilde{\mu}$ belongs to the convex hull of $\tilde{M} = j(\Sigma_2^k)$ and therefore is positive semi-definite and satisfies

$$\tilde{\mu}\mathbf{1}_k = 0, \quad \text{Trace}(\tilde{\mu}) = 1, \quad \text{rank}(\tilde{\mu}) \geq 1.$$

Let T be a matrix in $SU(k)$ such that

$$T\tilde{\mu}T^* = D = \text{Diag}(\lambda_1, \lambda_2, \ldots, \lambda_k),$$

where $\lambda_1 \leq \lambda_2 \leq \cdots \leq \lambda_k$ are the eigenvalues of $\tilde{\mu}$ in ascending order. Then, writing $v = Tu'$ with $u \in \mathbb{C}S^{k-1}$, we get

$$\|u'\bar{u} - \tilde{\mu}\|^2 = \|vv^* - D\|^2 = \sum_j \sum_i |v_j\bar{v}_i - \lambda_j\delta_{ji}|^2$$

$$= \sum_j (|v_j|^2 + \lambda_j^2 - 2\lambda_j|v_j|^2)$$

$$= \sum_j \lambda_j^2 + 1 - 2\sum_{j=1}^k \lambda_j|v_j|^2,$$

which is minimized on $j(\Sigma_2^k)$ by taking $v_j = 0$ for all $j \neq k$, and $|v_k| = 1$, that is, by taking u' to be a unit (column) eigenvector of $\tilde{\mu}$ having the largest eigenvalue λ_k. This implies that the projection set of $\tilde{\mu}$ on $\tilde{M}$, as defined in Section 4.1, consists of all $\mu'\bar{\mu}$, where μ' is a unit eigenvector of $\bar{\mu}$ corresponding to λ_k. The projection set is a singleton, in other words, $\bar{\mu}$ is a nonfocal point of $S(k,\mathbb{C})$, if and only if the eigenspace for the largest eigenvalue of $\bar{\mu}$ is (complex) one-dimensional, that is, when λ_k is a simple eigenvalue. In this case Q has a unique extrinsic mean μ_E, say, which is given by $\mu_E = j^{-1}(\mu'\bar{\mu}) = \pi(\mu)$.

The minimum squared distance between $\bar{\mu}$ and the (image of the) extrinsic mean is $\sum \lambda_j^2 + 1 - 2\lambda_k$.

If one writes $X_j = \pi(Z_j)$, $j = 1, 2, \ldots, n$, where Z_j is a preshape of X_j in $\mathbb{C}S^{k-1}$, then from Proposition 4.2 it follows that the extrinsic variation of Q has the expression

$$V = \mathrm{E}\left[\|Z_1'\bar{Z}_1 - \bar{\mu}\|^2\right] + \|\bar{\mu} - \mu'\bar{\mu}\|^2 = 2(1 - \lambda_k).$$

Therefore, we have the following consequence of Corollary 3.4 and Proposition 3.8.

Proposition 8.4 *Let μ_n' denote a unit eigenvector of $\frac{1}{n}\sum_{j=1}^{n} Z_j'\bar{Z}_j$ having the largest eigenvalue λ_{kn}. (a) If the largest eigenvalue λ_k of $\bar{\mu}$ is simple with a unit eigenvector μ', then the sample extrinsic mean $\pi(\mu_n)$ is a strongly consistent estimator of the extrinsic mean $\pi(\mu)$ of Q. (b) The sample extrinsic variation $V_n = 2(1 - \lambda_{kn})$ is a strongly consistent estimator of the extrinsic variation $V = 2(1 - \lambda_k)$ of Q.*

8.8 Asymptotic distribution of the sample extrinsic mean

In this section, we assume that Q has a unique extrinsic mean $\mu_E = \pi(\mu)$, where μ' is a unit eigenvector corresponding to the largest eigenvalue of the mean $\bar{\mu}$ of $Q \circ j^{-1}$. To get the asymptotic distribution of the sample extrinsic mean μ_{nE} using Proposition 4.3, we need to differentiate the projection map

$$P: S(k,\mathbb{C}) \to j(\Sigma_2^k), \quad P(\bar{\mu}) = \mu'\bar{\mu}$$

in a neighborhood of a nonfocal point such as $\bar{\mu}$. We consider $S(k,\mathbb{C})$ as a linear subspace of $\mathbb{C}^{k^2}$ (over $\mathbb{R}$) and, as such, a regular submanifold of $\mathbb{C}^{k^2}$ embedded by the inclusion map and inheriting the metric tensor

$$\langle A, B \rangle = \mathrm{Re}\left(\mathrm{Trace}(AB^*)\right).$$

The (real) dimension of $S(k,\mathbb{C})$ is k^2. An orthonormal basis for $S(k,\mathbb{C})$ is given by $\{v_b^a : 1 \le a \le b \le k\}$ and $\{w_b^a : 1 \le a < b \le k\}$, defined as

$$v_b^a = \begin{cases} \frac{1}{\sqrt{2}}(e_a e_b^t + e_b e_a^t) & a < b, \\ e_a e_a^t & a = b, \end{cases}$$

$$w_b^a = +\frac{i}{\sqrt{2}}(e_a e_b^t - e_b e_a^t) \quad a < b,$$

where $\{e_a : 1 \le a \le k\}$ is the standard canonical basis for $\mathbb{R}^k$. One can also take $\{v_b^a : 1 \le a \le b \le k\}$ and $\{w_b^a : 1 \le a < b \le k\}$ as the (constant) orthonormal frame for $S(k, \mathbb{C})$. For any $U \in SU(k)$ ($UU^* = U^*U = I$, $\det(U){=}{+}1$), $\{Uv_b^a U^* : 1 \le a \le b \le k\}$, $\{Uw_b^a U^* : 1 \le a < b \le k\}$ is also an orthonormal frame for $S(k, \mathbb{C})$. Because of its Euclidean structure one can identify $T_A S(k, \mathbb{C})$ with $S(k, \mathbb{C})$ for every $A \in S(k, \mathbb{C})$. We may then first consider P as a map on $S(k, \mathbb{C})$ into $S(k, \mathbb{C})$ and view the differential $d_{\tilde{\mu}} P$ as a map on $S(k, \mathbb{C})$ into $S(k, \mathbb{C})$. After identifying an orthonormal basis of $T_{P(\tilde{\mu})} j(\Sigma_2^k)$, we will finally view $d_{\tilde{\mu}} P$ as a map on $T_{\tilde{\mu}} S(k, \mathbb{C}) \approx S(k, \mathbb{C})$ into $T_{P(\tilde{\mu})} j(\Sigma_2^k)$, for the desired application of Proposition 4.3. Choose $U \in SU(k)$ such that $U^* \tilde{\mu} U = D$:

$$U = (U_1, \ldots, U_k) \text{ and } D = \text{Diag}(\lambda_1, \ldots, \lambda_k).$$

Here $\lambda_1 \le \cdots \le \lambda_{k-1} < \lambda_k$ are the eigenvalues of $\tilde{\mu}$ and $U_1, \ldots, U_k$ are the corresponding orthonormal eigenvectors. Choose the orthonormal basis frame $\{Uv_b^a U^*, Uw_b^a U^*\}$ for $S(k, \mathbb{C})$. Then it can be shown that

$$d_{\tilde{\mu}} P(Uv_b^a U^*) = \begin{cases} 0 & \text{if } 1 \le a \le b < k, a = b = k, \\ (\lambda_k - \lambda_a)^{-1} Uv_k^a U^* & \text{if } 1 \le a < k, b = k; \end{cases}$$

$$d_{\tilde{\mu}} P(Uw_b^a U^*) = \begin{cases} 0 & \text{if } 1 \le a < b < k, \\ (\lambda_k - \lambda_a)^{-1} Uw_k^a U^* & \text{if } 1 \le a < k, b = k. \end{cases} \tag{8.10}$$

The proof is similar to that for the real projective shape that is considered in Section 12.6. Let $\tilde{X}_j = J(X_j)$, $j = 1, 2, \ldots, n$, where $X_1, \ldots, X_n$ is an i.i.d. random sample from Q. Write

$$\tilde{X}_j - \tilde{\mu} = \sum_{1 \le a \le b \le k} \langle (\tilde{X}_j - \tilde{\mu}), Uv_b^a U^* \rangle Uv_b^a U^*$$

$$+ \sum_{1 \le a < b \le k} \langle (\tilde{X}_j - \tilde{\mu}), Uw_b^a U^* \rangle Uw_b^a U^*. \tag{8.11}$$

Because $\tilde{X}_j \mathbf{1}_k = \tilde{\mu} \mathbf{1}_k = 0$, hence $\lambda_1 = 0$ and one can choose $U_1 = \alpha \mathbf{1}_k$, where $|\alpha| = 1/\sqrt{k}$. Therefore,

$$\langle (\tilde{X}_j - \tilde{\mu}), Uv_b^1 U^* \rangle = \langle (\tilde{X}_j - \tilde{\mu}), Uw_b^1 U^* \rangle = 0, \quad 1 \le b \le k.$$

Then, from equations (8.10) and (8.11), it follows that

$$d_{\tilde{\mu}} P(\tilde{X}_j - \tilde{\mu})$$

$$= \sum_{a=2}^{k-1} \langle (\tilde{X}_j - \tilde{\mu}), U v_k^a U^* \rangle (\lambda_k - \lambda_a)^{-1} U v_k^a U^*$$

$$+ \sum_{a=2}^{k-1} \langle (\tilde{X}_j - \tilde{\mu}), U w_k^a U^* \rangle (\lambda_k - \lambda_a)^{-1} U w_k^a U^*$$

$$= \sum_{a=2}^{k-1} \sqrt{2} \, \mathrm{Re}(U_a^* \tilde{X}_j U_k)(\lambda_k - \lambda_a)^{-1} U v_k^a U^*$$

$$+ \sum_{a=2}^{k-1} \sqrt{2} \, \mathrm{Im}(U_a^* \tilde{X}_j U_k)(\lambda_k - \lambda_a)^{-1} U w_k^a U^*. \tag{8.12}$$

From equation (8.12) it is easy to check that the vectors

$$\{U v_k^a U^*, U w_k^a U^* : \ a = 2, \ldots, k-1\} \tag{8.13}$$

form an orthonormal basis for $T_{P(\tilde{\mu})} \tilde{M}$. Further, $d_{\tilde{\mu}} P(\tilde{X}_j - \tilde{\mu})$ has coordinates

$$T_j(\tilde{\mu}) \equiv (T_j^1(\tilde{\mu}), \ldots, T_j^{2k-4}(\tilde{\mu}))'$$

with respect to this orthonormal basis, where

$$T_j^a(\tilde{\mu}) = \begin{cases} \sqrt{2}(\lambda_k - \lambda_a)^{-1} \mathrm{Re}(U_{a+1}^* \tilde{X}_j U_k) & \text{if } 1 \le a \le k-2, \\ \sqrt{2}(\lambda_k - \lambda_a)^{-1} \mathrm{Im}(U_{a-k+3}^* \tilde{X}_j U_k) & \text{if } k-1 \le a \le 2k-4. \end{cases} \tag{8.14}$$

It follows from Proposition 4.3 that

$$\sqrt{n} \, \bar{T}(\tilde{\mu}) \xrightarrow{\mathcal{L}} N(0, \Sigma), \tag{8.15}$$

where $\bar{T}(\tilde{\mu}) = \frac{1}{n} \sum_{j=1}^n T_j(\tilde{\mu})$ and $\Sigma = \mathrm{Cov}(T_1(\tilde{\mu}))$.

8.9 Two-sample extrinsic tests on the planar shape space

Suppose Q_1 and Q_2 are two probability distributions on the planar shape space. Let $X_1, \ldots, X_{n1}$ and $Y_1, \ldots, Y_{n2}$ be two i.i.d. samples from Q_1 and Q_2, respectively, that are mutually independent. One many detect differences between Q_1 and Q_2 by comparing the sample extrinsic mean shapes or the sample extrinsic variations. This puts us in the same set-up as in Section 4.5.1.

To compare the extrinsic means, one may use the statistics T_1 or T_2 defined through equations (4.15) and (4.18), respectively. To get the expression for T_1, one needs to find the coordinates of $d_{\hat{\mu}} P(\tilde{X}_j - \hat{\mu})$ and $d_{\hat{\mu}} P(\tilde{Y}_j - \hat{\mu})$, which are obtained from equation (8.14) by replacing $\tilde{\mu}$ with $\hat{\mu}$. For the statistic T_2, which is

$$T_2 = L[P(\hat{\mu}_1) - P(\hat{\mu}_2)]'\hat{\Sigma}^{-1}L[P(\hat{\mu}_1) - P(\hat{\mu}_2)],$$

where

$$\hat{\Sigma} = \frac{1}{n_1}L\,\hat{\Sigma}_1 L' + \frac{1}{n_2}L\,\hat{\Sigma}_2 L', \tag{8.16}$$

we need expressions for the linear projections L. With respect to the orthonormal basis in equation (8.13) for $T_{P(\tilde{\mu})}j(\Sigma_2^k)$, the linear projection $L(A)$ of a matrix $A \in S(k, \mathbb{C})$ onto $T_{P(\tilde{\mu})}j(\Sigma_2^k)$ has the coordinates

$$L(A) = \{\langle A, Uv_k^a U^*\rangle, \langle A, Uw_k^a U^*\rangle : a = 2, \dots, k-1\}$$
$$= \sqrt{2}\,\{\mathrm{Re}(U_a^* A U_k), \mathrm{Im}(U_a^* A U_k) : a = 2, \dots, k-1\}.$$

For $A_1, A_2 \in S(k, \mathbb{R})$, if we label the bases for $T_{P(A_i)}j(\Sigma_2^k)$ as $\{v_1^i, \dots, v_d^i\}$, $i = 1, 2$, then it is easy to check that the linear projection matrix L_1 from $T_{P(A_1)}j(\Sigma_2^k)$ onto $T_{P(A_2)}j(\Sigma_2^k)$ is the $d \times d$ matrix, $d = 2k - 4$, with the coordinates

$$(L_1)_{ab} = \langle v_a^2, v_b^1\rangle, \quad 1 \le a, b \le d.$$

When the sample sizes are smaller than the dimension d (see Section 8.11.2), the standard error $\hat{\Sigma}$ in equation (8.16) may be singular or close to singular. Then it becomes more effective to estimate it from bootstrap simulations. When the sample sizes are small, we can also perform a bootstrap test using the bootstrap version T_2^* of the test statistic T_2 defined in equation (4.18), say,

$$T_2^* = v^{*'}\Sigma^{*-1}v^*.$$

However, Σ^* is even more likely to be singular or close to singular in most simulations. Then we may compare only a few principal scores of the coordinates of the means. If $d_1 < d$ is the number of principal scores that we want to compare, then the appropriate test statistic to be used is given by

$$T_{21} = (L[P(\hat{\mu}_1) - P(\hat{\mu}_2)])'\hat{\Sigma}_{11}^{-1}L[P(\hat{\mu}_1) - P(\hat{\mu}_2)], \tag{8.17}$$

where $\hat{\Sigma} = U\Lambda U'$, $U = (U_1, \dots, U_d) \in SO(d)$, $\Lambda = \mathrm{Diag}(\lambda_1, \cdots, \lambda_d)$, $\lambda_1 \ge \cdots \ge \lambda_d$ is a singular-value decomposition (s.v.d.) for $\hat{\Sigma}$, and

$$\hat{\Sigma}_{11}^{-1} \doteq \sum_{j=1}^{d_1} \lambda_j^{-1} U_j U_j'.$$

Then T_{21} has an asymptotic $\mathcal{X}^2_{d_1}$ distribution. We can construct its bootstrap analog, say T^*_{21}, and compare the first d_1 principal scores by a pivotal bootstrap test. Alternatively, we may use a nonpivotal bootstrap test statistic

$$T_2^{**} = w^{*\prime}\Sigma^{**-1}w^* \tag{8.18}$$

for comparing the mean shapes, where

$$w^* = L[\{P(\mu_1^*) - P(\hat{\mu}_1)\} - \{P(\mu_2^*) - P(\hat{\mu}_2)\}]$$

and Σ^{**} is the sample covariance of w^* values, estimated from the bootstrap resamples.

To compare the sample extrinsic variations, one may use the statistic T_3 defined in equation (4.20). If $\hat{\lambda}_i$ denotes the largest eigenvalue of $\hat{\mu}_i$, $i = 1, 2$, then

$$T_3 = 2\frac{\hat{\lambda}_2 - \hat{\lambda}_1}{\sqrt{\frac{s_1^2}{n_1} + \frac{s_2^2}{n_2}}}. \tag{8.19}$$

The bootstrap version of T_3 is given by

$$T_3^* = 2\frac{(\lambda_2^* - \hat{\lambda}_2) - (\lambda_1^* - \hat{\lambda}_1)}{\sqrt{\frac{s_1^{*2}}{n_1} + \frac{s_2^{*2}}{n_2}}},$$

where λ_i^* and s_i^* are the bootstrap analogs of $\hat{\lambda}_i$ and s_i, $i = 1, 2$, respectively.

8.10 Planar size-and-shape manifold

It is often important to study change in shape with size – a field known as *allometry*. For a number of applications we refer to Hopkins (1966), Sprent (1972), and the book by Dryden and Mardia (1998). In addition, a study of size and shape together may provide more efficient rules for classification and discrimination in certain cases.

Denote by ξ a centered k-ad, $\xi = z - \langle z \rangle$, using the notation in Sections 8.1 and 8.2. The *size-and-shape* $s\sigma(z)$ of the k-ad z (under Euclidean isometry) is defined as the orbit of ξ under all rotations:

$$s\sigma(z) = \{A\xi : A \in \text{SO}(2)\} \tag{8.20}$$

The *planar size-and-shape space* $S\Sigma_2^k$ comprising all those $s\sigma(z)$, such that not all k points of z are the same, is then a noncompact manifold of dimension $2k - 3$, which is diffeomorphic to $(0, \infty) \times \Sigma_2^k$: $s\sigma(z) \rightarrow (r(\xi), \sigma(z))$, $r(\xi) := \|\xi\|$. An intrinsic analysis can be based on the product metric on

$(0, \infty) \times \Sigma_2^k$. For extrinsic analysis consider the embedding j_s of $S\Sigma_2^k$ into $S(k, \mathbb{C})$ given by

$$j_s s\sigma(z) = r(\xi)u'\bar{u} = \xi'\bar{\xi}/\|\xi\|, \qquad (8.21)$$

where $u = \xi/\|\xi\|$ (as in Section 8.2). As in the case of the embedding j (see Section 8.6), j_s is equivariant under the action of the special unitary group $SU(k-1)$.

Let Q be a probability measure on $S\Sigma_2^k$ and $\tilde{Q} = Q \circ j_s^{-1}$. Let $\tilde{\mu}$ denote the (Euclidean) mean of $\tilde{Q}$.

Proposition 8.5 *Assume $\tilde{Q}$ has finite second moments. Then the extrinsic mean of Q exists if and only if the largest eigenvalue λ_k of $\tilde{\mu}$ is simple. In this case, if u_0 is a unit eigenvector with eigenvalue λ_k, then the extrinsic mean of Q is $\mu_E = j_s^{-1}(\lambda_k u_0'\bar{u}_0)$ and the sample extrinsic mean μ_{nE} is a consistent estimator of μ_E.*

Proof The proof is analogous to that of Proposition 8.4. Because $\tilde{\mu}$ is Hermitian and positive semi-definite, there exists an orthogonal matrix T such that $T\tilde{\mu}T^* = D \equiv \text{Diag}(\lambda_1, \cdots, \lambda_k)$, where $0 \leq \lambda_1 \leq \lambda_2 \leq \cdots \leq \lambda_k$ are the eigenvalues of $\tilde{\mu}$ (counting multiplicities). The squared distance between $\tilde{\mu}$ and an arbitrary element $r(\xi)u'\bar{u}$ of $\tilde{M} = j_s S\Sigma_2^k$ may be expressed in terms of $v = \sqrt{r(\xi)}Tu'$ as

$$\begin{aligned}
\text{Trace}(\tilde{\mu} - ru'\bar{u})^2 &= \text{Trace}(D - vv^*)^2 \\
&= \Sigma_j\lambda_j^2 + \Sigma_j|v_j|^4 - 2\Sigma_j\lambda_j|v_j|^2 + \Sigma_{j\neq j'}|v_j\bar{v}_{j'}|^2 \\
&= \Sigma_j\lambda_j^2 + \Sigma_j|v_j|^4 - 2\Sigma_j\lambda_j|v_j|^2 + \Sigma_j|v_j|^2 \cdot \Sigma_{j'}|v_{j'}|^2 - \Sigma_j|v_j|^4 \\
&= \Sigma_j\lambda_j^2 - 2\Sigma_j\lambda_j|v_j|^2 + r^2(\xi). \qquad (8.22)
\end{aligned}$$

For a given size $r(\xi)$, the minimum of equation (8.22) is achieved by $v = \sqrt{r(\xi)}e_k$, where e_k has 1 as its kth coordinate and zeros elsewhere. The corresponding u is given by $r(\xi)^{-1/2}v'\bar{T}$. With this choice, the last line of (8.22) equals

$$\Sigma_j\lambda_j^2 - 2r\lambda_k + r^2. \qquad (8.23)$$

The minimum of equation (8.23) over all $r > 0$ is attained with $r = \lambda_k$. Hence the minimum of equation (8.22) on $\tilde{M}$ is attained by $\lambda_k u_0'\bar{u}_0$, where u_0 is a unit eigenvector with eigenvalue λ_k. For this to be the unique minimizer, it is necessary and sufficient that λ_k is simple. $\square$

Note that the minimum value of equation (8.22) on $\tilde{M}$ is given by $\Sigma\lambda_j^2 - \lambda_k^2 = \Sigma_{1\leq j\leq k-1}\lambda_j^2$. Hence the extrinsic variation of Q is obtained by adding to this minimum value the trace of the covariance matrix of $\tilde{Q}$ (see Proposition 4.2).

8.11 Applications

In this section, we record the results of two-sample tests carried out in the two examples from Section 8.3.

8.11.1 Gorilla skulls

Consider the data on gorilla skull images from Section 8.3.1. There are 30 female and 29 male gorillas giving rise to two independent samples of sizes 30 and 29, respectively, on Σ_2^k, $k = 8$. To detect the differences in the shapes of the skulls between the two sexes, one may compare the sample mean shapes or variations in shape.

Figure 2.4 shows the plots of the sample extrinsic means for the two sexes along with the pooled sample extrinsic mean. In fact, the Procrustes coordinates for the two means with respect to a preshape of the pooled sample extrinsic mean have been plotted. The coordinates are

$$\hat{\mu}_1 = \begin{matrix} (-0.37, & -0.33; & 0.35, & 0.28; & 0.09, & 0.35; & -0.00, & 0.24; \\ -0.17, & 0.00; & -0.28, & -0.30; & 0.05, & -0.24; & 0.32, & -0.01) \end{matrix}$$

$$\hat{\mu}_2 = \begin{matrix} (-0.36, & -0.35; & 0.35, & 0.27; & 0.11, & 0.34; & 0.02, & 0.26; \\ -0.18, & 0.01; & -0.29, & -0.32; & 0.05, & -0.22; & 0.30, & 0.01) \end{matrix}$$

$$\hat{\mu} = \begin{matrix} (-0.36, & -0.34; & 0.35, & 0.28; & 0.10, & 0.34; & 0.01, & 0.25; \\ -0.17, & 0.01; & -0.29, & -0.31; & 0.05, & -0.23; & 0.31, & 0.00) \end{matrix}$$

where $\hat{\mu}_1$, $\hat{\mu}_2$, denote the Procrustes coordinates of the extrinsic mean shapes for the female and male samples, respectively, with respect to $\hat{\mu}$, and $\hat{\mu}$ is a preshape of the pooled sample extrinsic mean. The x and y coordinates for each landmark are separated by commas, while the different landmarks are separated by semicolons. The sample intrinsic means are very close to their extrinsic counterparts; the geodesic distances between the intrinsic and extrinsic means are 5.54×10^{-7} for the female sample and 1.96×10^{-6} for the male sample.

The value of the two-sample test statistic defined in equation (5.21) for comparing the intrinsic mean shapes and the asymptotic p-value for the chi-squared test are

$$T_{n1} = 391.63, \quad \text{p-value} = P(\mathcal{X}_{12}^2 > 391.63) < 10^{-16}.$$

Hence we reject the null hypothesis that the two sexes have the same intrinsic mean shape.

The two-sample test statistics defined in equations (4.15) and (4.18) for comparing the extrinsic mean shapes and the corresponding asymptotic p-values are

$$T_1 = 392.6, \quad \text{p-value} = P(\mathcal{X}_{12}^2 > 392.6) < 10^{-16},$$
$$T_2 = 392.0585, \quad \text{p-value} < 10^{-16}.$$

Hence we reject the null hypothesis that the two sexes have the same extrinsic mean shape. We can also compare the mean shapes by a pivotal bootstrap method using the t bootstrap version of the test statistic T_2 defined in equation (4.18). The p-value for the bootstrap test using 10^5 simulations turns out to be 0.

The sample extrinsic variations for the female and male samples are 0.0038 and 0.005, respectively. The value of the two-sample test statistic in equation (8.19) for testing the equality of the extrinsic variations is 0.923, and the asymptotic p-value is

$$P(|Z| > 0.923) = 0.356, \quad \text{where } Z \sim N(0, 1).$$

Hence we do not reject the null hypothesis that the two underlying distributions have the same extrinsic variation. However, because the mean shapes are different, it is possible to distinguish between the distribution of shapes for the two sexes.

8.11.2 Schizophrenia detection

In this example from Section 8.3.2, we have two independent random samples of size 14 each on Σ_2^k, $k = 13$. To distinguish between the underlying distributions, we compare the mean shapes and shape variations.

Figure 8.3 shows the Procrustes coordinates of the sample extrinsic means for the two groups of patients along with a preshape for the pooled sample extrinsic mean. The coordinates for the two sample means have been obtained with respect to the pooled sample mean's preshape. The coordinates for the three means are

$$\hat{\mu}_1 = (0.14, 0.01; -0.22, 0.22; 0.01, 0.21; 0.31, 0.30; 0.24, -0.28;$$
$$0.15, -0.06; 0.06, -0.19; -0.01, -0.33; -0.05, -0.04; -0.09, -0.19;$$
$$-0.20, 0.02; -0.39, 0.32; 0.04, -0.00)$$
$$\hat{\mu}_2 = (0.16, 0.02; -0.22, 0.22; 0.02, 0.22; 0.31, 0.31; 0.24, -0.28;$$
$$0.15, -0.07; 0.06, -0.18; -0.01, -0.33; -0.06, -0.04; -0.09, -0.20;$$
$$-0.19, 0.03; -0.39, 0.30; 0.03, 0.00)$$

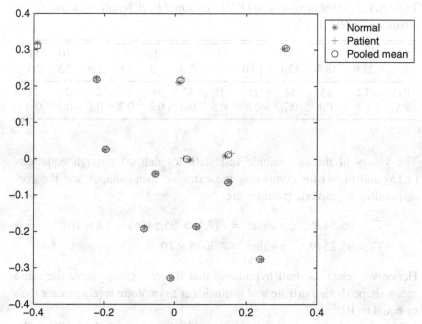

Figure 8.3 The sample extrinsic means for the two groups along with the pooled sample mean, corresponding to Figure 8.2.

$$\hat{\mu} = (0.15, 0.01; -0.22, 0.22; 0.02, 0.22; 0.31, 0.30; 0.24, -0.28;$$
$$0.15, -0.06; 0.06, -0.19; -0.01, -0.33; -0.05 - 0.04; -0.09, -0.19;$$
$$-0.20, 0.03; -0.39, 0.310.03, 0.00).$$

Here $\hat{\mu}_i$, $i = 1, 2$, denotes the Procrustes coordinates of the extrinsic mean shape for the sample of normal and schizophrenic patients, respectively, and $\hat{\mu}$ is the preshape of the pooled sample extrinsic mean.

As in case of the gorilla skull images from the last section, the sample intrinsic means are very close to their extrinsic counterparts; the geodesic distances between the intrinsic and extrinsic means are 1.65×10^{-5} for the normal patients sample and 4.29×10^{-5} for the sample of schizophrenic patients.

The values of the two-sample test statistic in equation (5.21) for testing the equality of the population intrinsic mean shapes along with the asymptotic p-values are

$$T_{n1} = 95.4587, \quad \text{p-value} = P(\mathcal{X}_{22}^2 > 95.4587) = 3.97 \times 10^{-11}.$$

Table 8.1 *Percent of variation (P.V.) explained by different principal components (P.C.) of* $\hat{\Sigma}$

P.C.	1	2	3	4	5	6	7	8	9	10	11
P.V.	21.6	18.4	12.1	10.0	9.9	6.3	5.3	3.6	3.0	2.5	2.1
P.C.	12	13	14	15	16	17	18	19	20	21	22
P.V.	1.5	1.0	0.7	0.5	0.5	0.3	0.2	0.2	0.1	0.1	0.0

The values of the two-sample test statistics defined through equations (4.15) and (4.18) for comparing the extrinsic mean shapes and the corresponding asymptotic p-values are

$$T_1 = 95.5476, \quad \text{p-value} = P(X_{22}^2 > 95.5476) = 3.8 \times 10^{-11},$$

$$T_2 = 95.2549, \quad \text{p-value} = 4.3048 \times 10^{-11}.$$

Hence we reject the null hypothesis that the two groups have the same mean shape (both extrinsic and intrinsic) at asymptotic levels greater than or equal to 10^{-10}.

Next we compare the extrinsic means by bootstrap methods. Since the dimension 22 of the underlying shape space is much higher than the sample sizes, it becomes difficult to construct a bootstrap test statistic as in the earlier section. That is because the bootstrap estimate of the standard error $\hat{\Sigma}$ defined in equation (8.16) tends to be singular in most simulations. Hence we only compare the first few principal scores of the coordinates of the sample extrinsic means. Table 8.1 displays the percentage of variation explained by each principal component of $\hat{\Sigma}$. The value of T_{21} from equation (8.17) for comparing the first five principal scores of $L[P(\hat{\mu}_1) - P(\hat{\mu}_2)]$ with $\mathbf{0}$ and the asymptotic p-value are

$$T_{21} = 12.1872, \quad \text{p-value} = P(X_5^2 > 12.1872) = 0.0323.$$

The bootstrap p-value from 10^4 simulations equals 0.0168, which is fairly small.

When we use the nonpivotal bootstrap test statistic T_2^{**} from equation (8.18), the p-value for testing the equality of the extrinsic mean shapes from 10^4 simulations equals 0. The value of T_2 with $\hat{\Sigma}$ replaced by its bootstrap estimate Σ^{**} equals 105.955 and the asymptotic p-value using X_{22}^2 approximation is 5.7798×10^{-13}. Hence we again reject H_0 and conclude that the extrinsic mean shapes are different. Once again, we caution the reader on applying principal component analysis to multi-sample problems (see Remarks 2.1 and 2.2).

Next we test for the equality of extrinsic variations for the two groups of patients. The sample extrinsic variations for the patient and normal samples turn out to be 0.0107 and 0.0093, respectively. The value of the two-sample test statistic in equation (8.19) for testing the equality of the population extrinsic variations is 0.9461 and the asymptotic p-value using standard a Normal approximation is 0.3441. The bootstrap p-value with 10^4 simulations equals 0.3564. Hence we conclude that the extrinsic variations in shapes for the two distributions are not significantly different

Because the mean shapes are different, we conclude that the probability distributions of the shapes of brain scans of normal and schizophrenic patients are distinct.

8.12 References

Theorems 8.1 and 8.3 and Corollary 8.2 are due to Bhattacharya and Bhattacharya (2008b). A general CLT for intrinsic sample means and the asymptotic distribution of the extrinsic sample means are essentially due to Bhattacharya and Patrangenaru (2003, 2005). Computation of the two-sample tests in this chapter are mostly taken from Bhattacharya (2008a) and Bhattacharya and Bhattacharya (2008a). Proposition 8.4 is due to Bandulasiri et al. (2009).

See Dryden and Mardia (1998, pp. 279–286), for a discussion of the problem of consistency of Procrustes estimators in parametric and semiparametric models. Kent and Mardia (1997) have shown that for consistency of such estimators in Kendall's planar shape spaces, one needs to assume that landmarks have *isotropic errors*. We point out that inconsistency aries because of a "disconnect" between the estimators and the population mean that they are meant to estimate. For example, Procrustes estimators often turn out to be the same as extrinsic sample means in our terminology, while the population mean in the semiparametric model used in common literature is not in general the extrinsic population mean.

For many applications in archaeology, astronomy, morphometrics, medical diagnostics, and so on, see Bookstein (1986, 1991), Kendall (1989), Dryden and Mardia (1998), Bhattacharya and Patrangenaru (2003, 2005), Bhattacharya and Bhattacharya (2008a, 2009), and Small (1996).

9

Reflection similarity shape spaces $R\Sigma_m^k$

As pointed out in Chapter 7, for $m > 2$ Kendall's similarity shape space Σ_m^k is not a manifold, and the manifold that results after removing a set of singular points is incomplete and has an unbounded sectional curvature in the usual Riemannian metric. In this chapter we consider a suitable open dense subset NS_m^k of the preshape sphere S_m^k, modulo the full orthogonal group $O(m)$. The resulting shape space is called the reflection similarity shape space. We derive an embedding of this space analogous to the Veronese–Whitney embedding and carry out extrinsic analysis on it.

9.1 Introduction

The *reflection similarity shape* of a k-ad, as defined in Section 6.2.2, is its orbit under translation, scaling, and all orthogonal transformations. Let $x = (x_1, \ldots, x_k)$ be a configuration of k points in $\mathbb{R}^m$, and let z denote its preshape in S_m^k as defined in equation (7.2) or, equivalently, equation (7.3) with Helmerized k-ads. Then the reflection similarity shape of the k-ad x is given by the orbit

$$\sigma(x) = \sigma(z) = \{Az : A \in O(m)\}, \tag{9.1}$$

where $O(m)$ is the group of all $m \times m$ orthogonal matrices (with determinants either $+1$ or -1). For the action of $O(m)$ on S_m^k to be free and the reflection shape space to be a (Riemannian) manifold, we consider only those shapes where the columns of z span $\mathbb{R}^m$. The set of all such z's is denoted by NS_m^k. Then the *reflection similarity shape space* is

$$R\Sigma_m^k = \{\sigma(z) : z \in S_m^k, \ \text{rank}(z) = m\} = NS_m^k / O(m), \tag{9.2}$$

which is a Riemannian manifold of dimension $km - m - 1 - m(m-1)/2$. Note that $NS_m^k \subset \tilde{N}S_m^k$, with the latter set comprising the incomplete manifold in Chapter 6. It is not difficult to see that NS_m^k is a dense open subset of S_m^k, and $R\Sigma_m^k$ is a dense open subset of Σ_m^k in the quotient topology. Throughout

110

this chapter we will regard Σ^k_{0m} as $NS^k_m/SO(m)$, that is, replace $\tilde{N}S^k_m$ by NS^k_m. Note that the calculations in Chapter 7 remain intact. As always, we assume that $k > m$.

9.2 Extrinsic analysis on the reflection shape space

It may be shown that the map

$$j: R\Sigma^k_m \to S(k, \mathbb{R}), \quad j(\sigma(z)) = z'z \qquad (9.3)$$

is an embedding of the reflection shape space into $S(k, \mathbb{R})$.

To prove that j is an embedding one needs to show that (i) j is one-to-one and (ii) its differential dj is one-to-one. To establish (i), we will first show that

$$j(R\Sigma^k_m) = \{A \in M^k_m : A\mathbf{1}_k = 0\}, \qquad (9.4)$$

where M^k_m is the set of all $k \times k$ symmetric positive semi-definite matrices of rank m and trace one. Clearly, the left side of equation (9.4) is contained in the right side. To prove the converse, let $A \in M^k_m$, with $\lambda_1 \leq \lambda_2 \leq \cdots \leq \lambda_m$ its positive eigenvalues and $Y_1, \ldots, Y_m$ the corresponding mutually orthogonal (k-dimensional) unit eigenvectors. Then $A = Y\Lambda Y'$, where Y is the $k \times m$ matrix $(Y_1, \ldots, Y_m)$ and $\Lambda = \text{Diag}(\lambda_1, \lambda_2, \ldots, \lambda_m)$. Write $z' = Y\Lambda^{1/2} = (\lambda_1^{1/2}Y_1, \ldots, \lambda_m^{1/2}Y_m)$, that is, $z = \Lambda^{1/2}Y' = (z_1, \ldots, z_k)$, say, with the z_i being m-dimensional vectors. Then $A = z'z = ((z'_i z_j))_{1 \leq i,j \leq k}$ is the $k \times k$ matrix of inner products (in $\mathbb{R}^m$) of the k vectors $z_1, \ldots, z_k$. Because the row sums of A are all zero, the z_i are centered, that is, $\langle z \rangle = (1/k)(z_1 + \cdots + z_k) = 0$. Also, Trace $(A) = 1$ implies $\sum \|z_i\|^2 = 1$. This establishes equation (9.4). Suppose now that there exists another k-ad $w = (w_1, \ldots, w_k)$ such that $w'w = A$. Then the inner product matrix of w is the same as that of z and, in particular, $\|w_i - w_j\|^2 = \|z_i - z_j\|^2 \ \forall i, j$. By Euclidean geometry, this means $w_i = Bz_i + \mathbf{c}$ for some $B \in O(m)$ and some $\mathbf{c} \in \mathbb{R}^m$ ($i = 1, \ldots, k$). Because the z_i and w_i are centered, $\mathbf{c} = 0$, so that $w_i = Bz_i$ ($i = 1, \ldots, k$). That is, $\sigma(z) = \sigma(w)$. We have shown that j is one-to-one into $S(k, \mathbb{R})$, and that its image $\tilde{M} = j(R\Sigma^k_m)$ is given by the right side of equation (9.4). We omit the proof of (ii), except to notice that j is clearly smooth. The fact that dj is one-to-one may be shown by using its expression derived in Proposition 9.1.

Note that j induces the extrinsic distance

$$\rho^2_E(\sigma(z_1), \sigma(z_2)) = \|j(\sigma(z_1)) - j(\sigma(z_2))\|^2 = \text{Trace}(z'_1 z_1 - z'_2 z_2)^2$$

$$= \text{Trace}(z_1 z'_1)^2 + \text{Trace}(z_2 z'_2)^2 - 2\text{Trace}(z_1 z'_2 z_2 z'_1),$$

$$z_1, z_2 \in S^k_m$$

on $R\Sigma_m^k$. The embedding is H-equivariant where H is isomorphic to $O(k - 1)$. To see this, consider the Helmertized and scaled landmarks $\tilde{z}$ as $m \times (k - 1)$ matrices (See p. 83, and (7.3)), and define the group action $A\sigma(\tilde{z}) = \sigma(\tilde{z}A')$, for $A \in O(k - 1)$. Then

$$j(A\sigma(\tilde{z})) = A\tilde{z}\tilde{z}'A' = \phi(A)j(\sigma(\tilde{z})),$$

where $\phi(A)$ defined by

$$\phi(A): S(k - 1, \mathbb{R}) \to S(k - 1, \mathbb{R}), \quad \phi(A)B = ABA',$$

is an isometry for each $A \in O(k - 1)$.

Using Helmertized $(k + 1)$-ads or restricting the $(k + 1)$-ads to lie in an $(m \times k)$-dimensional subspace (of $\mathbb{R}^{m \times (k+1)}$) isomorphic to $\mathbb{R}^{m \times k}$ (see equations (7.2), (7.3), and (9.4)), one may represent $R\Sigma_m^{k+1}$ as $\{z \in \mathbb{R}^{m \times k} :$ rank of $z = m$, Trace $(z'z) = 1\}$. The image of the latter set under the map $j: z \to z'z$ is precisely M_m^k. Hence one has the representation $M_m^k = j(R\Sigma_m^{k+1})$, and M_m^k is a submanifold (not complete) of $S(k, \mathbb{R})$ of dimension $km - 1 - m(m - 1)/2$. Proposition 9.1 below identifies the tangent and normal spaces of M_m^k.

Proposition 9.1 *Let $A \in M_m^k$. (a) The tangent space of M_m^k at A is given by*

$$T_A(M_m^k) = \left\{ U \begin{pmatrix} T & S \\ S' & 0 \end{pmatrix} U' : T \in S(m, \mathbb{R}), \text{ Trace}(T) = 0 \right\}, \qquad (9.5)$$

where $A = UDU'$ is a spectral decomposition of A, $U \in SO(m)$, and $D = \text{Diag}(\lambda_1, \dots, \lambda_k)$. (b) The orthocomplement of the tangent space in $S(k, \mathbb{R})$ or the normal space is given by

$$T_A(M_m^k)^\perp = \left\{ U \begin{pmatrix} \lambda I_m & 0 \\ 0 & T \end{pmatrix} U' : \lambda \in \mathbb{R}, \ T \in S(k - m, \mathbb{R}) \right\}. \qquad (9.6)$$

Proof As in equation (7.3), represent the preshape of a $(k + 1)$-ad x by the $m \times k$ matrix z, where $\|z\|^2 = \text{Trace}(zz') = 1$, and let S_m^{k+1} be the preshape sphere,

$$S_m^{k+1} = \{z \in \mathbb{R}^{m \times k} : \|z\| = 1\}.$$

Let NS_m^{k+1} be the nonsingular part of S_m^{k+1}, that is,

$$NS_m^{k+1} = \{z \in S_m^{k+1} : \text{rank}(z) = m\}.$$

Then $R\Sigma_m^{k+1} = NS_m^{k+1}/O(m)$ and $M_m^k = j(R\Sigma_m^{k+1})$. The map

$$j: R\Sigma_m^{k+1} \longrightarrow S(k, \mathbb{R}), \quad j(\sigma(z)) = z'z = A$$

is an embedding. Hence

$$T_A(M_m^k) = dj_{\sigma(z)}(T_{\sigma(z)}R\Sigma_m^{k+1}). \qquad (9.7)$$

Because $R\Sigma_m^{k+1}$ is locally like Σ_{0m}^{k+1}, $T_{\sigma(z)}R\Sigma_m^{k+1}$ can be identified with the horizontal subspace H_z of $T_z S_m^{k+1}$ obtained in Section 7.2, which is

$$H_z = \{v \in R^{m \times k} : \text{Trace}(zv') = 0, \; zv' = vz'\}. \tag{9.8}$$

Consider the map

$$\tilde{j} : NS_m^{k+1} \to S(k, \mathbb{R}), \quad \tilde{j}(z) = z'z. \tag{9.9}$$

Its derivative is an isomorphism between the horizontal subspace of $TNS_m^{k+1} \equiv TS_m^{k+1}$ and TM_m^k. The derivative is given by

$$d\tilde{j} : TS_m^{k+1} \to S(k, \mathbb{R}), \quad d\tilde{j}_z(v) = z'v + v'z. \tag{9.10}$$

Hence

$$T_A M_m^k = d\tilde{j}_z(H_z) = \{z'v + v'z : v \in H_z\}. \tag{9.11}$$

From the description of H_z in equation (9.8), and using the fact that z has full row rank, it follows that

$$H_z = \{zv : v \in \mathbb{R}^{k \times k}, \text{Trace}(z'zv) = 0, zvz' \in S(m, \mathbb{R})\}. \tag{9.12}$$

From equations (9.11) and (9.12), we get that

$$T_A M_m^k = \{Av + v'A : AvA \in S(k, \mathbb{R}), \text{Trace}(Av) = 0\}. \tag{9.13}$$

Let $A = UDU'$ be a spectral decomposition of A as in the statement of the proposition. Using the fact that A has rank m, equation (9.13) can be written as

$$\begin{aligned}
T_A M_m^k &= \{U(Dv + v'D)U' : DvD \in S(k, \mathbb{R}), \text{Trace}(Dv) = 0\} \\
&= \left\{ U \begin{pmatrix} T & S \\ S' & 0 \end{pmatrix} U' : T \in S(m, \mathbb{R}), \text{Trace}(T) = 0 \right\}.
\end{aligned} \tag{9.14}$$

This proves part (a). From the definition of orthocomplement and equation (9.14), we get that

$$\begin{aligned}
T_A M_m^{k\perp} &= \{v \in S(k, \mathbb{R}) : \text{Trace}(v'w) = 0 \; \forall w \in T_A M_m^k\} \\
&= \left\{ U \begin{pmatrix} \lambda I_m & 0 \\ 0 & R \end{pmatrix} U' : \lambda \in \mathbb{R}, R \in S(k - m, \mathbb{R}) \right\},
\end{aligned} \tag{9.15}$$

where I_m is the $m \times m$ identity matrix. This proves (b) and completes the proof. □

For a $k \times k$ positive semi-definite matrix μ with rank at least m, its projection on to M_m^k is defined as

$$P(\mu) = \{A \in M_m^k : \|\mu - A\|^2 = \arg \min_{x \in M_m^k} \|\mu - x\|^2\} \tag{9.16}$$

if this set is nonempty. The following theorem shows that the projection set is nonempty and derives a formula for the projection matrices.

Theorem 9.2 *$P(\mu)$ is nonempty and consists of*

$$A = \sum_{j=1}^{m} \left(\lambda_j - \bar\lambda + \frac{1}{m}\right) U_j U_j', \tag{9.17}$$

where $\lambda_1 \geq \lambda_2 \geq \cdots \geq \lambda_k$ are the ordered eigenvalues of μ; $U_1, U_2, \ldots, U_k$ are some corresponding orthonormal eigenvectors; and $\bar\lambda = \frac{1}{m}\sum_{j=1}^{m}\lambda_j$.

Proof Let

$$f(x) = \|\mu - x\|^2, \quad x \in S(k, \mathbb{R}). \tag{9.18}$$

If f has a minimizer A in M_m^k, then $(\mathrm{grad}\, f)(A) \in T_A(M_m^k)^\perp$, where grad denotes the Euclidean derivative operator. But $(\mathrm{grad}\, f)(A) = 2(A - \mu)$. Hence, if A minimizes f, then

$$A - \mu = U^A \begin{pmatrix} \lambda I_m & 0 \\ 0 & T \end{pmatrix} U^{A'}, \tag{9.19}$$

where $U^A = (U_1^A, U_2^A, \ldots, U_k^A)$ is a $k \times k$ matrix consisting of an orthonormal basis of eigenvectors of A corresponding to its ordered eigenvalues $\lambda_1^A \geq \lambda_2^A \geq \cdots \geq \lambda_m^A > 0 = \cdots = 0$. From equation (9.19) it follows that

$$\mu U_j^A = (\lambda_j^A - \lambda) U_j^A, \quad j = 1, 2, \ldots, m. \tag{9.20}$$

Hence $\{\lambda_j^A - \lambda\}_{j=1}^m$ are eigenvalues of μ with $\{U_j^A\}_{j=1}^m$ as the corresponding eigenvectors. Because these eigenvalues are ordered, this implies that there exists a singular value decomposition of μ: $\mu = \sum_{j=1}^{k}\lambda_j U_j U_j'$, and a set of indices $S = \{i_1, i_2, \ldots, i_m\}$, $1 \leq i_1 < i_2 < \cdots < i_m \leq k$ such that

$$\lambda_j^A - \lambda = \lambda_{i_j} \tag{9.21}$$

and

$$U_j^A = U_{i_j}, \quad j = 1, \ldots, m. \tag{9.22}$$

Add the equations in (9.21) to get $\lambda = \frac{1}{m} - \bar\lambda$, where $\bar\lambda = \frac{\sum_{j \in S}\lambda_j}{m}$. Hence

$$A = \sum_{j \in S}\left(\lambda_j - \bar\lambda + \frac{1}{m}\right) U_j U_j'. \tag{9.23}$$

Because $\sum_{j=1}^{k}\lambda_j = 1$, $\bar\lambda \leq \frac{1}{m}$ and $\lambda_j - \bar\lambda + \frac{1}{m} > 0$ for all $j \in S$. So A is positive semi-definite of rank m. It is easy to check that Trace(A)=1 and hence $A \in M_m^k$. It can be shown that among the matrices A of the form (9.23), the function f defined in equation (9.18) is minimized when

$$S = \{1, 2, \dots, m\}. \tag{9.24}$$

Define $M_{\leq m}^k$ as the set of all $k \times k$ positive semi-definite matrices of rank at most m and trace 1. This is a compact subset of $S(k, \mathbb{R})$. Hence f restricted to $M_{\leq m}^k$ attains a minimum value. Let A_0 be a corresponding minimizer. If rank$(A_0) < m$, say $= m_1$, then A_0 minimizes f restricted to $M_{m_1}^k$. Recall that $M_{m_1}^k$ is a Riemannian manifold (it is $j(R\Sigma_{m_1}^{k+1})$). Hence A_0 must have the form

$$A_0 = \sum_{j=1}^{m_1} \left(\lambda_j - \bar{\lambda} + \frac{1}{m_1} \right) U_j U_j', \tag{9.25}$$

where $\bar{\lambda} = \frac{\sum_{j=1}^{m_1} \lambda_j}{m_1}$. But if one defines

$$A = \sum_{j=1}^{m} \left(\lambda_j - \bar{\lambda} + \frac{1}{m} \right) U_j U_j' \tag{9.26}$$

with $\bar{\lambda} = \frac{\sum_{j=1}^{m} \lambda_j}{m}$, then it is easy to check that $f(A) < f(A_0)$. Hence A_0 cannot be a minimizer of f over $M_{\leq m}^k$; that is, a minimizer must have rank m. Thus it lies in M_m^k and, from equations (9.23) and (9.24), it follows that it has the form given in equation (9.26). This completes the proof. □

Let Q be a probability distribution on $R\Sigma_m^k$ and let $\tilde{\mu}$ be the mean of $\tilde{Q} \equiv Q \circ j^{-1}$ in $S(k, \mathbb{R})$. Then $\tilde{\mu}$ is positive semi-definite of rank at least m and $\tilde{\mu}\mathbf{1}_k = 0$. Theorem 9.2 can be used to get the formula for the extrinsic mean set of Q. This is obtained in the next corollary.

Corollary 9.3 *(a) The projection of $\tilde{\mu}$ into $j(R\Sigma_m^k)$ is given by*

$$P_{j(R\Sigma_m^k)}(\tilde{\mu}) = \left\{ A : A = \sum_{j=1}^{m} \left(\lambda_j - \bar{\lambda} + \frac{1}{m} \right) U_j U_j' \right\}, \tag{9.27}$$

where $\lambda_1 \geq \cdots \geq \lambda_k$ are the ordered eigenvalues of $\tilde{\mu}$, $U_1, \dots, U_k$ are the corresponding orthonormal eigenvectors, and $\bar{\lambda} = \frac{\sum_{j=1}^{m} \lambda_j}{m}$. (b) The projection set is a singleton and Q has a unique extrinsic mean μ_E if and only if $\lambda_m > \lambda_{m+1}$. Then $\mu_E = \sigma(F)$, where $F = (F_1, \dots, F_m)'$, $F_j = \sqrt{\lambda_j - \bar{\lambda} + \frac{1}{m}} U_j$.

Proof Because $\tilde{\mu}\mathbf{1}_k = 0$, therefore $U_j' \mathbf{1}_k = 0$ for all $j \leq m$. Hence any A in equation (9.27) lies in $j(R\Sigma_m^k)$. Now part (a) follows from Theorem 9.2

using the fact that $j(R\Sigma_m^k) \subseteq M_m^k$. For simplicity, let us denote $\lambda_j - \bar{\lambda} + \frac{1}{m}$ by λ_j^*, $j = 1, \ldots, m$. To prove part (b), note that if $\lambda_m = \lambda_{m+1}$, then clearly $A_1 = \sum_{j=1}^{m} \lambda_j^* U_j U_j'$ and $A_2 = \sum_{j=1}^{m-1} \lambda_j^* U_j U_j' + \lambda_m^* U_{m+1} U_{m+1}'$ are two distinct elements in the projection set of equation (9.27). Consider next the case when $\lambda_m > \lambda_{m+1}$. Let $\tilde{\mu} = U\Lambda U' = V\Lambda V'$ be two different spectral decompositions of $\tilde{\mu}$. Then $U'V$ consists of orthonormal eigenvectors of $\Lambda = \text{Diag}(\lambda_1, \ldots, \lambda_k)$. The condition $\lambda_m > \lambda_{m+1}$ implies that

$$U'V = \begin{pmatrix} V_{11} & 0 \\ 0 & V_{22} \end{pmatrix}, \tag{9.28}$$

where $V_{11} \in SO(m)$ and $V_{22} \in SO(k - m)$. Write

$$\Lambda = \begin{pmatrix} \Lambda_{11} & 0 \\ 0 & \Lambda_{22} \end{pmatrix}.$$

Then $\Lambda U'V = U'V\Lambda$ implies $\Lambda_{11} V_{11} = V_{11} \Lambda_{11}$ and $\Lambda_{22} V_{22} = V_{22} \Lambda_{22}$. Hence

$$\sum_{j=1}^{m} \lambda_j^* V_j V_j'$$

$$= U \sum_{j=1}^{m} \begin{pmatrix} \lambda_j^* (V_{11})_j (V_{11})_j' & 0 \\ 0 & 0 \end{pmatrix} U'$$

$$= U \begin{pmatrix} \Lambda_{11} + (\frac{1}{m} - \bar{\lambda})I_m & 0 \\ 0 & 0 \end{pmatrix} U'$$

$$= \sum_{j=1}^{m} \lambda_j^* U_j U_j'.$$

This proves that the projection set in equation (9.27) is a singleton when $\lambda_m > \lambda_{m+1}$. Then, for any F in part (b) and A in the projection set of equation (9.27), $A = F'F = j(\sigma(F))$. This proves part (b) and completes the proof. $\qquad \square$

From Proposition 4.2 and Corollary 9.3 it follows that the extrinsic variation of Q has the following expression:

$$V = \int_{j(R\Sigma_m^k)} \|x - \tilde{\mu}\|^2 \tilde{Q}(dx) + \|\tilde{\mu} - A\|^2, \quad A \in P_{j(R\Sigma_m^k)}(\tilde{\mu})$$

$$= \int_{j(R\Sigma_m^k)} \|x\|^2 \tilde{Q}(dx) + m\left(\frac{1}{m} - \bar{\lambda}\right)^2 - \sum_{j=1}^{m} \lambda_j^2. \tag{9.29}$$

Remark 9.4 From the proof of Theorem 9.2 and Corollary 9.3 it follows that the extrinsic mean set C_Q of Q is also the extrinsic mean set of Q

restricted to $M_{\leq m}^k$. Because $M_{\leq m}^k$ is a compact metric space, from Theorem 3.3 it follows that C_Q is compact. Let $X_1, X_2, \ldots, X_n$ be an i.i.d. sample from Q and let μ_{nE} and V_n be the sample extrinsic mean and variation. Then, from Proposition 3.8, it follows that V_n is a consistent estimator of V. From Corollary 3.4 it follows that if Q has a unique extrinsic mean μ_E, then μ_{nE} is a consistent estimator of μ_E.

9.3 Asymptotic distribution of the sample extrinsic mean

Let $X_1, \ldots, X_n$ be an i.i.d. sample from some probability distribution Q on $R\Sigma_m^k$ and let μ_{nE} be the sample extrinsic mean (any measurable selection from the sample extrinsic mean set). In the last section, we saw that if Q has a unique extrinsic mean μ_E, that is, if the mean $\tilde{\mu}$ of $\tilde{Q} = Q \circ j^{-1}$ is a nonfocal point of $S(k, \mathbb{R})$, then μ_{nE} converges a.s. to μ_E as $n \to \infty$. Also from Proposition 4.3 it follows that if the projection map $P \equiv P_{j(R\Sigma_m^k)}$ is continuously differentiable at $\tilde{\mu}$, then $\sqrt{n}[j(\mu_{nE}) - j(\mu_E)]$ has an asymptotic mean zero Gaussian distribution on $T_{j(\mu_E)}j(R\Sigma_m^k)$. To find the asymptotic covariance, we need to compute the differential of P at $\tilde{\mu}$.

Consider first the map $P \colon N(\tilde{\mu}) \to S(k, \mathbb{R})$, $P(\mu) = \sum_{j=1}^{m}(\lambda_j(\mu) - \bar{\lambda}(\mu) + 1/m)U_j(\mu)U_j(\mu)'$ as in Theorem 9.2. Here $N(\tilde{\mu})$ is an open neighborhood of $\tilde{\mu}$ in $S(k, \mathbb{R})$ where P is defined. Hence, for $\mu \in N(\tilde{\mu})$, $\lambda_m(\mu) > \lambda_{m+1}(\mu)$. It can be shown that P is smooth on $N(\tilde{\mu})$ (see Theorem 9.5). Let $\gamma(t) = \tilde{\mu} + tv$ be a curve in $N(\tilde{\mu})$ with $\gamma(0) = \tilde{\mu}$ and $\dot{\gamma}(0) = v \in S(k, \mathbb{R})$. Let $\tilde{\mu} = U\Lambda U'$, $U = (U_1, \ldots, U_k)$, $\Lambda = \mathrm{Diag}(\lambda_1, \ldots, \lambda_k)$ be a spectral decomposition of $\tilde{\mu}$. Then

$$\gamma(t) = U(\Lambda + tU'vU)U' = U\tilde{\gamma}(t)U', \tag{9.30}$$

where $\tilde{\gamma}(t) = \Lambda + tU'vU$. Note that $\tilde{\gamma}(t)$ is a curve in $S(k, \mathbb{R})$ starting at Λ. Write $\tilde{v} = \dot{\tilde{\gamma}}(0) = U'vU$. From equation (9.30) and from the definition of P, we obtain

$$P[\gamma(t)] = UP[\tilde{\gamma}(t)]U'. \tag{9.31}$$

Differentiate equation (9.31) at $t = 0$, noting that $\frac{d}{dt}P[\gamma(t)]|_{t=0} = d_{\tilde{\mu}}P(v)$ and $\frac{d}{dt}P[\tilde{\gamma}(t)]|_{t=0} = d_\Lambda P(\tilde{v})$, to get

$$d_{\tilde{\mu}}P(v) = Ud_\Lambda P(\tilde{v})U'. \tag{9.32}$$

Let us find $\frac{d}{dt}P[\tilde{\gamma}(t)]|_{t=0}$. For that, without loss of generality, we may assume that $\lambda_1 > \lambda_2 > \cdots > \lambda_k$. The set of all such matrices forms an open dense set of $S(k, \mathbb{R})$. We can choose a spectral decomposition for $\tilde{\gamma}(t)$: $\tilde{\gamma}(t) = \sum_{j=1}^{k} \lambda_j(t)e_j(t)e_j(t)'$ such that $\{e_j(t), \lambda_j(t)\}_{j=1}^{k}$ are some smooth functions of

t satisfying $e_j(0) = e_j$ and $\lambda_j(0) = \lambda_j$, where $\{e_j\}_{j=1}^k$ is the canonical basis for $\mathbb{R}^k$. Because $e_j(t)'e_j(t) = 1$, by differentiating we get

$$e_j'\dot{e}_j(0) = 0, \qquad j = 1,\ldots,k. \tag{9.33}$$

Also, because $\tilde{\gamma}(t)e_j(t) = \lambda_j(t)e_j(t)$, one has

$$\tilde{v}e_j + \Lambda\dot{e}_j(0) = \lambda_j\dot{e}_j(0) + \dot{\lambda}_j(0)e_j, \qquad j = 1,\ldots,k. \tag{9.34}$$

Consider the orthonormal basis (frame) for $S(k,\mathbb{R})$: $\{E_{ab} : 1 \le a \le b \le k\}$ defined as

$$E_{ab} = \begin{cases} \frac{1}{\sqrt{2}}(e_a e_b^t + e_b e_a^t) & \text{if } a < b, \\ e_a e_a^t & \text{if } a = b. \end{cases} \tag{9.35}$$

Let $\tilde{v} = E_{ab}$, $1 \le a \le b \le k$. From equations (9.33) and (9.34) we get

$$\dot{e}_j(0) = \begin{cases} 0 & \text{if } a = b \text{ or } j \notin \{a,b\}, \\ 2^{-1/2}(\lambda_a - \lambda_b)^{-1}e_b & \text{if } j = a < b, \\ 2^{-1/2}(\lambda_b - \lambda_a)^{-1}e_a & \text{if } j = b > a, \end{cases} \tag{9.36}$$

and

$$\dot{\lambda}_j(0) = \begin{cases} 1 & \text{if } j = a = b, \\ 0 & \text{otherwise.} \end{cases} \tag{9.37}$$

Because

$$P[\tilde{\gamma}(t)] = \sum_{j=1}^m \left(\lambda_j(t) - \bar{\lambda}(t) + \frac{1}{m}\right)e_j(t)e_j(t)',$$

where $\bar{\lambda}(t) = \frac{1}{m}\sum_{j=1}^m \lambda_j(t)$,

$$\dot{\bar{\lambda}}(0) = \frac{1}{m}\sum_{j=1}^m \dot{\lambda}_j(0),$$

$$\frac{d}{dt}P[\tilde{\gamma}(t)]|_{t=0} = \sum_{j=1}^m \left(\dot{\lambda}_j(0) - \dot{\bar{\lambda}}(0)\right)e_j e_j'$$

$$+ \sum_{j=1}^m \left(\lambda_j - \bar{\lambda} + \frac{1}{m}\right)\left(e_j\dot{e}_j(0)' + \dot{e}_j(0)e_j'\right). \tag{9.38}$$

Take $\dot{\tilde{\gamma}}(0) = \tilde{v} = E_{ab}$, $1 \le a \le b \le k$ in equation (9.38). From equations (9.36) and (9.37) we get

$$\frac{d}{dt}P[\tilde{\gamma}(t)]|_{t=0} = d_\Lambda P(E_{ab})$$

$$= \begin{cases} E_{ab} & \text{if } a < b \le m, \\ E_{aa} - \frac{1}{m}\sum_{j=1}^{m} E_{jj} & \text{if } a = b \le m, \\ (\lambda_a - \bar{\lambda} + \frac{1}{m})(\lambda_a - \lambda_b)^{-1} E_{ab} & \text{if } a \le m < b \le k, \\ 0 & \text{if } m < a \le b \le k. \end{cases} \quad (9.39)$$

Then from equations (9.32) and (9.39) one obtains

$$d_{\tilde{\mu}}P(UE_{ab}U') = \begin{cases} UE_{ab}U' & \text{if } a < b \le m, \\ U\left(E_{aa} - \frac{1}{m}\sum_{j=1}^{m} E_{jj}\right)U' & \text{if } a = b \le m, \\ (\lambda_a - \bar{\lambda} + \frac{1}{m})(\lambda_a - \lambda_b)^{-1} UE_{ab}U' & \text{if } a \le m < b \le k, \\ 0 & \text{if } m < a \le b \le k. \end{cases}$$

$$(9.40)$$

From the description of the tangent space $T_{P(\tilde{\mu})}M_m^k$ in Proposition 9.1 it is clear that

$$d_{\tilde{\mu}}P(UE_{ab}U') \in T_{P(\tilde{\mu})}M_m^k, \quad \forall a \le b.$$

Let us denote

$$F_{ab} = UE_{ab}U', \quad 1 \le a \le m, a < b \le k, \quad (9.41)$$

$$F_a = UE_{aa}U', \quad 1 \le a \le m. \quad (9.42)$$

Then from equation (9.40) we get

$$d_{\tilde{\mu}}P(UE_{ab}U') = \begin{cases} F_{ab} & \text{if } 1 \le a < b \le m, \\ F_a - \bar{F} & \text{if } a = b \le m, \\ \left(\lambda_a - \bar{\lambda} + \frac{1}{m}\right)(\lambda_a - \lambda_b)^{-1}F_{ab} & \text{if } 1 \le a \le m < b \le k, \\ 0 & \text{otherwise}, \end{cases}$$

$$(9.43)$$

where $\bar{F} = \frac{1}{m}\sum_{a=1}^{m} F_a$. Note that the vectors $\{F_{ab}, F_a\}$ in equations (9.41) and (9.42) are orthonormal and $\sum_{a=1}^{m} F_a - \bar{F} = 0$. Hence from equation (9.43) we conclude that the subspace spanned by

$$\{d_{\tilde{\mu}}P(UE_{ab}U') : 1 \le a \le b \le k\}$$

has dimension

$$\frac{m(m-1)}{2} + m - 1 + m(k-m) = km - m - \frac{m(m-1)}{2},$$

which is the dimension of M_m^k. This proves that

$$T_{P(\tilde\mu)}M_m^k = \text{Span}\{d_{\tilde\mu}P(UE_{ab}U')\}_{a\le b}.$$

Consider the orthonormal basis $\{UE_{ab}U' : 1 \le a \le b \le k\}$ of $S(k, \mathbb{R})$. Define

$$\tilde F_a = \sum_{j=1}^{m} H_{aj}F_j, \quad 1 \le a \le m-1, \tag{9.44}$$

where H is an $(m-1)\times m$ Helmert matrix, that is, $HH' = I_{m-1}$ and $H\mathbf{1}_m = 0$. Then the vectors $\{F_{ab}\}$ defined in equation (9.41) and $\{\tilde F_a\}$ defined in equation (9.44) together form an orthonormal basis of $T_{P\tilde\mu}M_m^k$. This is proved in the next theorem.

Theorem 9.5 *Let $\tilde\mu$ be a nonfocal point in $S(k, \mathbb{R})$, and let $\tilde\mu = U\Lambda U'$ be a spectral decomposition of $\tilde\mu$. (a) The projection map $P: N(\tilde\mu) \to S(k, \mathbb{R})$ is smooth and its derivative $dP: S(k, \mathbb{R}) \to TM_m^k$ is given by equation (9.40). (b) The vectors (matrices) $\{F_{ab}: 1 \le a \le m, a < b \le k\}$ defined in equation (9.41) and $\{\tilde F_a : 1 \le a \le (m-1)\}$ defined in equation (9.44) together form an orthonormal basis of $T_{P(\tilde\mu)}M_m^k$. (c) Let $A \in S(k, \mathbb{R}) \equiv T_{\tilde\mu}S(k, \mathbb{R})$ have coordinates $((a_{ij}))_{1\le i\le j\le k}$ with respect to the orthonormal basis $\{UE_{ij}U'\}$ of $S(k, \mathbb{R})$. That is,*

$$A = \sum_{1\le i\le j\le k}\sum a_{ij}UE_{ij}U',$$

$$a_{ij} = \langle A, UE_{ij}U'\rangle = \begin{cases} \sqrt{2}U_i'AU_j & \text{if } i < j, \\ U_i'AU_i & \text{if } i = j. \end{cases}$$

Then $d_{\tilde\mu}P(A)$ has coordinates

$$a_{ij}, \quad 1 \le i < j \le m,$$
$$\tilde a_i, \quad 1 \le i \le (m-1),$$
$$\left(\lambda_i - \bar\lambda + \frac{1}{m}\right)(\lambda_i - \lambda_j)^{-1}a_{ij}, \quad 1 \le i \le m < j \le k,$$

with respect to the orthonormal basis $\{F_{ij} : 1 \le i < j \le m\}$, $\{\tilde F_i : 1 \le i \le (m-1)\}$ and $\{F_{ij} : 1 \le i \le m < j \le k\}$ of $T_{P(\tilde\mu)}M_m^k$. Here

$$\mathbf{a} \equiv (a_{11}, a_{22}, \ldots, a_{mm})',$$
$$\tilde{\mathbf{a}} \equiv (\tilde a_1, \tilde a_2, \ldots, \tilde a_{m-1})' = H\mathbf{a}.$$

Proof Let $\mu \in N(\tilde\mu)$ have ordered eigenvalues $\lambda_1(\mu) \ge \lambda_2(\mu) \ge \cdots \ge \lambda_k(\mu)$ with corresponding orthonormal eigenvectors $U_1(\mu), U_2(\mu), \ldots, U_k(\mu)$. Then, from perturbation theory, it follows that if $\lambda_m(\mu) > \lambda_{m+1}(\mu)$, and

$$\mu \mapsto \mathrm{Span}(U_1(\mu), \ldots, U_m(\mu)), \quad \sum_{i=1}^{m} \lambda_i(\mu)$$

are smooth maps into their respective codomains (see Dunford and Schwartz, 1958, p. 598). Write

$$P(\mu) = \sum_{j=1}^{m} \lambda_j(\mu) U_j(\mu) U_j(\mu)' + \left(\frac{1}{m} - \bar\lambda(\mu)\right) \sum_{j=1}^{m} U_j(\mu) U_j(\mu)'.$$

Then $\sum_{j=1}^{m} U_j(\mu) U_j(\mu)'$ is the projection matrix of the subspace $\mathrm{Span}(U_1(\mu), \ldots, U_m(\mu))$, which is a smooth function of μ. Thus $\sum_{j=1}^{m} \lambda_j U_j(\mu) U_j(\mu)'$ is the projection of μ on the subspace $\mathrm{Span}(U_1(\mu) U_1(\mu)', \ldots, U_m(\mu) U_m(\mu)')$ and hence is a smooth function of μ, so that $\mu \mapsto P(\mu)$ is a smooth map on $N(\tilde\mu)$. This proves part (a).

From equation (9.43) we conclude that $\{F_{ab} : 1 \le a \le m, a < b \le k\}$ and $\{F_a - \bar F : 1 \le a \le m\}$ span $T_{P(\tilde\mu)} M_m^k$. It is easy to check from the definition of H that $\mathrm{Span}\{\tilde F_a : 1 \le a \le (m-1)\} = \mathrm{Span}\{F_a - \bar F : 1 \le a \le m\}$. Also, because $\{F_a\}$ are mutually orthogonal, so are $\{\tilde F_a\}$. This proves that $\{F_{ab} : 1 \le a \le m, a < b \le k\}$ and $\{\tilde F_a : 1 \le a \le (m-1)\}$ together form an orthonormal basis of $T_{P\tilde\mu} M_m^k$, as claimed in part (b).

If $A = \sum \sum_{1 \le i \le j \le k} a_{ij} U E_{ij} U'$, then

$$d_{\tilde\mu} P(A) = \sum \sum a_{ij} d_{\tilde\mu} P(U E_{ij} U') \tag{9.45}$$

$$= \sum_{1 \le i < j \le m} \sum a_{ij} F_{ij} + \sum_{i=1}^{m} a_{ii}(F_i - \bar F) + \sum_{i=1}^{m} \sum_{j=m+1}^{k} a_{ij}(\lambda_i - \bar\lambda + \frac{1}{m})(\lambda_i - \lambda_j)^{-1} F_{ij} \tag{9.46}$$

$$= \sum_{1 \le i < j \le m} \sum a_{ij} F_{ij} + \sum_{i=1}^{m-1} \tilde a_i \tilde F_i + \sum_{i=1}^{m} \sum_{j=m+1}^{k} (\lambda_i - \bar\lambda + \frac{1}{m})(\lambda_i - \lambda_j)^{-1} a_{ij} F_{ij}. \tag{9.47}$$

This proves part (c). To get equation (9.47) from equation (9.46), we use the fact that $\sum_{i=1}^{m} a_{ii}(F_i - \bar F) = \sum_{i=1}^{m-1} \tilde a_i \tilde F_i$. To show that denote by F the matrix $(F_1, \ldots, F_m)$, by $F - \bar F$ the matrix $(F_1 - \bar F, \ldots, F_m - \bar F)$, and by $\tilde F$ the matrix $(\tilde F_1, \ldots, \tilde F_{m-1})$. Then

$$\sum_{i=1}^{m-1} \tilde a_i \tilde F_i = \tilde F \tilde{\mathbf{a}}$$

$$= \tilde F H \mathbf{a} = F(I_m - \mathbf{1}_m \mathbf{1}_m') \mathbf{a}$$

$$= (F - \bar F)\mathbf{a} = \sum_{i=1}^{m} a_{ii}(F_i - \bar F).$$

This completes the proof. □

Corollary 9.6 *Consider the projection map restricted to $S_0(k, \mathbb{R}) \equiv \{A \in S(k, \mathbb{R}) : A\mathbf{1}_k = 0\}$. Then its derivative is given by*

$$dP \colon S_0(k, \mathbb{R}) \to T\,j(R\Sigma_m^k),$$

$$d_{\tilde{\mu}}P(A) = \sum_{1 \le i < j \le m} a_{ij}F_{ij} + \sum_{i=1}^{m-1} \tilde{a}_i \tilde{F}_i + \sum_{i=1}^{m} \sum_{j=m+1}^{k-1} (\lambda_i - \bar{\lambda} + \frac{1}{m})(\lambda_i - \lambda_j)^{-1} a_{ij}F_{ij}.$$

$$(9.48)$$

Hence $d_{\tilde{\mu}}P(A)$ has coordinates $\{a_{ij} : 1 \le i < j \le m\}$, $\{\tilde{a}_i : 1 \le i \le (m-1)\}$, $\{(\lambda_i - \bar{\lambda} + \frac{1}{m})(\lambda_i - \lambda_j)^{-1}a_{ij} : 1 \le i \le m < j < k\}$ with respect to the orthonormal basis $\{F_{ij} : 1 \le i < j \le m\}$, $\{\tilde{F}_i : 1 \le i \le (m-1)\}$, and $\{F_{ij} : 1 \le i \le m < j < k\}$ of $T_{P\tilde{\mu}}j(R\Sigma_m^k)$.

Proof Follows from the fact that

$$T_{P(\tilde{\mu})}j(R\Sigma_m^k) = \{v \in T_{P(\tilde{\mu})}M_m^k : v\mathbf{1}_k = 0\}$$

and $\{F_{ij} : j = k\}$ lie in $T_{P(\tilde{\mu})}j(R\Sigma_m^k)^{\perp}$. □

Consider the same set-up as in Section 4.2. Let $\tilde{X}_j = j(X_j)$, $j = 1, \ldots, n$, be the embedded sample in $j(R\Sigma_m^k)$. Let d be the dimension of $R\Sigma_m^k$. Let T_j, $j = 1, \ldots, n$, be the coordinates of $d_{\tilde{\mu}}P(\tilde{X}_j - \tilde{\mu})$ in $T_{P(\tilde{\mu})}j(R\Sigma_m^k) \approx \mathbb{R}^d$. Then from equation (4.5) and Proposition 4.3 it follows that

$$\sqrt{n}\,[P(\bar{\tilde{X}}) - P(\tilde{\mu})] = \sqrt{n}\bar{T} + o_P(1) \xrightarrow{\mathcal{L}} N(0, \mathrm{Cov}(T_1)).$$

We can get expressions for T_j and hence $\bar{T}$ from Corollary 9.6 as follows. Define

$$(Y_j)_{ab} = \begin{cases} \sqrt{2}U_a'Y_jU_b & \text{if } 1 \le a < b \le k, \\ U_a'Y_jU_a - \lambda_a & \text{if } a = b, \end{cases}$$

$$S_j = H((Y_j)_{11}, (Y_j)_{22}, \ldots, (Y_j)_{mm})',$$

$$(T_j)_{ab} = \begin{cases} (Y_j)_{ab} & \text{if } 1 \le a < b \le m, \\ (S_j)_a & \text{if } 1 \le a = b \le (m-1), \\ (\lambda_a - \bar{\lambda} + \frac{1}{m})(\lambda_a - \lambda_b)^{-1}(Y_j)_{ab} & \text{if } 1 \le a \le m < b < k. \end{cases}$$

$$(9.49)$$

Then $T_j \equiv ((T_j)_{ab})$ is the vector of coordinates of $d_{\tilde{\mu}}P(\tilde{X}_j - \tilde{\mu})$ in $\mathbb{R}^d$.

9.4 Two-sample tests on the reflection shape spaces

Now we are in the same set-up as in Section 4.5: there are two samples on $R\Sigma_m^k$ and we want to test if they come from the same distribution, by comparing their sample extrinsic means and variations. To use the test statistic T_1 from equation (4.15) to compare the extrinsic means, we need the coordinates of $\{d_{\hat\mu}P(\tilde X_j - \hat\mu)\}$ and $\{d_{\hat\mu}P(\tilde Y_j - \hat\mu)\}$ in $T_{P(\hat\mu)}(jR\Sigma_m^k)$. We get them from Corollary 9.6 as described in equation (9.49). To use the test statistic T_2 from equation (4.18), we need expressions for $L: S(k,\mathbb{R}) \to T_{P(\hat\mu)}(jR\Sigma_m^k)$ and $L_i: T_{P(\hat\mu_i)}(jR\Sigma_m^k) \to T_{P(\hat\mu)}(jR\Sigma_m^k)$, $i = 1, 2$. Let $\hat\mu = U\Lambda U'$ be a spectral decomposition of $\hat\mu$. Consider the orthonormal basis $\{UE_{ij}U' : 1 \le i \le j \le k\}$ of $S(k,\mathbb{R})$ and the orthonormal basis of $T_{P(\hat\mu)}(jR\Sigma_m^k)$ derived in Corollary 9.6. Then, if $A \in S(k,\mathbb{R})$ has coordinates $\{a_{ij} : 1 \le i \le j \le k\}$, it is easy to show that $L(A)$ has coordinates $\{a_{ij} : 1 \le i < j \le m\}$, $\{\tilde a_i : 1 \le i \le m - 1\}$ and $\{a_{ij} : 1 \le i \le m < j < k\}$ in $T_{P(\hat\mu)}(jR\Sigma_m^k)$. If we label the bases of $T_{P(\hat\mu_i)}(jR\Sigma_m^k)$ as $\{v_1^i, \ldots, v_d^i\}$, $i = 1, 2$, and that of $T_{P(\hat\mu)}(jR\Sigma_m^k)$ as $\{v_1, \ldots, v_d\}$, then one can show that L_i is the $d \times d$ matrix with coordinates

$$(L_i)_{ab} = \langle v_a, v_b^i \rangle, \quad 1 \le a, b \le d, \ i = 1, 2.$$

Here L_i is the restriction to $T_{P(\hat\mu_i)}(jR\Sigma_m^k)$ of the linear projection L (on $S(k,\mathbb{R})$) into $T_{P(\hat\mu)}(jR\Sigma_m^k)$ $(i = 1, 2)$.

9.5 Other distances on the reflection shape spaces

In this section we introduce some distances other than the extrinsic distance on $R\Sigma_m^k$ that can be used to construct appropriate Fréchet functions and hence Fréchet means and variations.

9.5.1 Full Procrustes distance

Given two k-ads X_1 and X_2 in $\mathbb{R}^{m\times k}$, we define the *full Procrustes distance* between their reflection shapes as

$$d_F(\sigma(X_1), \sigma(X_2)) = \inf_{\Gamma \in O(m), \beta \in \mathbb{R}^+} \|Z_2 - \beta\Gamma Z_1\|, \tag{9.50}$$

where Z_1 and Z_2 are the preshapes of X_1 and X_2, respectively. By a proof similar to that of result 4.1 in Dryden and Mardia (1998), it can be shown that

$$d_F(X_1, X_2) = \Big[1 - \Big(\sum_{i=1}^m \lambda_i\Big)^2\Big]^{1/2}$$

and the values of Γ and β for which the infimum in equation (9.50) is attained are

$$\hat\Gamma = VU', \quad \hat\beta = \sum_{i=1}^m \lambda_i.$$

Here $Z_1 Z_2' = U\Lambda V'$ is the singular-value decomposition of $Z_1 Z_2'$, that is, $U, V \in O(m)$ and

$$\Lambda = \text{Diag}(\lambda_1, \dots, \lambda_m), \quad \lambda_1 \geq \lambda_2 \geq \cdots \geq \lambda_m \geq 0.$$

The quantity $\hat{\beta}\hat{\Gamma}Z_1$ is called the *full Procrustes coordinates* of the shape of Z_1 with respect to that of Z_2.

9.5.2 Partial Procrustes distance

Now define the *partial Procrustes distance* between the shapes of X_1 and X_2 as

$$d_P(\sigma(X_1), \sigma(X_2)) = \inf_{\Gamma \in O(m)} \|Z_2 - \Gamma Z_1\|, \tag{9.51}$$

which is

$$d_P(X_1, X_2) = \sqrt{2}\Big(1 - \sum_{i=1}^m \lambda_i\Big)^{1/2}.$$

The value $\hat{\Gamma}$ of Γ for which the infimum in equation (9.51) is attained is the same as in Section 9.5.1. The quantity $\hat{\Gamma}Z_1$ is called the *partial Procrustes coordinates* of the shape of Z_1 with respect to that of Z_2.

9.5.3 Geodesic distance

We saw in Section 9.1 that $R\Sigma_m^k = NS_m^k/O(m)$. Therefore the geodesic distance between the shapes of two k-ads X_1 and X_2 is given by

$$d_g(\sigma(X_1), \sigma(X_2)) = d_g(\sigma(Z_1), \sigma(Z_2)) = \inf_{\Gamma \in O(m)} d_{gs}(Z_1, \Gamma Z_2). \tag{9.52}$$

Here Z_1 and Z_2 are the preshapes of X_1 and X_2, respectively, in the unit sphere S_m^k and $d_{gs}(.,.)$ denotes the geodesic distance on S_m^k that is given by

$$d_{gs}(Z_1, Z_2) = \arccos(\text{Trace}(Z_1 Z_2')).$$

Therefore

$$d_g(\sigma(X_1), \sigma(X_2))$$
$$= \inf_{\Gamma \in O(m)} \arccos(\text{Trace}(Z_1 Z_2')) = \arccos(\max_{\Gamma \in O(m)}(\text{Trace}(\Gamma Z_1 Z_2'))).$$

Let $Z_1 Z_2' = U\Lambda V$ be the singular-value decomposition of $Z_1 Z_2'$, that is, $U, V \in O(m)$ and

$$\Lambda = \text{Diag}(\lambda_1, \dots, \lambda_m), \quad \lambda_1 \geq \lambda_2 \geq \cdots \lambda_m \geq 0.$$

Then

$$\text{Trace}(\Gamma Z_1 Z_2') = \text{Trace}(\Gamma U\Lambda V) = \text{Trace}(V\Gamma U\Lambda)$$
$$= \sum_{j=1}^m \lambda_j (V\Gamma U)_{jj}.$$

This is maximized when $V\Gamma U = I_m$ or $\Gamma = V'U'$ and then

$$\text{Trace}(\Gamma Z_1 Z_2') = \sum_{j=1}^{m} \lambda_j.$$

Therefore the geodesic distance is

$$d_g(\sigma(X_1), \sigma(X_2)) = \arccos\left(\sum_{j=1}^{m} \lambda_j\right).$$

9.6 Application: glaucoma detection

This section is devoted to an application of three-dimensional similarity shape analysis in glaucoma detection. Glaucoma is a leading cause of eye blindness.

To detect any shape change due to glaucoma, three-dimensional images of the optic nerve head (ONH) of both eyes of 12 mature rhesus monkeys were collected. One of the eyes was treated to increase the intraocular pressure (IOP), which is often the case of glaucoma onset, while the other was left untreated. Five landmarks were recorded on each eye. The landmark coordinates can be found in Bhattacharya and Patrangenaru (2005). In this section, we consider the reflection shape of the k-ads in $R\Sigma_3^k$, $k = 5$. We want to test if there is any significant difference between the shapes of the treated and untreated eyes by comparing the extrinsic means and variations. The analysis is carried out in Bhattacharya (2008b).

Figure 9.1(a) shows the partial Procrustes coordinates of the untreated eye shapes along with a preshape of the untreated eye sample extrinsic mean. Figure 9.1(b) shows the coordinates for the treated eye shapes along with a preshape of the treated eye sample extrinsic mean. In both cases the Procrustes coordinates are obtained with respect to the respective sample means. Figure 9.2 shows the Procrustes coordinates of the mean shapes for the two eyes along with a preshape of the pooled sample extrinsic mean. Here the coordinates are with respect to the preshape of the pooled sample extrinsic mean. The sample extrinsic means have the coordinates

$L[P(\hat\mu_1) - P(\hat\mu)]$
$\quad = (0.003, -0.011, -0.04, 0.021, 0.001, -0.001, 0.007, -0.004),$
$L[P(\hat\mu_2) - P(\hat\mu)]$
$\quad = (-0.003, 0.011, 0.04, -0.021, -0.001, 0.001, -0.007, 0.005)$

in the tangent space of $P(\hat\mu)$. Here $P(\hat\mu_1)$ and $P(\hat\mu_2)$ are the embeddings of the sample extrinsic mean shapes of the untreated and treated eyes,

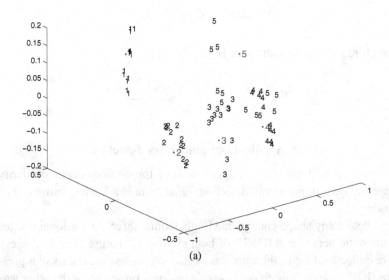

(a)

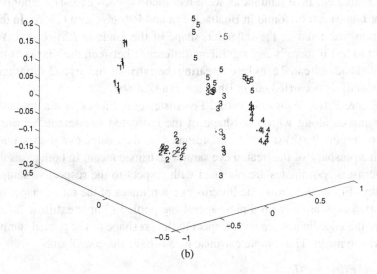

(b)

Figure 9.1 (a) and (b) show five landmarks from untreated and treated eyes of 12 monkeys, respectively, along with the mean shapes. The dots correspond to the mean shapes' landmarks.

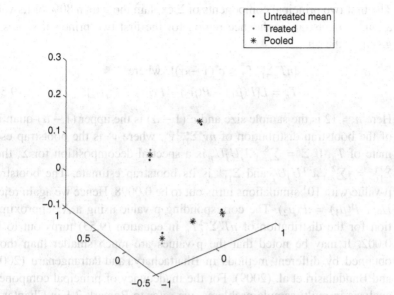

Figure 9.2 Extrinsic mean shapes for the two eyes along with the pooled sample extrinsic mean.

respectively; $P(\hat{\mu})$ is the embedded extrinsic mean shape for the pooled sample; and L denotes the linear projection onto $T_{P(\hat{\mu})} j(R\Sigma_3^5)$. The sample extrinsic variations for the untreated and treated eyes are 0.041 and 0.038, respectively.

This is an example of a matched paired sample. To compare the extrinsic means and variations, we use the methodology of Section 4.5.2. The value of the matched pair test statistic T_{1p} in equation (4.22) is 36.29 and the asymptotic p-value for testing if the shape distributions for the two eyes are the same is

$$P(X_8^2 > 36.29) = 1.55 \times 10^{-5}.$$

The value of the test statistic T_{2p} from equation (4.24) for testing whether the extrinsic means are the same is 36.56 and the p-value of the chi-squared test turns out to be 1.38×10^{-5}. Hence we conclude at an asymptotic level of 0.0001 or higher that the mean shapes of the two eyes are significantly different. Because of a lack of sufficient data and high dimension, the bootstrap estimates of the covariance matrix $\hat{\Sigma}$ in equation (4.25) turn out to be singular or close to singular in many simulations. To avoid that, we construct a pivotal bootstrap confidence region for the first few principal scores of $L_{\tilde{\mu}}[P(\mu_1) - P(\mu_2)]$ and see if it includes $\mathbf{0}$. Here $P(\mu_i)$ is the embedding of the extrinsic mean of Q_i, $i = 1, 2$ (see Section 4.5.2) and $\tilde{\mu} = (\mu_1 + \mu_2)/2$.

The first two principal components of $\hat{\Sigma}$ explain more than 80% of its variation. A bootstrap confidence region for the first two principal scores is given by the set

$$\{nT_n'\hat{\Sigma}_{11}^{-1}T_n \le c^*(1-\alpha)\}, \text{ where} \tag{9.53}$$

$$T_n = L[P(\hat{\mu}_1) - P(\hat{\mu}_2) - P(\mu_1) + P(\mu_2)]. \tag{9.54}$$

Here $n = 12$ is the sample size and $c^*(1-\alpha)$ is the upper $(1-\alpha)$-quantile of the bootstrap distribution of $nv^*\Sigma_{11}^{*-1}v^*$, where v^* is the bootstrap estimate of T_n. If $\hat{\Sigma} = \sum_{j=1}^8 \lambda_j U_j U_j'$ is a spectral decomposition for $\hat{\Sigma}$, then $\hat{\Sigma}_{11}^{-1} \doteq \sum_{j=1}^2 \lambda_j^{-1} U_j U_j'$ and Σ_{11}^{*-1} is its bootstrap estimate. The bootstrap p-value with 10^4 simulations turns out to be 0.0098. Hence we again reject $H_0 : P(\mu_1) = P(\mu_2)$. The corresponding p-value using a $\mathcal{X}_2^2$ approximation for the distribution of $nT_n'\hat{\Sigma}_{11}^{-1}T_n$ in equation (9.53) turns out to be 0.002. It may be noted that the p-values are much smaller than those obtained by different methods in Bhattacharya and Patrangenaru (2005) and Bandulasiri et al. (2009). For the inadequacy of principal component analysis in multi-sample problems, we refer to Remark 2.1 in Chapter 2. Also see Remark 2.2 about the use of sample covariances in bootstrapping.

Next we test if the two eye shapes have the same extrinsic variation. The value of the test statistic T_{3p} from equation (4.27) equals -0.5572 and the asymptotic p-value equals

$$P(|Z| > 0.5572) = 0.577, \quad Z \sim N(0,1).$$

The bootstrap p-value with 10^4 simulations equals 0.59. Hence we accept H_0 and conclude that the extrinsic variations are equal at levels 0.5 or lower.

Because the mean shapes for the two eyes are found to be different, we conclude that the underlying probability distributions are distinct and hence glaucoma indeed changes the shape of the eyes.

9.7 References

The reflection similarity shape space and the embedding (9.3), sometimes referred to as the *Schoenberg embedding*, was first introduced by Bandulasiri and Patrangenaru (2005) and further developed in Bandulasiri et al. (2009). Independently of this, Dryden et al. (2008) obtained the shape space and its embedding. Theorems 9.2 and 9.5, Corollaries 9.3 and 9.6, and Sections 9.3 and 9.4 are due to Bhattacharya (2008b). The analysis of the glaucoma detection problem presented here is an improvement over its treatments in Bhattacharya and Patrangenaru (2005) and Bandulasiri et al. (2009). In the latter article, a *size-and-shape (reflection-similarity)* space is introduced and extrinsic analysis carried out under the embedding $z \to z'z/\|z\|$.

Lele (1991), Lele (1993), and Lele and Cole (1995) use the inner point distance matrix of landmarks and *multi-dimensional scaling (MDS)* for inference for reflection similarity shape spaces. Lele and Cole (1995) in particular discuss the problem of consistency of (generalized) Procrustes estimators and point out that such estimators, as used, for example, by Goodall (1991) are not generally consistent under the parametric/semiparametric models used. As pointed out in Section 8.12, there is a basic problem in the use of the Procrustes estimator for the population mean defined by such models. Also see Stoyan (1990) and Lewis et al. (1980).

10

Stiefel manifolds $V_{k,m}$

Statistical modeling on Stiefel manifolds may be found in Chikuse (2003). In this monograph, its study aids intrinsic and extrinsic analyses on affine shape spaces.

10.1 Introduction

The *Stiefel manifold* $V_{k,m}$ is the space of all *k-frames* in $\mathbb{R}^m$, $k \leq m$. A *k*-frame is a set of *k* orthonormal vectors. The Stiefel manifold can be represented as

$$V_{k,m} = \{x \in M(m,k) : x'x = I_k\},$$

where $M(m,k)$ denotes the set of all $m \times k$ real matrices. This is a complete Riemannian manifold of dimension $k(2m - k - 1)/2$. The sphere S^{m-1} arises as a special case when $k = 1$. When $k = m$, $V_{k,m}$ is the orthogonal group $O(m)$ of all $m \times m$ orthogonal matrices. The Stiefel manifold is connected except when $k = m$. Note that $O(m)$ is disconnected with two connected components, namely those matrices with determinants equal to $+1$, which is the special orthogonal group $SO(m)$, and those with determent equal to -1.

10.2 Extrinsic analysis on $V_{k,m}$

The Stiefel manifold is embedded in $M(m,k)$ by the inclusion map. For a distribution Q on $V_{k,m}$ with Euclidean mean μ, Theorem 10.2 derives its set of projections on $V_{k,m}$ and hence the extrinsic mean of Q. We can decompose μ as

$$\mu = US$$

with $U \in V_{k,m}$ and S being positive semi-definite (we write $S \geq 0$) of rank same as that of μ. We will call this an *orthogonal decomposition* of μ. When μ has full rank, the decomposition is unique and we have $U = \mu(\mu'\mu)^{-1/2}$

and $S = (\mu'\mu)^{1/2}$. Hence U can be thought of as the *orientation of μ*; it is a k-frame of the column-space $C(\mu)$ of μ and S is a measure of the size of μ. Denote by Skew(k) the space of $k \times k$ skew-symmetric matrices, that is,

$$\text{Skew}(k) = \{A \in M(k,k) : A + A' = 0\}.$$

Also we denote by $M(m)$ the space all real $m \times m$ matrices, that is, $M(m,m)$.

Proposition 10.1 *The tangent space of $V_{k,m}$ at U is given by*

$$T_U V_{k,m} = \{v \in M(m,k) : U'v + v'U = 0\}. \tag{10.1}$$

Let $\tilde{U} \in M(m, m - k)$ be an orthonormal frame for the null space of U, that is, $\tilde{U}'U = 0$ and $\tilde{U}'\tilde{U} = I_{m-k}$. Then the tangent space can be re-expressed as

$$\{UA + \tilde{U}B : A \in \text{Skew}(k), B \in M(m - k, k)\}. \tag{10.2}$$

Proof Differentiate the identity $U'U = I_k$ to get the expression for $T_U V_{k,m}$ as in equation (10.1). Clearly the linear space in equation (10.2) is contained in $T_U V_{k,m}$. The dimensions of Skew(k) and $M(m - k)$ are $k(k - 1)/2$ and $k(m-k)$, respectively; add them to get the dimension of $V_{k,m}$ or $T_U V_{k,m}$. Hence the two spaces are identical. $\quad\square$

Theorem 10.2 *The projection set of $\mu \in M(m, k)$ is given by*

$$P(\mu) = \{U \in V_{k,m} : \mu = U(\mu'\mu)^{1/2}\}. \tag{10.3}$$

The above set is nonempty. It is a singleton and hence μ is nonfocal if and only if μ has full rank and then $U = \mu(\mu'\mu)^{-1/2}$.

Proof The projection set of μ is the set of minimizers of $f(U) = \|U - \mu\|^2$, $U \in V_{k,m}$. Then $f(U) = \|\mu\|^2 + k - 2\,\text{Trace}(\mu'U)$ and minimizing f is equivalent to maximizing

$$g : V_{k,m} \to \mathbb{R}, \quad g(U) = \text{Trace}(\mu'U).$$

For U to be a maximizer, the derivative of g must vanish at U. When viewed as a map from $M(k, m)$ to $\mathbb{R}$, the derivative matrix of g is a constant μ. It must therefore lie on the orthocomplement of $T_U V_{k,m}$, that is,

$$\text{Trace}(\mu'UA + \mu'\tilde{U}B) = 0, \quad \forall A \in \text{Skew}(k), B \in M(m - k, k),$$

for every orthonormal frame $\tilde{U}$ for the null space of U. This implies that

$$\text{Trace}(\mu'UA) = 0, \quad \forall A \in \text{Skew}(k), \tag{10.4}$$

$$\text{Trace}(\mu'\tilde{U}) = 0. \tag{10.5}$$

Now equation (10.4) implies that $\mu'U$ is symmetric and from equation (10.5) it follows that

$$\mu = US \quad \text{for some } S \in M(k).$$

Because $\mu'U$ is symmetric, it follows that P is symmetric. For such a U, $g(U) = \text{Trace}(S) \leq \text{Trace}((S^2)^{1/2}) = \text{Trace}((\mu'\mu)^{1/2})$. Hence g is maximized if and only if $S = (\mu'\mu)^{1/2}$, and from this the projection set as in equation (10.3) follows. When $\text{rank}(\mu) = k$, it follows that $U = \mu(\mu'\mu)^{-1/2}$ and hence the projection set is a singleton. If $\text{rank}(\mu) < k$, say, equal to k_1, then pick $U_0 \in P(\mu)$. Then $U_0 U_1 \in P(\mu)$ for any $U_1 \in O(k)$ satisfying $U_1 = O\Lambda O'$ with $O \in O(k)$ being the frame of eigenvectors of $(\mu'\mu)^{1/2}$ in decreasing order of eigenvalues and Λ being diagonal with the first k_1 diagonal entries equal to 1 and $\Lambda^2 = I_k$. This is because with such an O, $O(\mu'\mu)^{1/2} = (\mu'\mu)^{1/2}$. Hence the projection is unique if and only if $\text{rank}(\mu) = k$. This completes the proof. □

Next we find the derivative of the projection P at a nonfocal point μ. Consider the pseudo-singular-value decomposition of such a μ as

$$\mu = U\Lambda V'$$

with $U \in O(m)$, $V \in SO(k)$, and $\Lambda = \binom{\Lambda_1}{0}$ with $\Lambda_1 = \text{Diag}(\lambda_1, \ldots, \lambda_k)$, $\lambda_1 \geq \cdots \geq \lambda_k > 0$. Corresponding to such a decomposition, $\Lambda\Lambda'$ and $\Lambda'\Lambda$ consist of the ordered eigenvalues of $\mu\mu'$ and $\mu'\mu$, respectively, while U and V are corresponding orthonormal eigenframes. Denote by U_j and V_j the jth columns of U and V, respectively. Then the projection of μ becomes $\mu \sum_{j=1}^{k} \lambda_j^{-1} V_j V_j' = \sum_{j=1}^{k} U_j V_j'$. Hence this is a smooth map in some open neighborhood of μ. Its derivative at μ is a linear map from $T_\mu M(m,k) \equiv M(m,k)$ to $T_{P(\mu)} V_{k,m}$. It is determined by its values at some basis for $M(m,k)$. One such basis is $\{U_i V_j' : 1 \leq i \leq m, 1 \leq j \leq k\}$. This forms an orthonormal frame for $M(m,k)$. Theorem 10.3 evaluates the derivative at this frame.

Theorem 10.3 *The derivative of the projection P at $\mu \in M(m,k)$ with $\text{rank}(\mu) = k$ is given by*

$$d_\mu P : M(m,k) \to T_{P(\mu)} V_{k,m}, \quad d_\mu P(A) = \sum_{i=1}^{m} \sum_{j=1}^{k} U_i' A V_j d_\mu P(U_i V_j'),$$

$$d_\mu P(U_i V_j') = \begin{cases} \lambda_j^{-1} U_i V_j' & \text{if } i > k, \\ (\lambda_i + \lambda_j)^{-1}(U_i V_j' - U_j V_i') & \text{if } i \leq k. \end{cases} \tag{10.6}$$

Proof We can assume that the eigenvalues of the $\mu'\mu$ are strictly ordered, which means that $\lambda_1 > \lambda_2 > \cdots > \lambda_k$. The space of all such μ's is dense, and because the derivative is continuous in a neighborhood of μ, we can evaluate it at the remaining μ's as well. This implies that the λ_j are smooth functions of μ and we can choose the corresponding eigenvectors V_j smoothly as well. Write $P(\mu) = \mu \sum_{j=1}^{k} \lambda_j^{-1} V_j V_j'$. Then

$$d_\mu P(A) = \mu \sum_{j=1}^{k} \{\lambda_j^{-1}(V_j \dot{V}_j' + \dot{V}_j V_j') - \lambda_j^{-2}\dot{\lambda}_j V_j V_j'\} + A \sum_{j=1}^{k} \lambda_j^{-1} V_j V_j'. \quad (10.7)$$

Here $\dot{\lambda}_j$ and $\dot{V}_j$ refer to the derivatives of the jth eigenvalue and eigenvector functions at μ evaluated at A. The constraint $V_j' V_j = 1$ implies that $V_j' \dot{V}_j = 0 \; \forall j$. The derivative of the identity $(\mu'\mu)V_j = \lambda_j^2 V_j$ ($j \le k$) evaluated at $A = U_a V_b'$, $a \le m$, $b \le k$, gives

$$(\mu'\mu)\dot{V}_j + \lambda_a(V_a V_b' + V_b V_a')V_j = \lambda_j^2 \dot{V}_j + 2\lambda_j \dot{\lambda}_j V_j \quad \text{if } a \le k \quad (10.8)$$

and

$$(\mu'\mu)\dot{V}_j = \lambda_j^2 \dot{V}_j + 2\lambda_j \dot{\lambda}_j V_j \quad \text{if } a > k. \quad (10.9)$$

Consider first the case $A = U_a V_a'$, $a \le k$. Pre-multiply identity (10.8) by V_j' and use the facts $\mu'\mu V_j = \lambda_j^2 V_j^2$ and $V_j' \dot{V}_j = 0$ to get

$$\lambda_a (V_j' V_a)^2 = \lambda_j \dot{\lambda}_j, \quad 1 \le j \le k,$$

which implies that

$$\dot{\lambda}_j = \begin{cases} 0 & \text{if } j \ne a, \\ 1 & \text{if } j = a. \end{cases}$$

Plug this value into equation (10.8) to get

$$(\mu'\mu)\dot{V}_j = \lambda_j^2 \dot{V}_j,$$

which means that V_j and $\dot{V}_j$ are mutually perpendicular eigenvectors corresponding to the same eigenvalue of $\mu'\mu$. The assumption that all the eigenvalues of $\mu'\mu$ are simple implies that $\dot{V}_j = 0$. Plug these values into equation (10.7) to conclude that $d_\mu P(U_a V_a') = 0$.

Next let $A = U_a V_b'$, $a, b \le k$, $a \ne b$. Again pre-multiply equation (10.8) by V_j' to get that $\dot{\lambda}_j = 0 \; \forall j$. When j is neither a nor b, equation (10.8) will then imply that

$$(\mu'\mu)\dot{V}_j = \lambda_j^2 \dot{V}_j,$$

from which it follows that $\dot{V}_j = 0$. In case $j = a$, it follows that

$$(\mu'\mu)\dot{V}_a + \lambda_a V_b = \lambda_a^2 \dot{V}_a,$$

which implies that $V_j' \dot{V}_a = 0 \; \forall j \neq b$. Because the columns of V form an orthonormal basis for $\mathbb{R}^k$, this means that $\dot{V}_a$ must be a scalar multiple of V_b. Solve for that multiple to derive

$$\dot{V}_a = \lambda_a (\lambda_a^2 - \lambda_b^2)^{-1} V_b.$$

Similarly we can obtain

$$\dot{V}_b = \lambda_a (\lambda_b^2 - \lambda_a^2)^{-1} V_a.$$

Plug these values into equation (10.7) to conclude that

$$d_\mu P(U_a V_b') = (\lambda_a + \lambda_b)^{-1} (U_a V_b' - U_b V_a'), \quad \forall a, b \leq k, \; a \neq b.$$

Finally take $A = U_a V_b'$, $a > k$, $b \leq k$. Pre-multiply the identity (10.9) by V_j' to conclude that $\lambda_j = 0 \; \forall j \leq k$. Then identity (10.9) becomes $\mu' \mu \dot{V}_j = \lambda_j^2 \dot{V}_j$, which implies that $\dot{V}_j = 0 \; \forall j \leq k$. Plug these values into equation (10.7) to get

$$d_\mu P(U_a V_b') = \lambda_b^{-1} U_a V_b', \quad \forall a > k, b \leq k. \qquad \square$$

Note that the $\{d_\mu P(U_i V_j')\}$ values in equation (10.6) are nonzero and mutually orthogonal whenever $j < i$. The values corresponding to $j \geq i$ are negative of those obtained by interchanging i and j. Also, the size of the subset $S = \{(i, j) : i \leq m, j \leq k, j < i\}$ equals the dimension of $V_{k,m}$. This implies that $\{d_\mu P(U_i V_j')\}$, $(i, j) \in S$ determine an orthogonal basis for $T_{P(\mu)} V_{k,m}$.

10.3 References

The main results of this chapter, namely, Theorems 10.2 and 10.3, are new. For general information on Stiefel manifolds we refer to Chikuse (2003). Also see Lee and Ruymgaart (1996) for nonparametric curve estimation on Stiefel manifolds.

11

Affine shape spaces $A\Sigma_m^k$

Affine shape spaces may be used in place of similarity shape spaces in many situations, especially in analyzing images taken from far away. In such cases a square may appear as a parallelogram, for example, with nonperpendicular adjacent sides of unequal lengths.

11.1 Introduction

The *affine shape* of a k-ad x with landmarks in $\mathbb{R}^m$ may be defined as the orbit of this k-ad under the group of all *affine transformations*

$$x \mapsto F(x) = Ax + b,$$

where A is an arbitrary $m \times m$ nonsingular matrix and b is an arbitrary point in $\mathbb{R}^m$. Then the affine shape space may be defined as the collection of all affine shapes, that is,

$$\bar{A}\Sigma_m^k = \{\sigma(x) : x \in \mathbb{R}^{m \times k}\}, \quad \text{where}$$

$$\sigma(x) = \{Ax + b : A \in GL(m, \mathbb{R}), \, b \in \mathbb{R}^m\}$$

and $GL(m, \mathbb{R})$ is the general linear group on $\mathbb{R}^m$ of all $m \times m$ nonsingular matrices. Note that two k-ads $x = (x_1, \ldots, x_k)$ and $y = (y_1, \ldots, y_k)$ ($x_j, y_j \in \mathbb{R}^m$ for all j) have the same affine shape if and only if the corresponding centered k-ads $u = (u_1, u_2, \ldots, u_k) = (x_1 - \bar{x}, \ldots, x_k - \bar{x})$ and $v = (v_1, v_2, \ldots, v_k) = (y_1 - \bar{y}, \ldots, y_k - \bar{y})$ are related by the transformation

$$Au := (Au_1, \ldots, Au_k) = v, \quad A \in GL(m, \mathbb{R}).$$

The centered k-ads lie in a linear subspace of $\mathbb{R}^{m \times k}$, call it $H(m, k)$, which is

$$H(m, k) = \left\{ u \in \mathbb{R}^{m \times k} : \sum_{j=1}^{k} u_j = 0 \right\}.$$

135

Hence $\bar{A}\Sigma_m^k$ can be represented as the quotient of this subspace under all general linear transformations, that is,

$$\bar{A}\Sigma_m^k = H(m,k)/\mathrm{GL}(m,\mathbb{R}).$$

The subspace $H(m,k)$ is a Euclidean manifold of dimension $m(k-1)$. The group $\mathrm{GL}(m,\mathbb{R})$ has the relative topology (and distance) of $\mathbb{R}^{m^2}$ and hence is a manifold of dimension m^2. Assume $k > m+1$. For the action of $\mathrm{GL}(m,\mathbb{R})$ on $H(m,k)$ to be free and the affine shape space to be a manifold, we require that the columns of u ($u \in H(m,k)$) span $\mathbb{R}^m$. Indeed the condition

$$Au = u \Leftrightarrow A = I_m$$

holds if and only if $\mathrm{rank}(u) = m$. Hence we consider only such centered k-ads u, that is,

$$u \in H_0(m,k) := \{v \in H(m,k) : \mathrm{rank}(v) = m\},$$

and redefine the *affine shape space* as

$$A\Sigma_m^k = H_0(m,k)/\mathrm{GL}(m,\mathbb{R}).$$

Then it follows that $A\Sigma_m^k$ is a manifold of dimension $m(k-1) - m^2$. To get rid of the linear constraint $\sum_{j=1}^k u_j = 0$ on $H(m,k)$, one may postmultiply u by a Helmert matrix H and consider the Helmertized k-ad $\tilde{u} = uH$ as in Section 7.1. Then $H(m,k)$ can be identified with $\mathbb{R}^{m(k-1)}$ and $H_0(m,k)$ is an open dense subset of $H(m,k)$.

For $u, v \in H_0(m,k)$, the condition $Au = v$ holds if and only if $u'A' = v'$ and, as A varies over $\mathrm{GL}(m,\mathbb{R})$, $u'A'$ generates the linear subspace L of $\mathbb{R}^{k-1}$ spanned by the m rows of u. The affine shape of u (or of the original k-ad x) can be identified with this subspace. Thus $A\Sigma_m^k$ may be identified with the set of all m-dimensional subspaces of $\mathbb{R}^{k-1}$, namely, the *Grassmannian* $G_m(k-1)$, a result of Sparr (1992; also see Boothby, 1986). This identification enables us to give a Riemannian structure to $A\Sigma_m^k$ and carry out an intrinsic analysis. This is discussed in Section 11.2.

To carry out an extrinsic analysis on $A\Sigma_m^k$, we embed it into the space of all $k \times k$ symmetric matrices $S(k,\mathbb{R})$ via an equivariant embedding. Then analytic expressions for the extrinsic mean and variation are available. This is the subject of Section 11.3. To get the asymptotic distribution of the sample extrinsic mean and carry out nonparametric inference on affine shapes, we need to differentiate the projection map of Proposition 11.1, which requires perturbation theory arguments for eigenvalues and eigenvectors. This is carried out in Section 11.4.

Affine shape spaces arise in many problems of bioinformatics, cartography, machine vision, and pattern recognition (see Berthilsson and Heyden,

1999; Berthilsson and Astrom, 1999; Sepiashvili et al., 2003; Sparr, 1992). We will see such an application in Section 11.5. The tools developed in Sections 11.3 and 11.4 are applied to this example to carry out an extrinsic analysis.

11.2 Geometry of affine shape spaces

Consider a Helmertized k-ad x in $\mathbb{R}^{m \times (k-1)}$. Define its *special affine shape* as the orbit

$$s\sigma(x) = \{Ax : A \in \mathrm{GL}(m, \mathbb{R}), \ \det(A) > 0\}. \tag{11.1}$$

Any $A \in \mathrm{GL}(m, \mathbb{R})$ has a *singular-value decomposition* $A = U \Lambda V$, where $U, V \in \mathrm{SO}(m)$ and

$$\Lambda = \mathrm{Diag}(\lambda_1, \ldots, \lambda_m), \quad \lambda_1 \geq \ldots \geq \lambda_{m-1} \geq |\lambda_m|, \ \mathrm{sign}(\lambda_m) = \mathrm{sign}(\det(A)).$$

Therefore a linear transformation $x \mapsto Ax$ consists of a rotation and different amount of stretching in different directions followed by another rotation or reflection. When $\det(A) > 0$, that is, when we consider the special affine shape, we look at the affine shape without any reflections. We can get the affine shape $\sigma(x)$ of x from its special affine shape $s\sigma(x)$ by identifying $s\sigma(x)$ with $s\sigma(Tx)$, where $T \in \mathrm{O}(m)$, $\det(T) = -1$. This T can be chosen to be any reflection matrix. Let the *special affine shape space* $SA\Sigma_m^k$ be the collection of all special affine shapes, which is

$$SA\Sigma_m^k = \{s\sigma(x) : x \in \mathbb{R}^{m \times (k-1)}, \ \mathrm{rank}(x) = m\}.$$

We restrict to full rank k-ads so that the group action is free and $SA\Sigma_m^k$ is a manifold. From the expression of $s\sigma(x)$ in equation (11.1), it is clear that it is a function of the "oriented" span of the rows of x, which in turn is a function of an orthogonal m-frame for the rowspace of x. In fact, $SA\Sigma_m^k$ can be viewed as the quotient of $\mathrm{SO}(k-1)$ as follows. Denote by $V_{k,m}$ the *Stiefel manifold* of all orthogonal m-frames in $\mathbb{R}^k$. For $V \in \mathrm{SO}(k-1)$, write $V = \left(\begin{smallmatrix} V_1 \\ V_2 \end{smallmatrix} \right)$ with $V_1 \in V_{k-1,m}$, $V_2 \in V_{k-1,k-m-1}$. Then the oriented span of the rows of V_1, which is the special affine shape of V_1, can be identified with the orbit

$$\pi(V) = \left\{ \left(\begin{smallmatrix} AV_1 \\ BV_2 \end{smallmatrix} \right) : A \in \mathrm{SO}(m), \ B \in \mathrm{SO}(k-m-1) \right\} = \left\{ \left(\begin{smallmatrix} A & 0 \\ 0 & B \end{smallmatrix} \right) V \right\}. \tag{11.2}$$

This implies that

$$SA\Sigma_m^k = \mathrm{SO}(k-1)/\mathrm{SO}(m) \times \mathrm{SO}(k-m-1).$$

Then $A\Sigma_m^k = SA\Sigma_m^k/G$, where G is a finite group generated by any $T \in SO(k-1)$ that looks like

$$T = \begin{pmatrix} T_1 & 0 \\ 0 & T_2 \end{pmatrix}, \; T_1 \in O(m), \; T_2 \in O(k-m-1), \; \det(T_1) = \det(T_2) = -1.$$

This means that two elements V, W in $SO(k-1)$ have the same affine shape if and only if either $\pi(V) = \pi(W)$ or $\pi(TV) = \pi(W)$. Hence $A\Sigma_m^k$ is locally like $SA\Sigma_m^k$. Because $SO(m) \times SO(k-m-1)$ acts as isometries on $SO(k-1)$, the map $\pi: SO(k-1) \rightarrow SA\Sigma_m^k$ in equation (11.2) is a Riemannian submersion. Then $SA\Sigma_m^k$, and hence $A\Sigma_m^k$, inherit the Riemannian metric tensor from $SO(k-1)$, making it a Riemannian manifold.

To derive an explicit expression for the tangent space of $SA\Sigma_m^k$ (or of $A\Sigma_m^k$), we need to identify the horizontal subspace of the tangent space of $SO(k-1)$. Then $d\pi$ provides an isometry between the horizontal subspace and the tangent space of $SA\Sigma_m^k$. We saw in Section 7.2 that geodesics in $SO(k-1)$ starting at $V \in SO(k-1)$ look like

$$\gamma(t) = \exp(tA)V,$$

where $A \in \text{Skew}(k-1)$ $(A + A' = 0)$ and

$$\exp(B) = I + B + \frac{B^2}{2!} + \cdots.$$

This geodesic is vertical if it lies in the orbit $\pi(V)$, that is, when

$$\gamma(t) = \begin{pmatrix} \exp(tA) & 0 \\ 0 & \exp(tB) \end{pmatrix} V,$$

where $A \in \text{Skew}(m)$, $B \in \text{Skew}(k-m-1)$. Then

$$\dot{\gamma}(0) = \begin{pmatrix} A & 0 \\ 0 & B \end{pmatrix} V.$$

Therefore, the vertical subspace of the tangent space $T_V SO(k-1)$ of $SO(k-1)$ at V has the form

$$\text{Ver}_V = \left\{ \begin{pmatrix} A & 0 \\ 0 & B \end{pmatrix} V : A + A' = 0, \; B + B' = 0 \right\}.$$

The horizontal subspace H_V is its orthocomplement in $T_V SO(k-1)$, which is given by

$$H_V = \left\{ AV : A \in \text{Skew}(k-1), \; \text{Trace}\left(A \begin{pmatrix} B_1 & 0 \\ 0 & B_2 \end{pmatrix}\right) = 0 \; \forall B_1 \in \text{Skew}(m), \right.$$

$$\left. B_2 \in \text{Skew}(k-m-1) \right\}$$

$$= \left\{ AV : A = \begin{pmatrix} 0 & B \\ -B' & 0 \end{pmatrix}, \; B \in \mathbb{R}^{m \times (k-m-1)} \right\}.$$

Hence

$$T_{\pi(V)}SA\Sigma_m^k = d\pi_V(H_V).$$

11.3 Extrinsic analysis on affine shape spaces

Let u be a centered k-ad in $H_0(m,k)$ and let $\sigma(u)$ denote its affine shape, which is the orbit

$$\sigma(u) = \{Au : A \in GL(m,\mathbb{R})\}.$$

Consider the map

$$j: A\Sigma_m^k \to S(k,\mathbb{R}), \qquad j(\sigma(u)) \equiv A = FF', \tag{11.3}$$

where $F = (f_1, f_2, \ldots, f_m)$ is an orthonormal basis for the row space of u. It has been shown that j is an embedding of $A\Sigma_m^k$ into $S(k,\mathbb{R})$, equivariant under the action of a group isomorphic to $O(k-1)$ (see Dimitric, 1996). In equation (11.3), A is the projection (matrix) onto the subspace spanned by the rows of u. Hence, through the embedding j, we identify an m-dimensional subspace of $\mathbb{R}^{k-1}$ with the projection map (matrix) onto that subspace. Because A is a projection matrix, it is characterized by

$$A^2 = A, \ A = A' \ \text{and} \ \mathrm{Trace}(A) = \mathrm{rank}(A) = m.$$

Also, u is a centered k-ad, that is, the rows of u are orthogonal to $\mathbf{1}_k$, therefore $A\mathbf{1}_k = 0$. Hence the image of $A\Sigma_m^k$ into $S(k,\mathbb{R})$ under the embedding j is given by

$$j(A\Sigma_m^k) = \{A \in S(k,\mathbb{R}) : A^2 = A, \ \mathrm{Trace}(A) = m, \ A\mathbf{1}_k = 0\}, \tag{11.4}$$

which is a compact Riemannian submanifold of $S(k,\mathbb{R})$ of dimension $mk - m - m^2$. It is easy to show that the image $j(\sigma(\mu))$ equals $A = u'(uu')^{-1}u$.

Let Q be a probability distribution on $A\Sigma_m^k$ and let $\tilde{Q} = Q \circ j^{-1}$ be its image in $j(A\Sigma_m^k)$. Let $\tilde{\mu}$ be the mean of $\tilde{Q}$, that is, $\tilde{\mu} = \int_{j(A\Sigma_m^k)} x\tilde{Q}(dx)$. Then $\tilde{\mu}$ is a $k \times k$ positive semi-definite matrix satisfying

$$\mathrm{Trace}(\tilde{\mu}) = m, \ \mathrm{rank}(\tilde{\mu}) \geq m, \ \text{and} \ \tilde{\mu}\mathbf{1}_k = 0.$$

To carry out an extrinsic analysis on $A\Sigma_m^k$, we need to identify the extrinsic mean (set) of Q, which is the projection (set) of $\tilde{\mu}$ on $j(A\Sigma_m^k)$. Denote by $P(\tilde{\mu})$ the set of projections of $\tilde{\mu}$ on $j(A\Sigma_m^k)$, as defined in equation (4.3). Proposition 11.1 below gives an expression for $P(\tilde{\mu})$ and hence finds the extrinsic mean set of Q.

Proposition 11.1 *(a) The projection of $\tilde{\mu}$ into $j(A\Sigma_m^k)$ is given by*

$$P(\tilde{\mu}) = \Big\{ \sum_{j=1}^{m} U_j U_j' \Big\}, \tag{11.5}$$

where $U = (U_1, \ldots, U_k) \in SO(k)$ is such that $\tilde{\mu} = U\Lambda U'$, $\Lambda = \mathrm{Diag}(\lambda_1, \ldots, \lambda_k)$, $\lambda_1 \geq \cdots \geq \lambda_k = 0$. (b) $\tilde{\mu}$ is nonfocal and Q has a unique extrinsic mean μ_E if and only if $\lambda_m > \lambda_{m+1}$. Then $\mu_E = \sigma(F')$, where $F = (U_1, \ldots, U_m)$.

Proof From the definition of $P(\tilde{\mu})$ it follows that for any $A_0 \in P(\tilde{\mu})$,

$$\|\tilde{\mu} - A_0\|^2 = \min_{A \in j(A\Sigma_m^k)} \|\tilde{\mu} - A\|^2.$$

Here $\|.\|$ denotes the Euclidean norm, which is

$$\|A\|^2 = \mathrm{Trace}(AA'), \quad A \in \mathbb{R}^{k \times k}.$$

Then, for any $A \in j(A\Sigma_m^k)$,

$$\|\tilde{\mu} - A\|^2 = \mathrm{Trace}(\tilde{\mu} - A)^2 = \sum_{i=1}^{k} \lambda_i^2 + m - 2\,\mathrm{Trace}(\tilde{\mu}A), \tag{11.6}$$

where $\lambda_1, \ldots, \lambda_k$ are the eigenvalues of $\tilde{\mu}$ defined in the statement of the proposition. Because A is a projection matrix, it can be written as

$$A = FF', \quad \text{where } F \in \mathbb{R}^{k \times m}, \; F'F = I_m.$$

Also write $\tilde{\mu} = U\Lambda U'$ as in the proposition. Then

$$\|\tilde{\mu} - A\|^2 = \sum_{i=1}^{k} \lambda_i^2 + m - 2\,\mathrm{Trace}(F'U\Lambda U'F)$$

$$= \sum_{i=1}^{k} \lambda_i^2 + m - 2\,\mathrm{Trace}(E\Lambda E'), \quad E = F'U. \tag{11.7}$$

To minimize $\|\tilde{\mu} - A\|^2$, we need to maximize $\mathrm{Trace}(E\Lambda E')$ over $E \in \mathbb{R}^{m \times k}$, $EE' = I_m$. Note that

$$\mathrm{Trace}(E\Lambda E') = \sum_{i=1}^{m} \sum_{j=1}^{k} e_{ij}^2 \lambda_j = \sum_{j=1}^{k} w_j \lambda_j,$$

where the e_{ij} are the entries of E and $w_j = \sum_{i=1}^{m} e_{ij}^2$, $j = 1, 2, \ldots, k$. Then $0 \leq w_j \leq 1$ and $\sum_{j=1}^{k} w_j = m$. Therefore, the maximum value of $\mathrm{Trace}(E\Lambda E')$ equals $\sum_{j=1}^{m} \lambda_j$, which is attained if and only if

$$w_1 = w_2 = \cdots = w_m = 1, \quad w_i = 0 \text{ for } i > m,$$

that is, when

$$E = (E_{11}, 0)$$

for some $E_{11} \in O(m)$. Then, from equation (11.7), it follows that $F = UE'$ and the value of A that minimizes equation (11.6) is given by

$$A_0 = FF' = UE'EU' = U \begin{pmatrix} I_m & 0 \\ 0 & 0 \end{pmatrix} U' = \sum_{j=1}^{m} U_j U_j'. \tag{11.8}$$

This proves part (a) of the proposition.

To prove part (b), note that $\sum_{j=1}^{m} U_j U_j'$ is the projection matrix of the subspace spanned by $\{U_1, \ldots, U_m\}$, which is unique if and only if $\lambda_m > \lambda_{m+1}$. Then $\mu_E = \sigma(F')$ for any F satisfying $A_0 = FF'$, with A_0 being defined in equation (11.8). Hence one can choose $F = (U_1, \ldots, U_m)$. This completes the proof. □

We can use Proposition 11.1 and Proposition 4.2 to get an expression for the extrinsic variation V of Q as follows:

$$V = \|\tilde{\mu} - \mu\|^2 + \int_{j(A\Sigma_m^k)} \|\tilde{\mu} - x\|^2 \tilde{Q}(dx), \quad \mu \in P(\tilde{\mu})$$

$$= 2(m - \sum_{i=1}^{m} \lambda_i). \tag{11.9}$$

Let $X_1, \ldots, X_n$ be an i.i.d. sample from Q and let μ_{nE} be the sample extrinsic mean, which can be any measurable selection from the sample extrinsic mean set. It follows from Corollary 3.4 that if Q has a unique extrinsic mean μ_E, that is, if $\tilde{\mu}$ is a nonfocal point of $S(k, \mathbb{R})$, then μ_{nE} is a consistent estimator of μ_E.

11.4 Asymptotic distribution of the sample extrinsic mean

In this section we assume that $\tilde{\mu}$ is a nonfocal point of $S(k, \mathbb{R})$. Then the map $P(\tilde{\mu}) = \sum_{j=1}^{m} U_j U_j'$ is well defined and smooth in a neighborhood $N(\tilde{\mu})$ of $\tilde{\mu}$ in $S(k, \mathbb{R})$. That follows from perturbation theory, because if $\lambda_m > \lambda_{m+1}$, then the subspace spanned by $\{U_1, \ldots, U_m\}$ is a smooth map from $S(k, \mathbb{R})$ into the Grassmannian $G_m(k)$ and $P(\tilde{\mu})$ is the matrix of projection onto that subspace. Then it follows from the calculations of Section 4.2 that $\sqrt{n}(j(\mu_{nE}) - j(\mu_E))$ is asymptotically Normal in the tangent space of $j(A\Sigma_m^k)$ at $j(\mu_E) \equiv P(\tilde{\mu})$. To get the asymptotic coordinates and the covariance matrix as in Proposition 4.3, we need to find the derivative of P. Define

$$N_m^k = \{A \in S(k, \mathbb{R}): A^2 = A, \text{Trace}(A) = m\}. \tag{11.10}$$

Then $N_m^k = j(A\Sigma_m^{k+1})$, which is a Riemannian manifold of dimension $km - m^2$. It has been shown in Dimitric (1996) that the tangent and normal spaces to N_m^k are given by

$$T_A N_m^k = \{v \in S(k, \mathbb{R}) : vA + Av = v\}, \tag{11.11}$$

$$T_A N_m^{k\perp} = \{v \in S(k, \mathbb{R}) : vA = Av\}. \tag{11.12}$$

Consider the map

$$P: N(\tilde{\mu}) \to N_m^k, \quad P(A) = \sum_{j=1}^{m} U_j(A)U_j(A)', \tag{11.13}$$

where $A = \sum_{j=1}^{k} \lambda_j(A)U_j(A)U_j(A)'$ is a singular value decomposition of A as in Proposition 11.1.

Proposition 11.2 *The derivative of P is given by*

$$dP: S(k, \mathbb{R}) \to TN_m^k, \quad d_{\tilde{\mu}}P(A) = \sum_{i=1}^{m} \sum_{j=m+1}^{k} (\lambda_i - \lambda_j)^{-1} a_{ij} UE_{ij}U', \tag{11.14}$$

where $A = \sum\sum_{1 \le i \le j \le k} a_{ij}UE_{ij}U'$ and $\{UE_{ij}U' : 1 \le i \le j \le k\}$ is the orthogonal basis (frame) for $S(k, \mathbb{R})$ obtained in Section 9.3.

Proof Let $\gamma(t) = \tilde{\mu} + tv$ be a curve in $N(\tilde{\mu})$ with $\gamma(0) = \tilde{\mu}$ and $\dot{\gamma}(0) = v \in S(k, \mathbb{R})$. Then

$$\gamma(t) = U(\Lambda + tU'vU)U' = U\tilde{\gamma}(t)U', \tag{11.15}$$

where $\tilde{\gamma}(t) = \Lambda + tU'vU$, which is a curve in $S(k, \mathbb{R})$ satisfying $\tilde{\gamma}(0) = \Lambda$ and $\dot{\tilde{\gamma}}(0) = \tilde{v} = U'vU$. From equations (11.13) and (11.15) we get that

$$P[\gamma(t)] = UP[\tilde{\gamma}(t)]U'. \tag{11.16}$$

Differentiate equation (11.16) at $t = 0$ to get

$$d_{\tilde{\mu}}P(v) = Ud_\Lambda P(\tilde{v})U'. \tag{11.17}$$

To find $d_\Lambda P(\tilde{v}) \equiv \frac{d}{dt}P[\tilde{\gamma}(t)]|_{t=0}$, we may assume without loss of generality that $\lambda_1 > \lambda_2 > \cdots > \lambda_k$. Then we can choose a spectral decomposition for $\tilde{\gamma}(t)$ as $\tilde{\gamma}(t) = \sum_{j=1}^{k} \lambda_j(t)e_j(t)e_j(t)'$ such that $\{e_j(t), \lambda_j(t)\}_{j=1}^{k}$ are some smooth functions of t satisfying $e_j(0) = e_j$ and $\lambda_j(0) = \lambda_j$, where $\{e_j\}_{j=1}^{k}$ is the canonical basis for $\mathbb{R}^k$. Let $\tilde{v} = E_{ab}$, $1 \le a \le b \le k$ (see equation (9.35)). Then we can get expressions for $\dot{e}_j(0)$ from equation (9.36). Because

$$P[\tilde{\gamma}(t)] = \sum_{j=1}^{m} e_j(t)e_j(t)',$$

one has

$$\frac{d}{dt}P[\tilde{\gamma}(t)]|_{t=0} = \sum_{j=1}^{m}[e_j\dot{e}_j(0)' + \dot{e}_j(0)e_j'].\tag{11.18}$$

From equations (9.37) and (11.18) it follows that

$$d_\Lambda P(E_{ab}) = \begin{cases} (\lambda_a - \lambda_b)^{-1}E_{ab} & \text{if } a \le m < b \le k, \\ 0 & \text{otherwise.} \end{cases}\tag{11.19}$$

Then from equation (11.17) we obtain

$$d_{\tilde{\mu}}P(UE_{ab}U') = \begin{cases} (\lambda_a - \lambda_b)^{-1}UE_{ab}U' & \text{if } a \le m < b \le k, \\ 0 & \text{otherwise.} \end{cases}\tag{11.20}$$

Hence, if $A = \sum\sum_{1\le i\le j\le k} a_{ij}UE_{ij}U'$, from equation (11.20) we get

$$d_{\tilde{\mu}}P(A) = \sum_{i=1}^{m}\sum_{j=m+1}^{k}(\lambda_i - \lambda_j)^{-1}a_{ij}UE_{ij}U'.\tag{11.21}$$

This completes the proof. □

Corollary 11.3 *Consider the projection map of equation* (11.13) *restricted to*

$$S_0(k,\mathbb{R}) := \{A \in S(k,\mathbb{R}) : A\mathbf{1}_k = 0\}.$$

It has the derivative

$$dP: S_0(k,\mathbb{R}) \to Tj(A_m^k), \quad d_{\tilde{\mu}}P(A) = \sum_{i=1}^{m}\sum_{j=m+1}^{k-1}(\lambda_i - \lambda_j)^{-1}a_{ij}UE_{ij}U'.$$

Proof Follows from Proposition 11.2 and the fact that

$$T_{P(\tilde{\mu})}j(A_m^k) = \{v \in T_{P(\tilde{\mu})}N_m^k : v\mathbf{1}_k = 0\}.$$

□

From Corollary 11.3 it follows that

$$\{UE_{ij}U' : 1 \le i \le m < j < k\}\tag{11.22}$$

forms an orthonormal basis for $T_{P(\tilde{\mu})}j(A_m^k)$ and if $A \in S(k,\mathbb{R})$ has coordinates $\{a_{ij} : 1 \le i \le j \le k\}$ with respect to the orthonormal basis $\{UE_{ij}U' : 1 \le i \le j \le k\}$ of $S(k,\mathbb{R})$, then $d_{\tilde{\mu}}P(A)$ has coordinates $\{(\lambda_i - \lambda_j)^{-1}a_{ij} : 1 \le i \le m < j < k\}$ in $T_{P(\tilde{\mu})}j(A_m^k)$. Also, it is easy to show that the linear projection $L(A)$ of A into $T_{P(\tilde{\mu})}j(A_m^k)$ has coordinates $\{a_{ij} : 1 \le i \le m < j < k\}$. Therefore we have the following corollary to

Proposition 4.3. In the statement of Corollary 11.4, $\tilde{X}_i = j(X_i)$, $i = 1,\ldots,n$, denotes the embedded sample in $j(A\Sigma_m^k)$ and

$$T_j = \left((T_j)_{ab} : 1 \le a \le m < b < k\right)$$

denotes the coordinates of $d_{\tilde{\mu}}P(\tilde{X}_j - \tilde{\mu})$ in $\mathbb{R}^{km-m-m^2}$ with respect to the orthonormal basis of $T_{P(\tilde{\mu})}j(A_m^k)$ obtained in equation (11.22). Then T_j has the following expression:

$$(T_j)_{ab} = \sqrt{2}(\lambda_a - \lambda_b)^{-1}U_a'\tilde{X}_jU_b, \quad 1 \le a \le m < b < k.$$

Corollary 11.4 *If $\tilde{\mu} = E[\tilde{X}_1]$ is a nonfocal point of $S(k, \mathbb{R})$, then*

$$\sqrt{n}\,[j(\mu_{nE}) - j(\mu_E)] = \sqrt{n}\,d_{\tilde{\mu}}P(\overline{\tilde{X}} - \tilde{\mu}) + o_P(1)$$

$$\xrightarrow{\mathcal{L}} N(0, \Sigma),$$

where Σ denotes the covariance matrix of $T_1 = d_{\tilde{\mu}}P(\tilde{X}, -\tilde{\mu})$.

Using Corollary 11.4, we may construct confidence regions for μ_E as in Section 4.2 or perform two-sample tests to compare the extrinsic means from two populations on $A\Sigma_m^k$ as in Section 4.5.

The asymptotic distribution of the sample extrinsic variation follows from Theorem 4.5, from which we may construct confidence intervals for the extrinsic variation of Q or compare the extrinsic variations for two populations via the two-sample tests described in Section 4.5.

11.5 Application to handwritten digit recognition

Consider the landmark data recorded by Anderson (1997) of 30 handwritten digits "3" with 13 landmarks. These are used in Dryden and Mardia (1998) for principal component analysis and a one-sample parametric F test to check whether a particular template of the digit "3" significantly differs from the mean similarity shape of the population from which the 30 observations are drawn.

We analyze the affine shape of the sample points and estimate the mean shape and variation in shape. This can be used as a prior model for digit recognition from images of handwritten codes. Our observations lie on the affine shape space $A\Sigma_2^k$, $k = 13$. A representative of the sample extrinsic mean shape has the coordinates

$$u = (-0.53, -0.32, -0.26, -0.41, 0.14, -0.43, 0.38, -0.29, 0.29, -0.11,$$
$$0.06, 0, -0.22, 0.06, 0.02, 0.08, 0.19, 0.13, 0.30, 0.21, 0.18, 0.31, -0.13,$$
$$0.38, -0.42, 0.38).$$

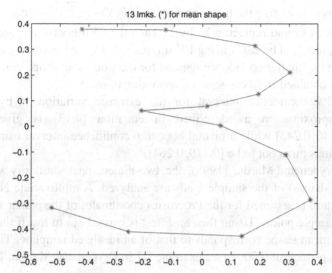

Figure 11.1 Extrinsic mean shape for handwritten digit "3" sample.

The coordinates are in pairs, with x coordinate followed by y. Figure 11.1 shows the plot of u.

The sample extrinsic variation turns out to be 0.27, which is fairly large. There seems to be a lot of variability in the data. Following are the extrinsic distances squared of the sample points from the mean affine shape:

$$(\rho^2(X_j, \mu_E), \; j = 1, \ldots, n) = (1.64, 0.28, 1.00, 0.14, 0.13, 0.07, 0.20, 0.09,$$
$$0.17, 0.15, 0.26, 0.17, 0.14, 0.20, 0.42, 0.31, 0.14, 0.12, 0.51, 0.10, 0.06,$$
$$0.15, 0.05, 0.31, 0.08, 0.08, 0.11, 0.18, 0.64, 0.12).$$

Here $n = 30$ is the sample size. From these distances it is clear that observations 1 and 3 are outliers. We remove them and recompute the sample extrinsic mean and variation. The sample variation now turns out to be 0.19.

An asymptotic 95% confidence region for the extrinsic mean μ_E as in equation (4.6) is given by

$$\{\mu_E : nL[P(\tilde{\mu}) - P(\overline{X})]'\hat{\Sigma}^{-1}L[P(\tilde{\mu}) - P(\overline{X})] \le \mathcal{X}_{20}^2(0.95) = 31.4104\}.$$

The dimension 20 of $A\Sigma_2^{13}$ is quite high compared to the sample size of 28. It is difficult to construct a pivotal bootstrap confidence region, because the bootstrap covariance estimates Σ^* tend to be singular or close to singular in most simulations. Instead, we construct a nonpivotal bootstrap confidence region by considering the linear projection $L[P(\overline{X}) - P(\overline{X}^*)]$ into the tangent space of $P(\overline{X})$ and replacing Σ^* by $\hat{\Sigma}$. Then the 95th bootstrap percentile $c^*(0.95)$ turns out be 1.077 using 10^5 simulations. Hence bootstrap methods yield much smaller confidence regions for the true mean shape compared to those obtained from chi-squared approximations.

A 95% confidence interval for the extrinsic variation V by normal approximation as described in equation (4.13) is given by $V \in [0.140, 0.243]$ while a pivotal bootstrap confidence interval using 10^5 simulations turns out to be $[0.119, 0.264]$.

In Dryden and Mardia (1998), the two-dimensional similarity shapes (planar shapes) of the sample k-ads are analyzed. A multivariate Normal distribution is assumed for the Procrustes coordinates of the planar shapes of the sample points. Using this, an F test is carried out to test if the population mean shape corresponds to that of an idealized template. The test yields a p-value of 0.0002 (see example 7.1 in Dryden and Mardia, 1998).

11.6 References

Apart from the work by Sparr (1992) and, especially, that of Dimitric (1996) mentioned in this chapter, we refer to Sugathadasa (2006), who derived the extrinsic mean (see Proposition 11.1). The proof given here is new. The asymptotic distribution theory presented here is taken from Bhattacharya (2008a).

12

Real projective spaces and projective shape spaces

The real projective space $\mathbb{R}P^m$ is the manifold of all lines through the origin in $\mathbb{R}^{m+1}$. It may be identified with the sphere S^m modulo the identification of a point p with its antipodal point $-p$. For $m = 1$ or 2, this is the *axial space*. Extrinsic and intrinsic inferences on $\mathbb{R}P^m$ are developed first in this chapter. This aids in the nonparametric analysis on the space of projective shapes, identified here, subject to a registration, with $(\mathbb{R}P^m)^{k-m-2}$.

12.1 Introduction

Consider a k-ad picked on a planar image of an object or a scene in three dimensions. If one thinks of images or photographs obtained through a central projection (a pinhole camera is an example of this), a ray is received as a landmark on the image plane (e.g., the film of the camera). Because axes in three dimensions comprise the projective space $\mathbb{R}P^2$, the k-ad in this view is valued in $\mathbb{R}P^2$. For a k-ad in three dimensions to represent a k-ad in $\mathbb{R}P^2$, the corresponding axes must all be distinct. To have invariance with regard to camera angles, one may first look at the original noncollinear three-dimensional k-ad u and achieve affine invariance by its affine shape (i.e., by the equivalence class Au, $A \in \mathrm{GL}(3, \mathbb{R})$), and finally take the corresponding equivalence class of axes in $\mathbb{R}P^2$ to define the projective shape of the k-ad as the equivalence class, or orbit, with respect to projective transformations on $\mathbb{R}P^2$. The projective shape of a k-ad is singular if the k axes lie on the vector plane $\mathbb{R}P^1$. For $k > 4$, the space of all nonsingular shapes is the *two-dimensional projective shape space*, denoted as $P_0\Sigma_2^k$.

In general, the *projective space* $\mathbb{R}P^m$ comprises axes or lines through the origin in $\mathbb{R}^{m+1}$. Thus elements of $\mathbb{R}P^m$ may be represented as equivalence classes:

$$[x] = [x^1 : x^2 : \cdots : x^{m+1}] = \{\lambda x : \lambda \neq 0\}, \quad x = (x^1, \ldots, x^{m+1})' \in \mathbb{R}^{m+1} \setminus \{0\}.$$

Then a *projective transformation* α on $\mathbb{R}P^m$ is defined in terms of an $(m+1) \times (m+1)$ nonsingular matrix $A \in GL(m+1, \mathbb{R})$ by

$$\alpha([x]) = [Ax].$$

The group of all projective transformations on $\mathbb{R}P^m$ is denoted by $PGL(m)$. Now consider a k-ad $y = (y_1, \ldots, y_k)$ in $(\mathbb{R}P^m)^k$, say $y_j = [x_j]$, $j = 1, \ldots, k$, $x_j \in \mathbb{R}^{m+1} \backslash \{0\}$ and $k > m + 2$. The *projective shape* of this k-ad is its orbit under $PGL(m)$, that is, $\alpha(y) \doteq \{(\alpha y_1, \ldots, \alpha y_k) : \alpha \in PGL(m)\}$. To exclude singular shapes, define a k-ad $y = (y_1, \ldots, y_k) = ([x_1], \ldots, [x_k])$ to be in a *general position* if there exists a subset of $m + 2$ landmarks, say $(y_{i_1}, \ldots, y_{i_{m+2}})$, such that the linear span of any $m + 1$ points from this set is $\mathbb{R}P^m$, that is, if the linear span of their representative points in $\mathbb{R}^{m+1}$ is $\mathbb{R}^{m+1}$. The space of shapes of all k-ads in the general position is the *projective shape space* $P_0\Sigma_m^k$.

12.2 Geometry of the real projective space $\mathbb{R}P^m$

Because any line through the origin in $\mathbb{R}^{m+1}$ is uniquely determined by its points of intersection with the unit sphere S^m, one may identify $\mathbb{R}P^m$ with S^m/G, with G comprising the identity map and the antipodal map $p \mapsto -p$. Its structure as a m-dimensional manifold (with quotient topology) and its Riemannian structure both derive from this identification. Among applications are observations on galaxies, on axes of crystals, on the line of a geological fissure (Watson, 1983; Mardia and Jupp, 2000; Fisher et al., 1987; Beran and Fisher, 1998; Kendall, 1989).

For $u, v \in S^m$, the geodesic distance between the corresponding elements $[u], [v] \in \mathbb{R}P^m$ is given by

$$d_g([u], [v]) = \min\{d_{gs}(u, v), d_{gs}(u, -v)\},$$

where $d_{gs}(u, v) = \arccos(u'v)$ is the geodesic distance on S^m. Therefore,

$$d_g([u], [v]) = \min\{\arccos(u'v), \arccos(-u'v)\} = \arccos(|u'v|).$$

The injectivity radius of $\mathbb{R}P^m$ is $\frac{\pi}{2}$. The map

$$\pi : S^m \to \mathbb{R}P^m, \quad u \mapsto [u]$$

is a Riemannian submersion. The exponential map of $\mathbb{R}P^m$ at $[u]$ is $\exp_{[u]} = \pi \circ \exp_u \circ d\pi_u^{-1}$ where $\exp_u : T_u S^m \to S^m$ is the exponential map of the sphere, which is

$$\exp_u(v) = \cos(\|v\|)u + \sin(\|v\|)\frac{v}{\|v\|}, \quad v \in T_u S^m.$$

The cut-locus of $[u]$ is

$$C([u]) = \{[v] \in \mathbb{R}P^m : d_g([u],[v]) = \frac{\pi}{2}\}$$
$$= \{[v] \in \mathbb{R}P^m : u'v = 0\}.$$

The inverse exponential map $\exp_{[u]}^{-1}$ is invertible on $\mathbb{R}P^m \setminus C([u])$ and is given by

$$\exp_{[u]}^{-1}([v]) = \frac{\arccos(|u'v|)}{\sqrt{1 - (u'v)^2}} d\pi_u \left(\frac{u'v}{|u'v|} v - |u'v|u \right), \quad u'v \neq 0.$$

The exponential map $\exp_{[u]}$ is a diffeomorphism on $\exp_{[u]}^{-1}(\mathbb{R}P^m \setminus C([u]))$ onto $\mathbb{R}P^m \setminus C([u])$. The projective space has a constant sectional curvature of 1 (see Gallot et al., 1990, p. 123).

12.3 Geometry of the projective shape space $P_0\Sigma_m^k$

Assume $k > m + 2$. Recall that the projective shape of a k-ad $y \in (\mathbb{R}P^m)^k$ is given by the orbit

$$\sigma(y) = \{\alpha y : \alpha \in \mathrm{PGL}(m)\}.$$

This orbit has full rank if y is in the general position. Then we define the projective shape space $P_0\Sigma_m^k$ to be the set of all shapes of k-ads in the general position. Define a *projective frame* in $\mathbb{R}P^m$ to be an ordered system of $m + 2$ points in the general position, that is, the linear span of any $m + 1$ points from this set is $\mathbb{R}P^m$. Let $I = i_1 < \cdots < i_{m+2}$ be an ordered subset of $\{1, \ldots, k\}$. A manifold structure on the set $P_I\Sigma_m^k$, of projective shapes of all k-ads $(y_1, \ldots, y_k)$, for which $(y_{i_1}, \ldots, y_{i_{m+2}})$ is a projective frame in $\mathbb{R}P^m$, is derived as follows. The *standard frame* is defined to be $([e_1], \ldots, [e_{m+1}], [e_1 + e_2 + \cdots + e_{m+1}])$, where $e_j \in \mathbb{R}^{m+1}$ has 1 in the jth coordinate and zeros elsewhere. One can check that, given two projective frames $(p_1, \ldots, p_{m+2})$ and $(q_1, \ldots, q_{m+2})$, there exists a unique $\alpha \in \mathrm{PGL}(m)$ such that $\alpha(p_j) = q_j$ $(j = 1, \ldots, m+2)$. By ordering the points in a k-ad such that the first $m + 2$ points are in the general position, one may bring this ordered set, say $(p_1, \ldots, p_{m+2})$, to the standard form by a unique $\alpha \in \mathrm{PGL}(m)$. Then the ordered set of remaining $k - m - 2$ points is transformed to a point in $(\mathbb{R}P^m)^{k-m-2}$. This provides a diffeomorphism between $P_I\Sigma_m^k$ and the product of $k - m - 2$ copies of the real projective space $\mathbb{R}P^m$. Hence, by developing corresponding inference tools on $\mathbb{R}P^m$, one can perform statistical inference in a dense open subset of $P_0\Sigma_m^k$. In the subsequent sections, we develop intrinsic and extrinsic analysis tools on

$\mathbb{R}P^m$. One may note that $P_I\Sigma_m^k$ is an open dense subset of the full projective shape space $P_0\Sigma_m^k$.

12.4 Intrinsic analysis on $\mathbb{R}P^m$

Let Q be a probability distribution on $\mathbb{R}P^m$ and let $X_1, \ldots, X_n$ be an i.i.d. random sample from Q. The value of r_* on $\mathbb{R}P^m$ as defined in Chapter 5 turns out to be the minimum of its injectivity radius of $\frac{\pi}{2}$ and $\frac{\pi}{\sqrt{C}}$, where C is its constant sectional curvature 4. Hence $r_* = \frac{\pi}{2}$ and therefore, if the support of Q is contained in an open geodesic ball of radius $\frac{\pi}{4}$, then it has a unique intrinsic mean in that ball (see Proposition 5.2). In this section we assume that $\text{supp}(Q) \subseteq B(p, \frac{\pi}{4})$, $p \in \mathbb{R}P^m$. Let $\mu_I = [\mu]$ ($\mu \in S^m$) be the intrinsic mean of Q in the ball. Choose an orthonormal basis $v_1, \ldots, v_d$ for $T_\mu S^m$ so that $\{d\pi_\mu(v_j)\}$ forms an orthonormal basis for $T_{\mu_I}\mathbb{R}P^m$. For $[x] \in B(p, \frac{\pi}{4})$ ($x \in S^m$), let $\phi([x])$ be the coordinates of $\exp_{\mu_I}^{-1}([x])$ with respect to this basis, which are

$$\phi([x]) = (x^1, \ldots, x^m),$$

$$x^j = \frac{x'\mu}{|x'\mu|} \frac{\arccos(|x'\mu|)}{\sqrt{1 - (x'\mu)^2}}(x'v_j), \quad j = 1, 2, \ldots, m.$$

Let $X_j = [Y_j]$ ($Y_j \in S^m$) and $\tilde{X}_j = \phi(X_j)$, $j = 1, 2, \ldots, n$. Let μ_{nI} be the sample intrinsic mean in $B(p, \frac{\pi}{4})$ and let $\mu_n = \phi(\mu_{nI})$. Then, from Theorem 5.3 and Corollary 5.4, it follows that if $\text{supp}(Q) \subseteq B(\mu_I, \frac{\pi}{4})$, then

$$\sqrt{n}\,\mu_n \xrightarrow{\mathcal{L}} N(0, \Lambda^{-1}\Sigma\Lambda^{-1}),$$

where $\Sigma = 4E(\tilde{X}_1\tilde{X}_1')$ and $\Lambda = ((\Lambda_{rs}))_{1 \leq r,s \leq d}$, where

$$\Lambda_{rs} = \Lambda_{sr} = 2E\left[\frac{1}{(1 - |\tilde{X}_1'\mu|^2)}\left(1 - \frac{\arccos(|\tilde{X}_1'\mu|)(2|\tilde{X}_1'\mu|^2 - 1)}{|\tilde{X}_1'\mu|\sqrt{1 - (\tilde{X}_1'\mu)^2}}\right)(\tilde{X}_1'v_r)(\tilde{X}_1'v_s)\right.$$
$$\left. + \frac{\arccos(|\tilde{X}_1'\mu|)(2|\tilde{X}_1'\mu|^2 - 1)}{|\tilde{X}_1'\mu|\sqrt{1 - (\tilde{X}_1'\mu)^2}}\delta_{rs}\right], \quad 1 \leq r \leq s \leq d.$$

A confidence region for μ_I of asymptotic confidence level $1 - \alpha$ is given by

$$\{\mu_I : n\mu_n'\hat{\Lambda}\hat{\Sigma}^{-1}\hat{\Lambda}\mu_n \leq X_m^2(1 - \alpha)\},$$

where $\hat{\Lambda}$ and $\hat{\Sigma}$ are sample estimates of Λ and Σ, respectively. One can also construct a pivotal bootstrap confidence region.

To compare the intrinsic means or variations of two probability distribution on $\mathbb{R}P^m$ and hence distinguish between them, we can use the methods developed in Section 5.4.

12.5 Extrinsic analysis on $\mathbb{R}P^m$

Another representation of $\mathbb{R}P^m$ is via the *Veronese–Whitney embedding* j of $\mathbb{R}P^m$ into the space of all $(m + 1) \times (m + 1)$ symmetric matrices $S(m + 1, \mathbb{R})$, which is a real vector space of dimension $\frac{(m+1)(m+2)}{2}$. This embedding is given by

$$j([u]) = uu' = ((u_i u_j))_{1 \le i, j \le m+1}, \quad u = (u_1, .., u_{m+1})' \in S^m.$$

It induces the extrinsic distance

$$\rho^2([u], [v]) = \|uu' - vv'\|^2 = \text{Trace}(uu' - vv')^2 = 2(1 - (u'v)^2).$$

If one denotes the space of all $(m + 1) \times (m + 1)$ positive semi-definite matrices as $S^+(m + 1, \mathbb{R})$, then

$$j(\mathbb{R}P^m) = \{A \in S^+(m + 1, \mathbb{R}) : \text{rank}(A) = 1, \ \text{Trace}(A) = 1\},$$

which is a compact Riemannian submanifold of $S(m + 1, \mathbb{R})$ of dimension m. The embedding j is equivariant under the action of the orthogonal group $O(m + 1)$, which acts on $\mathbb{R}P^m$ as $A[u] = [Au]$.

Let Q be a probability measure on $\mathbb{R}P^m$, and let $\tilde{\mu}$ be the mean of $\tilde{Q} := Q \circ j^{-1}$ considered as a probability distribution on $S(m + 1, \mathbb{R})$. To find the extrinsic mean set of Q, we need to find the projection of $\tilde{\mu}$ on $\tilde{M} := j(\mathbb{R}P^m)$, say $P_{\tilde{M}}(\tilde{\mu})$, as in Proposition 4.2. Because $\tilde{\mu}$ belongs to the convex hull of $\tilde{M}$, it lies in the space $S^+(m + 1, \mathbb{R})$ of $(m + 1) \times (m + 1)$ nonnegative definite matrices and satisfies

$$\text{rank}(\tilde{\mu}) \ge 1, \quad \text{Trace}(\tilde{\mu}) = 1.$$

There exists an orthogonal $(m + 1) \times (m + 1)$ matrix U such that $\tilde{\mu} = UDU'$, $D \equiv \text{Diag}(\lambda_1, \ldots, \lambda_{m+1})$, where the eigenvalues may be taken to be ordered: $0 \le \lambda_1 \le \cdots \le \lambda_{m+1}$. To find $P_{\tilde{M}}(\tilde{\mu})$, note first that, writing $v = U'u$, we get

$$\|\tilde{\mu} - uu'\|^2 = \text{Trace}[(\tilde{\mu} - uu')^2]$$
$$= \text{Trace}[\{U'(\tilde{\mu} - uu')U\}\{U'(\tilde{\mu} - uu')U\}] = \text{Trace}[(D - vv')^2].$$

Write $v = (v_1, \ldots, v_{m+1})$, so that

$$
\begin{aligned}
\|\tilde{\mu} - uu'\|^2 &= \sum_{i=1}^{m+1}(\lambda_i - v_i^2)^2 + \sum_{j \neq j'}(v_i v_j)^2 \\
&= \sum_{i=1}^{m+1}\lambda_i^2 + \sum_{i=1}^{m+1}v_i^4 - 2\sum_{i=1}^{m+1}\lambda_i v_i^2 + (\sum_j v_j^2)(\sum_{j'} v_{j'}^2) - \sum_{j=1}^{m+1}v_j^4 \\
&= \sum_{i=1}^{m+1}\lambda_i^2 - 2\sum_{i=1}^{m+1}\lambda_i v_i^2 + 1.
\end{aligned}
\tag{12.1}
$$

The minimum of equation (12.1) is achieved when $v = (0, 0, \ldots, 0, 1)' = e_{m+1}$, that is, when $u = Uv = Ue_{m+1}$ is an unit eigenvector of $\tilde{\mu}$ having the eigenvalue λ_{m+1}. Hence the minimum distance between $\tilde{\mu}$ and $\tilde{M}$ is attained by $\mu\mu'$, where μ is a unit vector in the eigenspace of the largest eigenvalue of $\tilde{\mu}$. There is a unique minimizer if and only if the largest eigenvalue of $\tilde{\mu}$ is simple, that is, if the eigenspace corresponding to the largest eigenvalue is one-dimensional. In that case, one says that $\tilde{\mu}$ is a nonfocal point of $S^+(m + 1, \mathbb{R})$ and then from Proposition 4.2 it follows that the extrinsic mean μ_E of Q is $[\mu]$. Also, the extrinsic variation of Q has the expression

$$
V = \mathrm{E}[\|J(X_1) - \tilde{\mu}\|^2] + \|\tilde{\mu} - uu'\|^2 = 2(1 - \lambda_{m+1}),
$$

where $X_1 \sim Q$. Therefore we have the following result.

Proposition 12.1 *Let Q be a probability distribution on $\mathbb{R}P^m$ and let $\tilde{Q} = Q \circ j^{-1}$ be its image in $S(m + 1, \mathbb{R})$. Let $\tilde{\mu} = \int_{S(m+1,\mathbb{R})} x\tilde{Q}(dx)$ denote the mean of $\tilde{Q}$. Then (a) the extrinsic mean set of Q consists of all $[\mu]$, where μ is a unit eigenvector of $\tilde{\mu}$ corresponding to its largest eigenvalue λ_{m+1}; (b) this set is a singleton and Q has a unique extrinsic mean if and only if $\tilde{\mu}$ is nonfocal, that is, λ_{m+1} is a simple eigenvalue; and (c) the extrinsic variation of Q has the expression $V = 2(1 - \lambda_{m+1})$.*

Consider a random sample $X_1, \ldots, X_n$ i.i.d. with distribution Q. Let μ_n denote a measurable unit eigenvector of $\tilde{\mu}_n = \frac{1}{n}\sum_{i=1}^n j(X_i)$ corresponding to its largest eigenvalue $\lambda_{m+1,n}$. Then it follows from Theorem 3.3 and Proposition 12.1 that if $\tilde{\mu}$ is nonfocal, then the sample extrinsic mean $\mu_{nE} = [\mu_n]$ is a strongly consistent estimator of the extrinsic mean of Q. Proposition 3.8 now implies that the sample extrinsic variation $2(1 - \lambda_{m+1,n})$ is a strongly consistent estimator of the extrinsic variation of Q.

12.6 Asymptotic distribution of the sample extrinsic mean

In this section we assume that $\tilde{\mu}$ is a nonfocal point of $S(m+1,\mathbb{R})$. Let $\tilde{X}_j = j(X_j)$, $j = 1,\ldots,n$ be the image of the sample in $\tilde{M}$ $(= j(\mathbb{R}P^m))$. Consider the projection map $P: S(m+1,\mathbb{R}) \to J(\mathbb{R}P^m)$, given by $P(A) = vv'$, with v being a unit eigenvector from the eigenspace of the largest eigenvalue of A. It follows from Proposition 4.3 that if P is continuously differentiable in a neighborhood of $\tilde{\mu}$, then $\sqrt{n}\,[j(\mu_{nE}) - j(\mu_E)]$ has an asymptotic mean zero Gaussian distribution on $T_{j(\mu_E)}\tilde{M}$. It has asymptotic coordinates $\sqrt{n}\,\bar{T}$, where T_j is the coordinate vector of $d_{\tilde{\mu}}P(\tilde{X}_j - \tilde{\mu})$ with respect to some orthonormal basis for $T_{j(\mu_E)}\tilde{M}$. To get these coordinates, and hence derive analytic expressions for the parameters in the asymptotic distribution, we need to compute the differential of P at $\tilde{\mu}$.

Let $\gamma(t) = \tilde{\mu} + tv$ be a curve in $S(m+1,\mathbb{R})$ with $\gamma(0) = \tilde{\mu}$ and $\dot{\gamma}(0) = v \in S(m+1,\mathbb{R})$. Let $\tilde{\mu} = UDU'$, $U = (U_1,\ldots,U_{m+1})$, $D = \text{Diag}(\lambda_1,\ldots,\lambda_{m+1})$ be a spectral decomposition of $\tilde{\mu}$ as in Section 12.5. Then

$$\gamma(t) = U(D + tU'vU)U' = U\tilde{\gamma}(t)U',$$

where $\tilde{\gamma}(t) = D + tU'vU$ is a curve in $S(m+1,\mathbb{R})$ starting at D with initial velocity $\dot{\tilde{\gamma}}(0) = \tilde{v} \equiv U'vU$. Because D has a simple largest eigenvalue, for t sufficiently small $\tilde{\gamma}(t)$ is nonfocal. Choose $e_{m+1}(t)$ to be a unit eigenvector corresponding to the largest (simple) eigenvalue $\lambda_{m+1}(t)$ of $\tilde{\gamma}(t)$, such that $t \mapsto e_{m+1}(t)$, $t \mapsto \lambda_{m+1}(t)$ are smooth (near $t = 0$) with $e_{m+1}(0) = e_{m+1}$, $\lambda_{m+1}(0) = \lambda_{m+1}$. Such a choice is possible by perturbation theory of matrices because $\lambda_{m+1} > \lambda_m$ (see Dunford and Schwartz, 1958). Then

$$\tilde{\gamma}(t)e_{m+1}(t) = \lambda_{m+1}(t)e_{m+1}(t), \qquad (12.2)$$

$$e'_{m+1}(t)e_{m+1}(t) = 1. \qquad (12.3)$$

Differentiate equations (12.2) and (12.3) with respect to t at $t = 0$ to get

$$(\lambda_{m+1}I_{m+1} - D)\dot{e}_{m+1}(0) = -\dot{\lambda}_{m+1}(0)e_{m+1} + \tilde{v}e_{m+1}, \qquad (12.4)$$

$$e'_{m+1}\dot{e}_{m+1}(0) = 0, \qquad (12.5)$$

where $\dot{e}_{m+1}(0)$ and $\dot{\lambda}_{m+1}(0)$ refer to $\frac{d}{dt}e_{m+1}(t)|_{t=0}$ and $\frac{d}{dt}\lambda_{m+1}(t)|_{t=0}$, respectively. Consider the orthonormal basis (frame) $\{E_{ab} : 1 \le a \le b \le m+1\}$ for $S(m+1,\mathbb{R})$ as defined in Section 9.3. Choose $\tilde{v} = E_{ab}$ for $1 \le a \le b \le m+1$. From equations (12.4) and (12.5) we get

$$\dot{e}_{m+1}(0) = \begin{cases} 0 & \text{if } 1 \le a \le b \le m \text{ or } a = b = m+1, \\ 2^{-1/2}(\lambda_{m+1} - \lambda_a)^{-1}e_a & \text{if } 1 \le a < b = m+1. \end{cases}$$

$$(12.6)$$

Because $P(\tilde{\gamma}(t)) = e_{m+1}(t)e'_{m+1}(t)$, one has

$$\frac{\mathrm{d}}{\mathrm{d}t}P(\tilde{\gamma}(t))|_{t=0} = d_D P(\tilde{v}) = e_{m+1}\dot{e}'_{m+1}(0) + \dot{e}_{m+1}(0)e'_{m+1}. \tag{12.7}$$

From equations (12.6) and (12.7) it follows that

$$d_D P(E_{ab}) = \begin{cases} 0 & \text{if } 1 \le a \le b \le m \text{ or } a = b = m+1, \\ (\lambda_{m+1} - \lambda_a)^{-1}E_{ab} & \text{if } 1 \le a < b = m+1. \end{cases} \tag{12.8}$$

Because P commutes with isometries $A \mapsto UAU'$, that is, $P(UAU') = UP(A)U'$ and $\gamma(t) = U\tilde{\gamma}(t)U'$,

$$\frac{\mathrm{d}}{\mathrm{d}t}P(\gamma(t))|_{t=0} = U\frac{\mathrm{d}}{\mathrm{d}t}P(\tilde{\gamma}(t))|_{t=0}U'$$

or

$$d_{\tilde{\mu}}P(v) = Ud_D P(\tilde{v})U'.$$

Hence from equation (12.8) it follows that

$$d_{\tilde{\mu}}P(UE_{ab}U') = \begin{cases} 0 & \text{if } 1 \le a \le b \le m \text{ or } a = b = m+1, \\ (\lambda_{m+1} - \lambda_a)^{-1}UE_{ab}U' & \text{if } 1 \le a < b = m+1. \end{cases} \tag{12.9}$$

Note that for all $U \in SO(m+1)$, $\{UE_{ab}U' : 1 \le a \le b \le m+1\}$ is also an orthonormal frame for $S(m+1, \mathbb{R})$. Further, from equation (12.9), it is clear that

$$\{UE_{ab}U' : 1 \le a < b = m+1\} \tag{12.10}$$

forms an orthonormal frame for $T_{P(\tilde{\mu})}\tilde{M}$. If $A \in S(m+1, \mathbb{R})$ has coordinates $\{a_{ij} : 1 \le i \le j \le m+1\}$ with respect to the basis $\{UE_{ab}U' : 1 \le a \le b \le m+1\}$, that is,

$$A = \sum_{1 \le i \le j \le m+1} a_{ij}UE_{ij}U',$$

$$a_{ij} = \langle A, UE_{ij}U'\rangle = \begin{cases} \sqrt{2}U'_i AU_j & \text{if } i < j, \\ U'_i AU_i & \text{if } i = j, \end{cases}$$

then from equation (12.9) it follows that

$$d_{\tilde{\mu}}P(A) = \sum_{1 \le i \le j \le m+1} a_{ij}d_{\tilde{\mu}}P(UE_{ij}U')$$

$$= \sum_{i=1}^{m} a_{im+1}(\lambda_{m+1} - \lambda_i)^{-1}UE_{im+1}U'.$$

Hence $d_{\tilde{\mu}} P(A)$ has the coordinates

$$\{ \sqrt{2}(\lambda_{m+1} - \lambda_i)^{-1} U_i' A U_{m+1} : 1 \leq i \leq m \} \tag{12.11}$$

with respect to the orthonormal basis in equation (12.10) for $T_{P(\tilde{\mu})}\tilde{M}$. This proves the following proposition.

Proposition 12.2 Let Q be a probability distribution on $\mathbb{R}P^m$ with unique extrinsic mean μ_E. Let $\tilde{\mu}$ be the mean of $\tilde{Q} := Q \circ j^{-1}$ regarded as a probability distribution on $S(m+1, \mathbb{R})$. Let μ_{nE} be the sample extrinsic mean from an i.i.d. sample $X_1, \ldots, X_n$. Let $\tilde{X}_j = j(X_j)$, $j = 1, \ldots, n$, and $\overline{\tilde{X}} = \frac{1}{n} \sum_{j=1}^n \tilde{X}_j$. (a) The projection map P is twice continuously differentiable in a neighborhood of $\tilde{\mu}$ and

$$\sqrt{n}\,[j(\mu_{nE}) - j(\mu_E)] = \sqrt{n}\, d_{\tilde{\mu}} P(\overline{\tilde{X}} - \tilde{\mu}) + o_P(1)$$

$$\xrightarrow{\mathcal{L}} N(0, \Sigma),$$

where Σ is the covariance of the coordinates of $d_{\tilde{\mu}} P(\tilde{X}_1 - \tilde{\mu})$.
(b) If $T_j = (T_j^1, \ldots, T_j^m)$ denotes the coordinates of $d_{\tilde{\mu}} P(\tilde{X}_j - \tilde{\mu})$ with respect to the orthonormal basis of $T_{P(\tilde{\mu})}\tilde{M}$ as in equation (12.10), then

$$T_j^a = \sqrt{2}(\lambda_{m+1} - \lambda_a)^{-1} U_a' \tilde{X}_j U_{m+1}, \quad a = 1, \ldots, m.$$

Proposition 12.2 can be used to construct an asymptotic or bootstrap confidence region for μ_E as in Section 4.2.

Given two random samples on $\mathbb{R}P^m$, we can distinguish between the underlying probability distributions by comparing the sample extrinsic means and variations by the methods developed in Section 4.5.

12.7 References

Nonparametric inferences on axial spaces $\mathbb{R}P^m$ were carried out in Prentice (1984) and Beran and Fisher (1998). The present treatment in terms of extrinsic means may be found in Bhattacharya and Patrangenaru (2003, 2005). Pivotal confidence regions for directional and axial means are provided in Fisher et al. (1996).

Section 12.3 on the geometry of the projective shape space $P_l\Sigma_m^k$ is taken from Mardia and Patrangenaru (2005) and Patrangenaru et al. (2010), where one can find detailed treatments.

13

Nonparametric Bayes inference on manifolds

In this chapter we adapt and extend the nonparametric density estimation procedures on Euclidean spaces to general (Riemannian) manifolds.

13.1 Introduction

So far in this book we have used notions of center and spread of distributions on manifolds to identify them or to distinguish between two or more distributions. However, in certain applications, other aspects of the distribution may also be important. The reader is referred to the data in Section 14.5.3 for such an example. Also, our inference method so far has been frequentist.

In this chapter and the next, we pursue different goals and a different route. Our approach here and in the next chapter will be nonparametric Bayesian, which involves modeling the full data distribution in a flexible way that is easy to work with. The basic idea will be to represent the unknown distribution as an infinite mixture of some known parametric distribution on the manifold of interest and then set a full support prior on the mixing distribution. Hence the parameters defining the distribution are no longer finite-dimensional but reside in the infinite-dimensional space of all probabilities. By making the parameter space infinite-dimensional, we ensure a flexible model for the unknown distribution and consistency of its estimate under mild assumptions. All these will be made rigorous through the various theorems we will encounter in the subsequent sections.

For a prior on the mixing distribution, a common choice can be the Dirichlet process prior (see Ferguson, 1973, 1974). We present a simple algorithm for posterior computations in Section 13.4.

For the sake of illustration, we apply our methods to two specific manifolds, namely, the unit sphere and the planar shape space. In such cases and on other Riemannian manifolds, when we have continuous data, it is natural to assume that they come from a density with respect to the invariant volume form, which we are modeling via the above method. We prove that the assumptions for full support for the density prior and consistency of the estimated density are satisfied.

Nonparametric density estimation by itself may not be very special and may be carried out by a kernel method. However, the ideas of this chapter provide the framework for nonparametric Bayes regression and hypothesis testing on manifolds discussed in Chapter 14. Further, by using a countably infinite mixture of kernels model, one can do clustering of data on manifolds, which does not require us to know or fix the number of clusters in advance. Also, when we have observations from two or more categories, modeling the density for each category data, we can predict the category to which a new subject belongs and hence classify it. Such an application with data on shapes is presented in Section 13.7. There we predict the gender of a gorilla based on the shape of its skull.

When the space of interest is Euclidean, a similar approach for density modeling can be found in Lo (1984) and Escobar and West (1995). The methods have been generalized to general compact metric spaces and hence manifolds in Bhattacharya and Dunson (2010a) and Bhattacharya and Dunson (2011).

To maintain continuity of flow, we will present all proofs at the very end, in Section 13.8.

13.2 Density estimation on metric spaces

Let (M, ρ) be a separable metric space and let X be a random variable on M. We assume that the distribution of X has a density, say f_0, with respect to some fixed base measure λ on M and we are interested in modeling this unknown density via a flexible model. Let $K(m; \mu, \kappa)$ be a probability density (kernel) on M (with respect to λ) with a known parametric form. It has variable $m \in M$ and parameters $\mu \in M$ and $\kappa \in N$, with N being a *Polish space*. A Polish space is a topological space that is homeomorphic to a complete separable metric space. The kernel satisfies $\int_M K(m; \mu, \kappa) \lambda(dm) = 1$ for all values of μ and κ in their respective domains. In most interesting examples, μ will turn out to be the Fréchet mean of the probability corresponding to kernel K. Hence we will call μ the kernel location. The parameter κ is comprised of all other parameters in appropriate spaces determining the kernel shape, spread, and so on. In the kernels considered in Sections 13.5 and 13.6, N is $(0, \infty)$ and κ is a decreasing function of the Fréchet variation of K. Such kernels are called *location-scale kernels*.

Given a probability P on M, one can define a *location mixture* probability density model for X at $m \in M$ as

$$f(m; P, \kappa) = \int_M K(m; \mu, \kappa) P(d\mu). \tag{13.1}$$

We denote by $\mathcal{D}(M)$ the space of all probability densities on M with respect to the base measure λ. For a pre-specified kernel K, a prior on $\mathcal{D}(M)$ is induced through a prior Π_1 on (P, κ) in equation (13.1).

For example, on $M = \mathbb{R}^d$ one can use the multivariate Gaussian kernel as K. Then κ is the kernel covariance matrix and hence $N = M^+(d)$, the space of all order d positive-definite matrices. For prior and posterior properties of the mixture model in this case, the reader is referred to Wu and Ghosal (2010) and the references cited therein. In case M is the unit sphere, we may use any of the densities introduced in Appendix D as the kernel. Similarly on different shape spaces, we have various choices for K. In each case λ is the volume form on M, which is the standard choice on a Riemannian manifold.

A common choice for prior on $\mathcal{M}(M)$, the space of all probabilities on M, is the *Dirichlet process (DP)* prior, which was introduced by Ferguson (1973) (see Appendix C). We put such a prior on P and an independent parametric prior on κ in equation (13.1).

13.3 Full support and posterior consistency

To justify the use of any specific kernel and prior on parameters and call our inference nonparametric, we would like to verify that the prior Π induced on density f has support as $\mathcal{D}(M)$ and that the posterior distribution of f, given a random realization of X, concentrates in arbitrarily small neighborhoods of the true data generating distribution as the sample size gets larger. The former property will be referred to as Π having full support while the latter as posterior consistency. These properties make sure that our inference results are independent of prior and kernel choices (for large samples).

To talk about a neighborhood of a probability and support of a prior on probabilities, we need to introduce a topology on $\mathcal{M}(M)$. In this chapter, we will use three, namely *weak, strong*, and *Kullback–Leibler* neighborhoods. A sequence of probabilities $\{P_n\}$ is said to converge weakly to P if $\int_M \phi dP_n \longrightarrow \int_M \phi dP$ for any continuous $\phi: M \to [-1, 1]$. The strong or total variation or L^1 distance between P and Q in $\mathcal{M}(M)$ is given by $\sup\{|\int_M \phi dP - \int_M \phi dQ|\}$, with the supremum being taken over all continuous $\phi: M \to [-1, 1]$. The Kullback–Leibler (KL) divergence from P to Q is defined as $d_{KL}(P|Q) = \int_M p \log \frac{p}{q} d\lambda$, with λ being any measure on M, with respect to which both P and Q have densities p and q, respectively, the definition being independent of the choice of λ. Then KL convergence implies strong convergence, which in turn implies weak convergence. We will also come across the uniform or L^∞ distance between p and q, which is simply $\sup_{m\in M} |p(m) - q(m)|$. When M is separable, so is $\mathcal{M}(M)$ under the weak topology, and hence it makes sense to talk of support of priors on $\mathcal{M}(M)$. We can also talk about a probability being in the KL or strong support of a prior, which simply refers to arbitrarily small neighborhoods of that probability under the respective topology receiving positive prior mass. Unless specified otherwise, by "support" or "supp" we will always refer to weak support. Because most of the non-Euclidean manifolds arising in this book are compact, we derive consistency results on such spaces. If M were Euclidean, for similar theorems, the reader may refer to Wu and Ghosal (2008) or any other work on Bayesian density estimation.

Under the following assumptions on kernel K and prior Π_1, Theorem 13.1 establishes full L^∞ and KL support for the prior Π induced on $\mathcal{D}(M)$ through the mixture density model in equation (13.1).

A1 M is compact.

A2 The kernel K is continuous in its arguments.

For any continuous function $f\colon M \to \mathbb{R}$, for any $\epsilon > 0$, there exists a compact subset N_ϵ of N with a nonempty interior N_ϵ^o, such that

A3 $\sup_{m\in M, \kappa\in N_\epsilon} \left| f(m) - \int_M K(m; \mu, \kappa) f(\mu)\lambda(d\mu) \right| < \epsilon$.

A4 For any $\epsilon > 0$, the set $\{F_0\} \times N_\epsilon^o$ intersects with the (weak) support of Π_1. Here F_0 is the probability distribution with density f_0.

A5 The true density f_0 is continuous everywhere.

Assumptions A2 and A3 place minor regularity conditions on the kernel. If K is symmetric in m and μ, A3 implies that as a probability distribution on M, $K(\cdot; \mu, \kappa)$ can be made arbitrarily close in a weak sense to the degenerate point mass at μ, uniformly in μ, for appropriate κ choice, thereby further justifying the name "location" for μ. We will verify it for the vMF (von Mises–Fisher) kernel on the sphere and for the complex Watson on the planar shape space. A common choice for Π_1 satisfying A4 can be a full support product prior. The support of a Dirichlet process (DP) prior consists of all probabilities whose supports are subsets of the support of its base (see, e.g., Ghosh and Ramamoorthi, 2003, theorem 3.2.4). Hence a DP prior on P with base having M as its support and an independent full support prior on κ satisfies A4.

Theorem 13.1 *Define $f \in \mathcal{D}(M)$ as in equation (13.1). Let Π be the prior on f induced from a prior Π_1 on parameters (P, κ). Under assumptions A1–A5, given any $\epsilon > 0$,*

$$\Pi\left\{ f \, : \, \sup_{m \in M} |f_0(m) - f(m)| < \epsilon \right\} > 0,$$

which implies that f_0 is in the KL support of Π.

Theorem 13.1 shows that the density prior Π assigns positive probability to arbitrarily small L^∞ neighborhoods of the true density f_0 under mild assumptions. This, in turn, implies that f_0 is in the KL support of Π (and hence in its strong and weak support). Then we say that Π *satisfies the KL condition at f_0.*

To achieve more flexibility, while defining the mixture density model we may mix across κ as well, that is, replace $P(d\mu)$ in equation (13.1) with $P(d\mu \, d\kappa)$. With such a mixture, the KL condition is verified in Bhattacharya and Dunson (2010a).

13.3.1 Weak posterior consistency

Let $X_1, \dots, X_n$ be an i.i.d. realization of X. The Schwartz (1965) theorem stated below provides a useful tool in proving posterior consistency as sample size $n \to \infty$.

Proposition 13.2 *If (1) f_0 is in the KL support of Π and (2) $U \subset \mathcal{D}(M)$ is such that there exists a uniformly exponentially consistent sequence of test functions for testing $H_0 : f = f_0$ versus $H_1 : f \in U^c$, then $\Pi(U|X_1, \dots, X_n) \to 1$ as $n \to \infty$ a.s. F_0^∞, where U^c is the complement of U.*

The posterior probability of U^c can be expressed as

$$\Pi(U^c|X_1, \dots, X_n) = \frac{\int_{U^c} \prod_{i=1}^n \frac{f(X_i)}{f_0(X_i)} \Pi(df)}{\int \prod_1^n \frac{f(X_i)}{f_0(X_i)} \Pi(df)}. \tag{13.2}$$

Condition (1) ensures that for any $\beta > 0$,

$$\liminf_{n \to \infty} \exp(n\beta) \int \prod_{i=1}^n \frac{f(X_i)}{f_0(X_i)} \Pi(df) = \infty \text{ a.s.} \tag{13.3}$$

while condition (2) implies that

$$\lim_{n \to \infty} \exp(n\beta_0) \int_{U^c} \prod_{i=1}^n \frac{f(X_i)}{f_0(X_i)} \Pi(df) = 0 \text{ a.s.}$$

for some $\beta_0 > 0$ and therefore

$$\lim_{n\to\infty} \exp(n\beta_0/2)\Pi(U^c|X_1,\ldots,X_n) = 0 \text{ a.s.}$$

Hence Proposition 13.2 provides conditions for posterior consistency at an exponential rate. When U is a weakly open neighborhood of f_0, condition (2) is always satisfied from the definition of such a neighborhood. Hence, from Theorem 13.1, weak posterior consistency at an exponential rate follows for the location mixture density model.

13.3.2 Strong posterior consistency

When U is a total variation neighborhood of f_0, LeCam (1973) and Barron (1989) show that condition (2) of Proposition 13.2 will not be satisfied in most cases. In Barron et al. (1999), a sieve method is considered to obtain sufficient conditions for the numerator in equation (13.2) to decay at an exponential rate and hence get strong posterior consistency at an exponential rate. This is stated in Proposition 13.3. In its statement, for $\mathcal{F} \subseteq \mathcal{D}(M)$ and $\epsilon > 0$, the L_1-metric entropy $N(\epsilon, \mathcal{F})$ is defined as the logarithm of the minimum number of ϵ-sized (or smaller) L_1 subsets needed to cover $\mathcal{F}$.

Proposition 13.3 *If there exists a $\mathcal{D}_n \subseteq \mathcal{D}(M)$ such that (1) for n sufficiently large, $\Pi(\mathcal{D}_n^c) < \exp(-n\beta)$ for some $\beta > 0$, and (2) $N(\epsilon, \mathcal{D}_n)/n \to 0$ as $n \to \infty$ for any $\epsilon > 0$, then for any total variation neighborhood U of f_0, there exists a $\beta_0 > 0$ such that $\lim \sup_{n\to\infty} \exp(n\beta_0) \int_{U^c} \prod_1^n \frac{f(X_i)}{f_0(X_i)} \Pi(df) = 0$ a.s. F_0^∞. Hence, if f_0 is in the KL support of Π, the posterior probability of any total variation neighborhood of f_0 converges to 1 almost surely.*

Theorem 13.4 describes a $\mathcal{D}_n$ that satisfies condition (2). We assume that there exists a continuous function $\phi\colon N \to [0, \infty)$ for which the following assumptions hold:

A6 There exist positive constants $\mathcal{K}_1, a_1, A_1$ such that for all $\mathcal{K} \geq \mathcal{K}_1$, $\mu, \nu \in M$,

$$\sup_{m\in M, \kappa\in\phi^{-1}[0,\mathcal{K}]} \left| K(m; \mu, \kappa) - K(m; \nu, \kappa) \right| \leq A_1 \mathcal{K}^{a_1} \rho(\mu, \nu).$$

A7 There exist positive constants a_2, A_2 such that for all $\kappa_1, \kappa_2 \in \phi^{-1}[0, \mathcal{K}]$, $\mathcal{K} \geq \mathcal{K}_1$,

$$\sup_{m,\mu\in M} \left| K(m; \mu, \kappa_1) - K(m; \mu, \kappa_2) \right| \leq A_2 \mathcal{K}^{a_2} \rho_2(\kappa_1, \kappa_2),$$

with ρ_2 metrizing the topology of N.

A8 For any $\mathcal{K} \geq \mathcal{K}_1$, the subset $\phi^{-1}[0, \mathcal{K}]$ is compact and, given $\epsilon > 0$, the minimum number of ϵ (or smaller) radius balls covering it (known as the ϵ-*covering number*) can be bounded by $(\mathcal{K}\epsilon^{-1})^{b_2}$ for an appropriate positive constant b_2 (independent of $\mathcal{K}$ and ϵ).

A9 There exist positive constants a_3, A_3 such that the ϵ-covering number of M is bounded by $A_3 \epsilon^{-a_3}$ for any $\epsilon > 0$.

Theorem 13.4 *For a positive sequence $\{\mathcal{K}_n\}$ diverging to ∞, define*

$$\mathcal{D}_n = \{f(\cdot\,; P, \kappa) : P \in \mathcal{M}(M),\ \kappa \in \phi^{-1}[0, \mathcal{K}_n]\}$$

with f as in equation (13.1). *Under assumptions A6–A9, given any $\epsilon > 0$, for n sufficiently large, $N(\epsilon, \mathcal{D}_n) \leq C(\epsilon)\mathcal{K}_n^{a_1 a_3}$ for some $C(\epsilon) > 0$. Hence $N(\epsilon, \mathcal{D}_n)$ is $o(n)$, that is, $\lim_{n \to \infty} N(\epsilon, \mathcal{D}_n)/n = 0$, whenever $\mathcal{K}_n = o(n^{(a_1 a_3)^{-1}})$.*

As a corollary, we derive conditions on the prior on the location mixture density model under which strong posterior consistency at an exponential rate follows.

Corollary 13.5 *Let Π_1 be the prior on (P, κ) for density model* (13.1). *Under assumptions A1–A9 and*

A10 $\Pi_1(\mathcal{M}(M) \times \phi^{-1}(n^a, \infty)) < \exp(-n\beta)$ *for some $a < (a_1 a_3)^{-1}$ and $\beta > 0$,*

the posterior probability of any total variation neighborhood of f_0 converges to 1 a.s. F_0^∞.

Theorem 13.1 ensures that f_0 is in the KL support of Π. Theorem 13.4 and assumption A10 ensure that $\mathcal{D}_n$ satisfies conditions (1) and (2) of Proposition 13.3. Hence, from the proposition, the proof follows.

When we use a location–scale kernel, that is, when $N = (0, \infty)$, choose a prior $\Pi_1 = \Pi_{11} \otimes \pi_1$ having full support, set ϕ to be the identity map, then a choice for π_1 for which assumptions A4 and A10 are satisfied is a Weibull density $\mathrm{Weib}(\kappa; a, b) \propto \kappa^{a-1} \exp(-b\kappa^a)$, whenever the shape parameter $a > a_1 a_3$. A Gamma prior on κ does not satisfy A10 (unless $a_1 a_3 < 1$). However, that does not prove that it is not eligible for strong consistency because Corollary 13.5 provides only sufficient conditions.

When the underlying space is noncompact such as $\mathbb{R}^d$, Corollary 13.5 applies to any true density f_0 supported on a compact set, say M. Then the kernel can be chosen to have noncompact support, such as multivariate Gaussian; but to apply Theorem 13.4, we need to restrict the prior on the location mixing distribution to have support in $\mathcal{M}(M)$. In that case, we are modeling a compactly supported density by a mixture density having full support but with locations drawn from a compact domain. While using a Gaussian mixture, strong posterior consistency is achieved with a suitably truncated and transformed Wishart prior on the covariance inverse. We do not go into the details here, but the interested reader is referred to Bhattacharya and Dunson (2011).

When the dimension of the manifold is large, as is the case in shape analysis with a large number of landmarks, the constraints on the shape parameter a in the proposed Weibull prior on the inverse-scale parameter κ become overly restrictive. For strong consistency, a needs to be very large, implying a prior on bandwidth $1/\kappa$ that places very small probability in neighborhoods close to zero, which is undesirable in many applications. Bhattacharya and Dunson (2011) propose an alternative by allowing the prior Π_1 to depend on the sample size n. For example, a DP prior on P and an independent Gamma prior on κ for which the scale parameter is of order $\log(n)/n$ is shown to satisfy the requirements for weak and strong consistency.

13.4 Posterior computations

In this section we describe methods for sampling from the posterior of the density f and obtaining its Bayes estimate given observations $X_1, \ldots, X_n$ i.i.d. f when using the mixture density model (13.1) for f. For the choice of prior Π, as recommended in earlier sections, we set a $\mathrm{DP}(w_0 P_0)$ prior on P and an independent prior π_1 on κ. Then, using the stick-breaking representation of Sethuraman (1994) for the Dirichlet process, a random draw from Π can be expressed as a countably infinite mixture density:

$$f = \sum_{j=1}^{\infty} w_j K(\cdot; \mu_j, \kappa),$$

with $\kappa \sim \pi_1$, μ_j i.i.d. P_0 and $w_j = V_j \prod_{l<j}(1 - V_l)$, for V_j i.i.d. $\mathrm{Beta}(1, w_0)$, $j = 1, \ldots, \infty$. Here Beta stands for the Beta distribution. Hence the parameters explaining f are κ, $\{\mu_j, V_j\}_{j=1}^{\infty}$. An exact sampling procedure for the infinitely many parameters from their joint posterior is impossible. We instead use a Gibbs sampler proposed by Yau et al. (2011) to get approximate draws from the posterior of f. We introduce cluster labels S_i denoting the mixture component for observation i, $i = 1, \ldots, n$, such that $\Pr(S_i = j) = w_j$, $j = 1, \ldots, \infty$. The complete data likelihood is then $\prod_{i=1}^{n}\{w_{S_i} K(X_i; \mu_{S_i}, \kappa)\}$ and the prior is $\pi_1(\kappa) \prod_{j=1}^{\infty}\{\mathrm{Beta}(V_j, 1, w_0)P_0(d\mu_j)\}$. We also introduce uniformly distributed slice sampling latent variables $u = \{u_i\}_{i=1}^{n}$ and rewrite the likelihood as

$$\prod_{i=1}^{n} \{I(u_i < w_{S_i})K(X_i; \mu_{S_i}, \kappa)\}.$$

Then the likelihood depends on only finitely many parameters, namely, those with their index in the set $\{j: w_j > \min(u)\}$, which can be given the upper bound J, which is the smallest index j satisfying $\sum_1^j w_l > 1 - \min(u)$. Hence, after setting some guessed values for the labels (which may be obtained, for example, using the k-mean algorithm), the Gibbs sampler iterates through the following steps.

Step 1. Update S_i, for $i = 1, \dots, n$, by sampling from their multinomial conditional posterior distributions given by $\Pr(S_i = j) \propto K(X_i; \mu_j, \kappa)$ for $j \in A_i$, where $A_i = \{j : 1 \le j \le J, w_j > u_i\}$ and J is the smallest index j satisfying $\sum_1^j w_l > 1 - \min(u)$. In implementing this step, draw $V_j \sim \text{Beta}(1, w_0)$ and $\mu_j \sim P_0$ for $\max(S) < j \le J$, with $\max(S)$ denoting the largest of the cluster labels from the previous iteration.

Step 2. Update the kernel locations μ_j, $j = 1, \dots, \max(S)$, by sampling from their conditional posteriors, which are proportional to

$$P_0(d\mu_j) \prod_{i:S_i=j} K(X_i; \mu_j, \kappa).$$

An appropriate choice of P_0 results in conjugacy while implementing this step.

Step 3. Draw κ from its full conditional posterior proportional to

$$\pi_1(d\kappa) \prod_{i=1}^{n} K(X_i; \mu_{S_i}, \kappa).$$

Step 4. Update the stick-breaking random variables V_j, $j = 1, \dots, \max(S)$, from their conditional posterior distributions given the cluster allocation but marginalizing out the slice sampling variables,

$$V_j \sim \text{Beta}\left(1 + \sum_i I(S_i = j), w_0 + \sum_i I(S_i > j)\right).$$

Step 5. Update the slice sampling latent variables $\{u_i\}_{i=1}^n$ from their conditional posteriors by letting $u_i \sim \text{Unif}(0, w_{S_i})$.

A draw from the posterior for f can be obtained using

$$f(\cdot; P, \kappa) = \sum_{j=1}^{\max(S)} w_j K(\cdot; \mu_j, \kappa) + \left(1 - \sum_{j=1}^{\max(S)} w_j\right) \int K(\cdot; \mu, \kappa) P_0(d\mu), \quad (13.4)$$

with κ and w_j, μ_j $(j = 1, \ldots, \max(S))$ a Markov chain Monte Carlo draw from the joint posterior of the parameters up to the maximum occupied. A Bayes estimate of f can then be obtained by averaging these draws across many iterations after discarding a suitable burn-in. If it is difficult to evaluate the integral in equation (13.4) in closed form, we replace the integral by $K(\cdot; \mu_1, \kappa)$, with μ_1 being a draw from P_0, or just ignore it if $1 - \sum_1^{\max(S)} w_j$ is negligible.

13.5 Application to unit sphere S^d

Let M be S^d endowed with the extrinsic distance d_E. To define a probability density model as in Section 13.2 with respect to the volume form V, we need a suitable kernel that satisfies the assumptions in Section 13.3. One of the most commonly used probability densities on this space is the Fisher or von Mises–Fisher (vMF) density

$$\text{vMF}(m; \mu, \kappa) = c^{-1}(\kappa) \exp(\kappa m' \mu) \text{ with}$$

$$m, \mu \in S^d, \ \kappa \in \mathbb{R}^+, \ c(\kappa) = \frac{2\pi^{d/2}}{\Gamma(\frac{d}{2})} \int_{-1}^1 \exp(\kappa t)(1 - t^2)^{d/2-1} dt.$$

It turns out that the parameter μ is the kernel extrinsic mean. Since besides μ the only other parameter is the scalar κ, this is an example of a location–scale kernel. Further, as κ diverges to ∞, the vMF distribution converges to a point mass at μ in a weak sense uniformly in μ, as shown in Theorem 13.6.

Theorem 13.6 *The vMF kernel satisfies assumptions A2 and A3.*

Hence, from Theorem 13.1, when using the location mixture density model (13.1) with a full support prior on (P, κ), the density prior includes all continuous densities in its L^∞ and hence KL support.

An appropriate prior choice is $\Pi_1 = \text{DP}(w_0 P_0) \otimes \pi_1$ with $P_0 = \text{vMF}(\cdot; \mu_0, \kappa_0)$. Then, sampling from the posterior distribution of the density using the algorithm in Section 13.4, we have conjugacy while updating the cluster locations in step 2. In particular,

$$\mu_j |- \ \sim \text{vMF}(\bar{\mu}_j / \|\bar{\mu}_j\|, \|\bar{\mu}_j\|), \quad j = 1, \ldots, \infty,$$

where the left side denotes the *full posterior* of u_j and $\bar{\mu}_j = \kappa \sum_{i:S_i=j} X_i + \kappa_0 \mu_0$. The posterior of κ in step 3 is proportional to

$$\pi_1(d\kappa)\{\kappa^{d/2} \exp(-\kappa)c(\kappa)\}^{-n} \kappa^{nd/2} \exp\left\{ -\kappa\left(n - \sum_i X_i' \mu_{S_i}\right)\right\}.$$

Hence if we choose π_1 to be the density proportional to

$$\{\kappa^{d/2} \exp(-\kappa)c(\kappa)\}^n \exp(-b\kappa)\kappa^{a-1} \tag{13.5}$$

for some $a, b > 0$, then the posterior becomes Gamma. The reason why this is a valid density is because $\kappa^{d/2} \exp(-\kappa)c(\kappa)$ is bounded (both below and above). Alternatively, one may choose a Gamma prior and the posterior becomes very close to Gamma under high concentrations. This is because $\kappa^{d/2} \exp(-\kappa)c(\kappa)$ has a finite limit as $\kappa \to \infty$. Hence one can implement a Metropolis–Hasting step with a Gamma proposal when updating κ.

Theorem 13.7 verifies the assumptions for strong consistency when using the Fisher mixture density model.

Theorem 13.7 *With ϕ being the identity map, the vMF kernel on S^d satisfies assumption A6 with $a_1 = d/2 + 1$ and A7 with $a_2 = d/2$. The compact metric space (S^d, d_E) satisfies assumption A9 with $a_3 = d$.*

As a result, a $\text{Weib}(\cdot\,; a, b)$ prior on κ with $a > (d + d^2/2)^{-1}$ satisfies the condition of Corollary 13.5, and strong posterior consistency follows.

When d is large, as is often the case for spherical data, a more appropriate prior on κ for which weak and strong consistencies hold can be a sample size dependent Gamma as mentioned at the end of Section 13.3.2.

The proofs of Theorems 13.6 and 13.7 use the following lemma, which establishes certain properties of the normalizing constant.

Lemma 13.8 *Define $\tilde{c}(\kappa) = \exp(-\kappa)c(\kappa)$, $\kappa \geq 0$. Then $\tilde{c}$ is decreasing and, for $\kappa \geq 1$,*

$$\tilde{c}(\kappa) \geq C\kappa^{-d/2}$$

for some appropriate positive constant C.

13.6 Application to the planar shape space Σ_2^k

We view the planar shape space Σ_2^k as a compact metric space endowed with the extrinsic distance d_E. To model an unknown density on Σ_2^k, we use a mixture density as in equation (13.1) with K corresponding to the complex Watson density:

$$\text{CW}(m; \mu, \kappa) = c^{-1}(\kappa) \exp(\kappa |x^* v|^2), \quad (m = [x],\ \mu = [v]),$$

$$m, \mu \in \Sigma_2^k,\ \kappa \in \mathbb{R}^+,\ c(\kappa) = (\pi \kappa^{-1})^{k-2}\Big\{ \exp(\kappa) - \sum_{r=0}^{k-3} \kappa^r/r! \Big\}.$$

The following theorem justifies its use.

Theorem 13.9 *For the complex Watson kernel, assumptions A2 and A3 of Section 13.3 are satisfied.*

Hence, if we choose a full support prior Π_1 on the parameters (P, κ), we induce a prior with L^∞ and KL support including all continuous densities over Σ_2^k. As a result, weak posterior consistency follows from Proposition 13.2 if the data generating density is continuous everywhere.

To specify a prior that satisfies the assumptions and leads to simplifications in implementing posterior computation when using model (13.1), we let $P \sim DP(w_0 P_0)$, with $P_0 = CW(\mu_0, \kappa_0)$, independently of $\kappa \sim \text{Gamma}(a, b)$. This prior choice leads to conditional conjugacy so that posterior computation given a random sample $X_1, \ldots, X_n$ can proceed via the Gibbs sampling algorithm developed in Section 13.4. For instance, in step 1, the full posterior is given by

$$\mu_j| - \sim CB(m_j \kappa \overline{X}_j + A_0),$$

where CB stands for the complex Bingham density introduced in Appendix D, $m_j = \sum_{i=1}^n I(S_i = j)$, $\overline{X}_j = \sum_{i:S_i=j} x_i x_i^* / m_j$ ($X_i = [x_i]$), $A_0 = \kappa_0 v_0 v_0^*$, and $\mu_0 = [v_0]$. We use a Metropolis–Hastings step to draw μ_j. In step 3, the full conditional posterior of κ is proportional to

$$\kappa^{n(k-2)+a-1} \exp\left\{ -\kappa\left(n + b - \sum_{j=1}^{\max(S)} m_j v_j^* \overline{X}_j v_j\right)\right\}\left\{1 - \exp(-\kappa)\sum_{r=0}^{k-3}\kappa^r/r!\right\}^{-n},$$

where $\mu_j = [v_j]$. For κ high, this conditional density is approximately equivalent to

$$\text{Gamma}\left\{a + n(k-2), b + \sum_{j=1}^{\max(S)} m_j(1 - v_j^* \overline{X}_j v_j)\right\}.$$

Hence we get approximate conjugacy for the conditional distribution of κ under a Gamma prior.

To show that strong posterior consistency holds for the complex Watson mixture density, we need to verify assumptions A6 and A7 for the kernel and A9 on Σ_2^k. These are shown in Theorems 13.10 and 13.12. The map ϕ is taken to be identity.

Theorem 13.10 *The complex Watson kernel on Σ_2^k satisfies assumption A6 with $a_1 = k - 1$ and assumption A7 with $a_2 = 3k - 8$.*

The proof uses Lemma 13.11, which verifies certain properties of the normalizing constant c.

Lemma 13.11 *Define* $c_1(\kappa) = \exp(-\kappa)c(\kappa)$ *and* $c_2(\kappa) = (\pi^{-1}\kappa)^{k-2}$ $\exp(-\kappa)c(\kappa)$. *Then* c_1 *is decreasing on* $[0, \infty)$ *with*

$$\lim_{\kappa \to 0} c_1(\kappa) = \frac{\pi^{k-2}}{(k-2)!} \text{ and } \lim_{\kappa \to \infty} c_1(\kappa) = 0,$$

while c_2 *is increasing with*

$$\lim_{\kappa \to 0} c_2(\kappa) = 0, \ \lim_{\kappa \to \infty} c_2(\kappa) = 1 \text{ and}$$
$$c_2(\kappa) \geq (k-2)!^{-1} \exp(-\kappa)\kappa^{k-2}.$$

The proof follows from direct computations.

Theorem 13.12 *The metric space* (Σ_2^k, d_E) *satisfies assumption A9 with* $a_3 = 2k - 3$.

As a result, Corollary 13.5 implies that strong posterior consistency holds with $\Pi_1 = DP(w_0 P_0) \otimes \pi_1$, for $\pi_1 = \text{Weib}(\cdot\,; a, b)$ whenever $a > (2k - 3)(k - 1)$. Alternatively, one may use a Gamma prior on κ with scale decreasing with n at a suitable rate and we have consistency from Bhattacharya and Dunson (2011).

13.7 Application to morphometrics: classification of gorilla skulls

We apply the method of density estimation to the data on shapes of 29 male and 30 female gorilla skulls, with eight landmarks chosen on the midline plane of two-dimensional images of each skull (Dryden and Mardia, 1998). The goal is to study how the shapes of the skulls vary between males and females, and build a classifier to predict gender. The shape samples lie on Σ_2^k, $k = 8$. We randomly pick 25 individuals of each gender as a training sample, with the remaining 9 used as test data. As Figure 2.4 shows, most of the landmarks corresponding to the preshapes of the sample extrinsic means are close for females and males even after rotation based alignment, but there is a larger difference in landmarks three and eight.

Applying nonparametric discriminant analysis, we assume that the unconditional probability of being female is 0.5 and use a separate Dirichlet process location mixture of complex Watson kernels for the shape density in the male and female groups. Letting $f_1(m)$ and $f_2(m)$ denote the female and male shape densities, the conditional probability of being female given shape data $[z]$ is simply $p([z]) = 1/\{1 + f_2([z])/f_1([z])\}$. To estimate the posterior probability, we average $p([z])$ across Markov

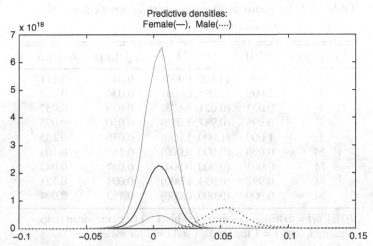

Figure 13.1 Estimated shape densities of gorillas: female (*solid*) and male (*dot*). Estimate (*black*), 95% confidence region (*gray*). Densities evaluated at a dense grid of points drawn from the unit speed geodesic starting at female extrinsic mean in direction of male extrinsic mean.

chain Monte Carlo iterations to obtain $\hat{p}([z])$. For simplicity, we choose the same prior form for both subsamples, namely, $\Pi_1 = DP(w_0 CW(\kappa_0, \mu_0)) \otimes$ Gamma(a, b) with hyperparameters $w_0 = 1$, $\kappa_0 = 1000$, $a = 1.01$, $b = 0.001$, and μ_0 as the corresponding subsample extrinsic mean. These choices are elicited based on our prior expectation for the gorilla example. Figure 13.1 displays the estimated shape densities (on the geodesic joining the male and female extrinsic sample means) for the two groups. It reveals some differences, which were also identified by nonparametric frequentist tests earlier.

Table 13.1 presents the estimated posterior probabilities of being female for each of the gorillas in the test sample along with a 95% credible interval for $p([z])$. In addition, we show the extrinsic distance between the shape for each gorilla and the female and male sample extrinsic means. For most of the gorillas, there is a high posterior probability of assigning the correct gender. There is misclassification only in the third female and third male. There is some uncertainty in predicting the gender of that female gorilla because the credible interval includes 0.5, but the corresponding male is surely misclassified.

Potentially we could define a distance-based classifier, which allocates a test subject to the group having the mean shape closest to that subject's

Table 13.1 *Posterior probability of being female for each gorilla in the test sample*

True gender	$\hat{p}([z])$	95% CI	$d_E([z], \hat{\mu}_1)$	$d_E([z], \hat{\mu}_2)$
F	1.000	(1.000,1.000)	0.041	0.111
F	1.000	(0.999,1.000)	0.036	0.093
F	0.023	(0.021, 0.678)	0.056	0.052
F	0.998	(0.987, 1.000)	0.050	0.095
F	1.000	(1.000, 1.000)	0.076	0.135
M	0.000	(0.000, 0.000)	0.167	0.103
M	0.001	(0.000, 0.004)	0.087	0.042
M	0.992	(0.934, 1.000)	0.091	0.121
M	0.000	(0.000, 0.000)	0.152	0.094

$d_E([z], \hat{\mu}_i)$ = extrinsic distance of subject $[z]$ from the mean shape in group i, with $i = 1$ for females and $i = 2$ for males

shape. Based on Table 13.1, such a classifier gives results consistent with the former approach. Indeed, the shape for the third female gorilla was closer to the mean shape for the male gorillas than that for the females, while the shape for the third male was closer to the mean for the females. Perhaps there is something unusual about the shapes for these individuals that was not represented in the training data, or, alternatively, they were labeled incorrectly. This is also revealed in Figure 13.2, where we plot these two sample preshapes. However, such a distance-based classifier may be suboptimal in not taking into account the variability within each group. In addition, the approach is deterministic and there is no measure of uncertainty in classification.

It is possible that classification performance could be improved in this application by also taking into account skull size. The proposed method can be easily extended to this case by using a Dirichlet process mixture density with the kernel being the product of a complex Watson kernel for the shape component and a log-Gaussian kernel for the size. Such a model induces a prior with support on the space of densities on the manifold $\Sigma_2^k \times \mathbb{R}^+$.

13.8 Proofs of theorems 13.1, 13.4, 13.6, 13.7, 13.8, 13.9, 13.10, and 13.12

Proof of Theorem 13.1 First we show that the set

$$\left\{ P \in \mathcal{M}(M) : \sup_{m \in M, \kappa \in N_\epsilon} |f(m; P, \kappa) - f(m; F_0, \kappa)| < \epsilon \right\} \tag{13.6}$$

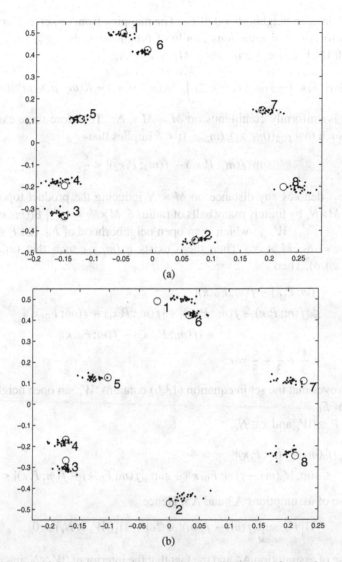

Figure 13.2 Landmarks from preshapes of training (*dot*) and misclassified test samples (*circle*) for females (a) and males (b). From Bhattacharya and Dunson (2010a).

contains a weakly open neighborhood of F_0, with F_0 being the distribution corresponding to f_0. The kernel K, being continuous from assumption A2, for any $(m, \kappa) \in M \times N_\epsilon$,

$$\mathcal{W}_{m,\kappa} = \left\{ P : |f(m; P, \kappa) - f(m; F_0, \kappa)| < \frac{\epsilon}{3} \right\},$$

defines an open neighborhood of F_0. The mapping from (m, κ) to $f(m; P, \kappa)$ is a uniformly equicontinuous family of functions on $M \times N_\epsilon$, labeled by $P \in \mathcal{M}(M)$, because, for $m_1, m_2 \in M$; $\kappa_1, \kappa_2 \in N_\epsilon$,

$$|f(m_1; P, \kappa_1) - f(m_2; P, \kappa_2)| \leq \int_M |K(m_1; \mu, \kappa_1) - K(m_2; \mu, \kappa_2)| P(d\mu)$$

and K is uniformly continuous on $M \times M \times N_\epsilon$. Therefore there exists a $\delta > 0$ such that $\rho_{12}((m_1, \kappa_1), (m_2, \kappa_2)) < \delta$ implies that

$$\sup_P |f(m_1; P, \kappa_1) - f(m_2; P, \kappa_2)| < \frac{\epsilon}{3}.$$

Here ρ_{12} denotes any distance on $M \times N$ inducing the product topology. Cover $M \times N_\epsilon$ by finitely many balls of radius δ: $M \times N_\epsilon = \bigcup_{i=1}^J B((m_i, \kappa_i), \delta)$. Let $\mathcal{W}_\epsilon = \bigcap_{i=1}^J \mathcal{W}_{m_i, \kappa_i}$, which is an open neighborhood of F_0. Let $P \in \mathcal{W}_\epsilon$ and $(m, \kappa) \in M \times N_\epsilon$. Then there exists a (m_i, κ_i) such that $(m, \kappa) \in B((m_i, \kappa_i), \delta)$. Then

$$|f(m; P, \kappa) - f(m; F_0, \kappa)|$$
$$\leq |f(m; P, \kappa) - f(m_i; P, \kappa_i)| + |f(m_i; P, \kappa_i) - f(m_i; F_0, \kappa_i)|$$
$$+ |f(m_i; F_0, \kappa_i) - f(m; F_0, \kappa)|$$
$$< \frac{\epsilon}{3} + \frac{\epsilon}{3} + \frac{\epsilon}{3} = \epsilon.$$

This proves that the set in equation (13.6) contains $\mathcal{W}_\epsilon$, an open neighborhood of F_0.

For $P \in \mathcal{W}_\epsilon$ and $\kappa \in N_\epsilon$,

$$\sup_m |f_0(m) - f(m; P, \kappa)|$$
$$\leq \sup_m |f_0(m) - f(m; F_0, \kappa)| + \sup_m |f(m; F_0, \kappa) - f(m; P, \kappa)| < 2\epsilon$$

because of assumptions A3 and A5. Hence

$$\Pi\left\{ f : \sup_{m \in M} |f(m) - f_0(m)| < 2\epsilon \right\} \geq \Pi_1(\mathcal{W}_\epsilon \times N_\epsilon) > 0$$

because of assumption A4 and the fact that the interior of $\mathcal{W}_\epsilon \times N_\epsilon$ intersects with supp(Π_1). This proves that the L^∞ support of Π includes all continuous densities.

Next we show that $\Pi(K_\epsilon(f_0)) > 0 \; \forall \epsilon > 0$, with $K_\epsilon(f_0)$ being the set $\{f : d_{KL}(f_0|f) < \epsilon\}$. If f_0 is strictly positive everywhere, with M being compact, f_0 must be bounded below by, say, $m_1 > 0$. For $\delta > 0$, denote by $\mathcal{W}_\delta(f_0)$ the set of densities $\{f : \sup_m |f(m) - f_0(m)| < \delta\}$ that have positive Π probability for any $\delta > 0$ and any continuous f_0. For δ sufficiently small,

$$\inf_{m \in M} f(m) \geq m_1/2, \quad \forall f \in \mathcal{W}_\delta(f_0).$$

Then, for $\epsilon > 0$, $f \in \mathcal{W}_\delta(f_0)$,

$$\int f_0 \log(f_0/f) \leq \sup_M |f_0(m)/f(m) - 1| \leq \frac{2\delta}{m_1} < \epsilon$$

if we choose $\delta < \frac{m_1\epsilon}{2}$. This proves $\Pi(K_\epsilon(f_0)) > 0$ when f_0 is bounded below.

In case f_0 is not bounded below, as shown in lemma 4 in Wu and Ghosal (2008), we can find an f_1 (depending on f_0 and ϵ) that is bounded below and

$$\Pi(K_{2\epsilon + \sqrt{\epsilon}}(f_0)) \geq \Pi(K_\epsilon(f_1)) > 0$$

as shown earlier. Because ϵ was arbitrary, this completes the proof. $\quad\square$

Proof of Theorem 13.4 In this proof and the subsequent ones, we use a general symbol C for any constant not depending on n (but possibly on ϵ).

Given $\delta_1 > 0$ ($\equiv \delta_1(\epsilon, n)$), cover M by N_1 ($\equiv N_1(\delta_1)$) many disjoint subsets of diameter at most δ_1: $M = \cup_{i=1}^{N_1} E_i$. Assumption A9 implies that for δ_1 sufficiently small, $N_1 \leq C\delta_1^{-a_3}$. Pick $\mu_i \in E_i$, $i = 1, \ldots, N_1$, and define for a probability P,

$$P_n = \sum_{i=1}^{N_1} P(E_i)\delta_{\mu_i}, \quad P_n(\mathbf{E}) = (P(E_1), \ldots, P(E_{N_1}))'. \quad (13.7)$$

Denoting the L_1-norm as $\|.\|$, for any κ with $\phi(\kappa) \leq \mathcal{K}_n$,

$$\|f(P, \kappa) - f(P_n, \kappa)\| \leq \sum_{i=1}^{N_1} \int_{E_i} \|K(\mu, \kappa) - K(\mu_i, \kappa)\| P(d\mu)$$

$$\leq C \sum_i \int_{E_i} \sup_{m \in M} |K(m; \mu, \kappa) - K(m; \mu_i, \kappa)| P(d\mu)$$

$$(13.8)$$

$$\leq C\mathcal{K}_n^{a_1}\delta_1. \quad (13.9)$$

The inequality in (13.9) follows from inequality (13.8) using assumption A6.

For $\kappa, \tilde{\kappa} \in \phi^{-1}[0, \mathcal{K}_n]$, $P \in \mathcal{M}(M)$,

$$\|f(P, \kappa) - f(P, \tilde{\kappa})\| \leq C \sup_{m, \mu \in M} |K(m; \mu, \kappa) - K(m; \mu, \tilde{\kappa})|$$

$$\leq C\mathcal{K}_n^{a_2}\rho_2(\kappa, \tilde{\kappa}), \quad (13.10)$$

which follows from from assumption A7. Given $\delta_2 > 0$ ($\equiv \delta_2(\epsilon, n)$), cover $\phi^{-1}[0, \mathcal{K}_n]$ by finitely many subsets of length at most δ_2, the number of such subsets required being at most $(\mathcal{K}_n\delta_2^{-1})^{b_2}$ from assumption A8. Call the collection of these subsets $W(\delta_2, n)$.

Letting $S_d = \{x \in [0, 1]^d : \sum x_i \leq 1\}$ be the $(d-1)$-dimensional simplex, S_d is compact under the L^1-metric ($\|x\|_{L^1} = \sum |x_i|$, $x \in \mathbb{R}^d$) and hence, given any $\delta_3 > 0$ ($\equiv \delta_3(\epsilon)$), can be covered by finitely many subsets of the cube $[0, 1]^d$ each of diameter at most δ_3, in particular, cover S_d with cubes of side length δ_3/d lying partially or totally in S_d. Then an upper bound on the number N_2 ($\equiv N_2(\delta_3, d)$) of such cubes can be shown to be $\frac{\lambda(S_d(1+\delta_3))}{(\delta_3/d)^d}$, with λ denoting the Lebesgue measure on $\mathbb{R}^d$ and $S_d(r) = \{x \in [0, \infty)^d : \sum x_i \leq r\}$. Because $\lambda(S_d(r)) = r^d/d!$, we have

$$N_2(\delta_3, d) \leq \frac{d^d}{d!} \left(\frac{1 + \delta_3}{\delta_3} \right)^d.$$

Let $\mathcal{W}(\delta_3, d)$ denote the partition of S_d as constructed above.

Let $d_n = N_1(\delta_1)$. For $1 \leq i \leq N_2(\delta_3, d_n)$, $1 \leq j \leq (\mathcal{K}_n \delta_2^{-1})^{b_2}$, define

$$\mathcal{D}_{ij} = \{f(P, \kappa) : P_n(\mathbf{E}) \in \mathcal{W}_i, \ \kappa \in W_j\},$$

with $\mathcal{W}_i$ and W_j being elements of $\mathcal{W}(\delta_3, d_n)$ and $W(\delta_2, n)$, respectively. We claim that this subset of $\mathcal{D}_n$ has L^1 diameter of at most ϵ. For $f(P, \kappa)$, $f(\tilde{P}, \tilde{\kappa})$ in this set,

$$\|f(P, \kappa) - f(\tilde{P}, \tilde{\kappa})\| \leq \|f(P, \kappa) - f(P_n, \kappa)\| + \|f(P_n, \kappa) - f(\tilde{P}_n, \kappa)\|$$
$$+ \|f(\tilde{P}_n, \kappa) - f(\tilde{P}, \kappa)\| + \|f(\tilde{P}, \kappa) - f(\tilde{P}, \tilde{\kappa})\|. \quad (13.11)$$

From inequality (13.9) it follows that the first and third terms in inequality (13.11) are at most $C\mathcal{K}_n^{a_1}\delta_1$. The second term can be bounded by

$$\sum_{i=1}^{d_n} |P(E_i) - \tilde{P}(E_i)| < \delta_3,$$

and from the inequality in (13.10) the fourth term is bounded by $C\mathcal{K}_n^{a_2}\delta_2$. Hence the claim holds if we choose $\delta_1 = C\mathcal{K}_n^{-a_1}$, $\delta_2 = C\mathcal{K}_n^{-a_2}$, and $\delta_3 = C$. The number of such subsets covering $\mathcal{D}_n$ is at most $N_2(\delta_3, d_n)(\mathcal{K}_n\delta_2^{-1})^{b_2}$. From assumption A9, it follows that for n sufficiently large,

$$d_n = N_1(\delta_1) \leq C\mathcal{K}_n^{a_1 a_3}.$$

Using Stirling's formula, we can bound $\log(N_2(\delta_3, d_n))$ by Cd_n. Also, $\mathcal{K}_n\delta_2^{-1}$ is bounded by $C\mathcal{K}_n^{a_2+1}$, so that

$$N(\epsilon, \mathcal{D}_n) \leq C + C\log(\mathcal{K}_n) + Cd_n \leq C\mathcal{K}_n^{a_1 a_3}$$

for n sufficiently large. This completes the proof. $\qquad \square$

Proof of Lemma 13.8 Express $\tilde{c}(\kappa)$ as

$$C \int_{-1}^{1} \exp\{-\kappa(1-t)\}(1-t^2)^{d/2-1} dt$$

and it is clear that it is decreasing. This expression suggests that

$$
\begin{aligned}
\tilde{c}(\kappa) &\geq C \int_{0}^{1} \exp\{-\kappa(1-t)\}(1-t^2)^{d/2-1} dt \\
&\geq C \int_{0}^{1} \exp\{-\kappa(1-t^2)\}(1-t^2)^{d/2-1} dt \\
&= C \int_{0}^{1} \exp(-\kappa u) u^{d/2-1}(1-u)^{-1/2} du \\
&\geq C \int_{0}^{1} \exp(-\kappa u) u^{d/2-1} du \\
&= C \kappa^{-d/2} \int_{0}^{\kappa} \exp(-v) v^{d/2-1} dv \\
&\geq C \Big\{ \int_{0}^{1} \exp(-v) v^{d/2-1} dv \Big\} \kappa^{-d/2}
\end{aligned}
$$

if $\kappa \geq 1$. This completes the proof. $\qquad \square$

Proof of Theorem 13.6 Denote by M the unit sphere S^d and by ρ the extrinsic distance on it. Express the vMF kernel as

$$K(m; \mu, \kappa) = c^{-1}(\kappa) \exp\left[\kappa\{1 - \rho^2(m, \mu)/2\}\right] \quad (m, \mu \in M; \kappa \in [0, \infty)).$$

Because ρ is continuous on the product space $M \times M$ and c is continuous and nonvanishing on $[0, \infty)$, K is continuous on $M \times M \times [0, \infty)$ and assumption A2 follows.

For a given continuous function f on M, $m \in M$, $\kappa \geq 0$, define

$$
\begin{aligned}
I(m, \kappa) &= f(m) - \int_{M} K(m; \mu, \kappa) f(\mu) V(d\mu) \\
&= \int_{M} K(m; \mu, \kappa)\{f(m) - f(\mu)\} V(d\mu).
\end{aligned}
$$

Then assumption A3 follows once we show that

$$\lim_{\kappa \to \infty} (\sup_{m \in M} |I(m, \kappa)|) = 0.$$

To simplify $I(m, \kappa)$, make a change of coordinates $\mu \mapsto \tilde{\mu} = U(m)'\mu$, $\tilde{\mu} \mapsto \theta \in \Theta_d \equiv (0, \pi)^{d-1} \times (0, 2\pi)$, where $U(m)$ is an orthogonal matrix with the first column equal to m and $\theta = (\theta_1, \ldots, \theta_d)'$ are the spherical coordinates of $\tilde{\mu} \equiv \tilde{\mu}(\theta)$ that are given by

$$\tilde{\mu}_j = \cos\theta_j \prod_{h<j} \sin\theta_h, \quad j = 1, \ldots, d, \quad \tilde{\mu}_{d+1} = \prod_{j=1}^{d} \sin\theta_j.$$

Using these coordinates, the volume form can be written as

$$V(d\mu) = V(d\tilde{\mu}) = \sin^{d-1}(\theta_1)\sin^{d-2}(\theta_2)\cdots\sin(\theta_{d-1})d\theta_1\cdots d\theta_d$$

and hence $I(m, \kappa)$ equals

$$\int_{\Theta_d} c^{-1}(\kappa)\exp\{\kappa\cos(\theta_1)\}\{f(m) - f(U(m)\tilde{\mu})\}\sin^{d-1}(\theta_1)\cdots\sin(\theta_{d-1})d\theta_1\cdots d\theta_d$$

$$= c^{-1}(\kappa)\int_{\Theta_{d-1}\times(-1,1)}\exp(\kappa t)\{f(m) - f(U(m)\tilde{\mu})\}(1 - t^2)^{d/2-1}$$

$$\sin^{d-2}(\theta_2)\cdots\sin(\theta_{d-1})d\theta_2\cdots d\theta_d dt, \quad (13.12)$$

where $t = \cos(\theta_1)$, $\tilde{\mu} = \tilde{\mu}(\theta(t))$, and $\theta(t) = (\arccos(t), \theta_2, \ldots, \theta_d)^T$. In the integrand in equation (13.12), the distance between m and $U(m)\tilde{\mu}$ is $\sqrt{2(1 - t)}$. Substitute $t = 1 - \kappa^{-1}s$ in the integral with $s \in (0, 2\kappa)$. Define

$$\Phi(s, \kappa) = \sup\{|f(m) - f(\tilde{m})| : m, \tilde{m} \in M, \rho(m, \tilde{m}) \leq \sqrt{2\kappa^{-1}s}\}.$$

Then

$$\left|f(m) - f(U(m)\tilde{\mu})\right| \leq \Phi(s, \kappa).$$

Because f is uniformly continuous on (M, ρ), Φ is therefore bounded on $(\mathbb{R}^+)^2$ and $\lim_{\kappa\to\infty}\Phi(s, \kappa) = 0$. Hence, from inequality (13.12) we deduce that

$$\sup_{m\in M}|I(m, \kappa)| \leq c^{-1}(\kappa)\kappa^{-1}\int_{\Theta_{d-1}\times(0,2\kappa)}\exp(\kappa - s)\Phi(s, \kappa)(\kappa^{-1}s(2 - \kappa^{-1}s))^{d/2-1}$$

$$\sin^{d-2}(\theta_2)\cdots\sin(\theta_{d-1})d\theta_2\cdots d\theta_d ds$$

$$\leq C\kappa^{-d/2}\tilde{c}^{-1}(\kappa)\int_0^\infty \Phi(s, \kappa)e^{-s}s^{d/2-1}ds.$$

$$(13.13)$$

From Lemma 13.8, it follows that

$$\limsup_{\kappa\to\infty}\kappa^{-d/2}\tilde{c}^{-1}(\kappa) < \infty.$$

This, in turn, using the Lebesgue DCT, implies that the expression in inequality (13.13) converges to 0 as $\kappa \to \infty$. This verifies assumption A3 and completes the proof. □

Proof of Theorem 13.7 In the proof,

$$B_d(r) = \{x \in \mathbb{R}^d : \|x\|_2 \le r\}$$

and B_d refers to $B_d(1)$.

It is clear from the vMF kernel expression that it is continuously differentiable on $\mathbb{R}^{d+1} \times \mathbb{R}^{d+1} \times [0, \infty)$. Hence

$$\sup_{m \in S^d, \kappa \in [0, \mathcal{K}]} \left| K(m; \mu, \kappa) - K(m; \nu, \kappa) \right| \le \sup_{m \in S^d, x \in B_{d+1}, \kappa \in [0, \mathcal{K}]} \left\| \frac{\partial}{\partial x} K(m; x, \kappa) \right\|_2 \|\mu - \nu\|_2.$$

Because

$$\frac{\partial}{\partial x} K(m; x, \kappa) = \kappa \tilde{c}^{-1}(\kappa) \exp\{-\kappa(1 - m'x)\} m,$$

its norm is bounded by $\kappa \tilde{c}^{-1}(\kappa)$. Lemma 13.8 implies that this in turn is bounded by

$$\mathcal{K} \tilde{c}^{-1}(\mathcal{K}) \le C \mathcal{K}^{d/2+1}$$

for $\kappa \le \mathcal{K}$ and $\mathcal{K} \ge 1$. This proves assumption A6 with $a_1 = \frac{d}{2} + 1$.

To verify A7, given $\kappa_1, \kappa_2 \le \mathcal{K}$, use the inequality

$$\sup_{m, \mu \in S^d} \left| K(m; \mu, \kappa_1) - K(m; \mu, \kappa_2) \right| \le \sup_{m, \mu \in S^d, \kappa \le \mathcal{K}} \left| \frac{\partial}{\partial \kappa} K(m; \mu, \kappa) \right| |\kappa_1 - \kappa_2|.$$

By direct computation one can show that

$$\frac{\partial}{\partial \kappa} K(m; \mu, \kappa) = -\frac{\partial}{\partial \kappa} \tilde{c}(\kappa) \tilde{c}^{-2}(\kappa) \exp\{-\kappa(1 - m'\mu)\}$$

$$- \tilde{c}^{-1}(\kappa) \exp\{-\kappa(1 - m'\mu)\}(1 - m'\mu),$$

$$\frac{\partial}{\partial \kappa} \tilde{c}(\kappa) = -C \int_{-1}^{1} \exp\{-\kappa(1 - t)\}(1 - t)(1 - t^2)^{d/2-1} dt,$$

$$\left| \frac{\partial}{\partial \kappa} \tilde{c}(\kappa) \right| \le C \tilde{c}(\kappa).$$

Therefore, using Lemma 13.8,

$$\left| \frac{\partial}{\partial \kappa} K(m; \mu, \kappa) \right| \le C \tilde{c}^{-1}(\kappa) \le C \tilde{c}^{-1}(\mathcal{K}) \le C \mathcal{K}^{d/2}$$

for any $\kappa \le \mathcal{K}$ and $\mathcal{K} \ge 1$. Hence assumption A7 is verified with $a_2 = d/2$.

Finally, to verify assumption A9, note that $S^d \subset B_{d+1} \subset [-1, 1]^{d+1}$, which can be covered by finitely many cubes of side length $\epsilon/(d + 1)$. Each such cube has L_2 diameter ϵ. Hence their intersections with S^d provide a finite ϵ-cover for this manifold. If $\epsilon < 1$, such a cube intersects with S^d only if it lies entirely in $B_{d+1}(1 + \epsilon) \cap B_{d+1}(1 - \epsilon)^c$. The number of such cubes, and hence the ϵ-cover size, can be bounded by

$$C\epsilon^{-(d+1)}\{(1 + \epsilon)^{d+1} - (1 - \epsilon)^{d+1}\} \le C\epsilon^{-d}$$

for some $C > 0$ not depending on ϵ. This verifies assumption A9 for appropriate positive constant A_3 and $a_3 = d$ and completes the proof.   □

Proof of Theorem 13.9   Express the complex Watson kernel as

$$K(m; \mu, \kappa) = c^{-1}(\kappa) \exp\{\kappa(1 - 1/2d_E^2(m, \mu))\},$$

where $c(\kappa) = (\pi\kappa^{-1})^{(k-2)}\{\exp(\kappa) - \sum_{r=0}^{k-3} \kappa^r/r!\}$ and d_E denotes the extrinsic distance and it is clear that assumption A2 holds on $\Sigma_2^k \times \Sigma_2^k \times (0, \infty)$.

Because the kernel is symmetric in m and μ, for any continuous f, define $I: \Sigma_2^k \to \mathbb{R}$ as

$$I(m) = f(m) - \int_{\Sigma_2^k} K(m; \mu, \kappa)f(\mu)V(d\mu) = \int_{\Sigma_2^k} \{f(m) - f(\mu)\}K(m; \mu, \kappa)V(d\mu).$$

$$(13.14)$$

Choose preshapes z and v for m and μ, respectively, in the complex sphere $\mathbb{C}S^{k-2}$, so that $m = [z]$ and $\mu = [v]$. Let V_1 denote the volume form on $\mathbb{C}S^{k-2}$. Then, for any integrable function $\phi: \Sigma_2^k \to \mathbb{R}$,

$$\int_{\Sigma_2^k} \phi(\mu)V(d\mu) = \frac{1}{2\pi} \int_{\mathbb{C}S^{k-2}} \phi([v])V_1(dv).$$

Hence the integral in equation (13.14) can be written as

$$I(m) = \frac{c^{-1}(\kappa)}{2\pi} \int_{\mathbb{C}S^{k-2}} \{f([z]) - f([v])\} \exp(\kappa v^* zz^* v)V_1(dv). \quad (13.15)$$

Consider a singular-value decomposition of zz^* as $zz^* = U\Lambda U^*$, where we take $\Lambda = \text{Diag}(1, 0, \ldots, 0)$ and $U = [U_1, \ldots, U_{k-1}]$ with $U_1 = z$. Then $v^* zz^* v = x^* \Lambda x = |x_1|^2$, where $x = U^* v = (x_1, \ldots, x_{k-1})'$. Make a change of variables from v to x in equation (13.15). This being an orthogonal transformation does not change the volume form. Then equation (13.15) becomes

$$I(m) = \frac{\exp(\kappa)}{2\pi c(\kappa)} \int_{\mathbb{C}S^{k-2}} \{f([z]) - f([Ux])\} \exp\{\kappa(|x_1|^2 - 1)\}V_1(dx). \quad (13.16)$$

Write $x_j = r_j^{1/2} \exp(i\theta_j)$, $j = 1, \ldots, k-1$, with $r = (r_1, \ldots, r_{k-1})'$ in the unit simplex S_{k-1} and $\theta = (\theta_1, \ldots, \theta_{k-1})' \in (-\pi, \pi)^{k-1}$, so that $V_1(dx) = 2^{2-k}dr_1 \cdots dr_{k-2}\, d\theta_1 \cdots d\theta_{k-1}$. Hence equation (13.16) can be written as

$$I(m) = 2^{1-k}\pi^{-1}e^\kappa c^{-1}(\kappa) \int_{S_{k-1}\times(-\pi,\pi)^{k-1}} \{f([z]) - f([y(r,\theta,z)])\}$$

$$\times \exp\{\kappa(r_1 - 1)\}dr\, d\theta, \quad (13.17)$$

with $y \equiv y(r,\theta,z) = \sum_{j=1}^{k-1} r_j^{1/2} \exp(i\theta_j)U_j$. Then $d_E^2([y],[z]) = 2(1 - r_1)$. For $d \in \mathbb{R}^+$, define

$$\psi(d) = \sup\{|f(m_1) - f(m_2)| : m_1, m_2 \in \Sigma_2^k, d_E^2(m_1, m_2) \le d\}.$$

Then the absolute value of $f([z]) - f([y(r,\theta,z)])$ in equation (13.17) is at most $\psi(2(1 - r_1))$, so that

$$\sup_{m\in\Sigma_2^k} |I(m)| \le \pi^{k-2}e^\kappa c^{-1}(\kappa) \int_{S_{k-1}} \psi(2(1 - r_1)) \exp\{\kappa(r_1 - 1)\}dr_1 \cdots dr_{k-2}$$

$$= \pi^{k-2}(k-3)!^{-1}e^\kappa c^{-1}(\kappa) \int_0^1 \psi(2(1 - r_1)) \exp\{\kappa(r_1 - 1)\}(1 - r_1)^{k-3}dr_1.$$

$$(13.18)$$

Make a change of variable $s = \kappa(1 - r_1)$ to rewrite equation (13.18) as

$$\sup_{m\in\Sigma_2^k} |I(m)| \le \pi^{k-2}(k-3)!^{-1}\kappa^{2-k}e^\kappa c^{-1}(\kappa) \int_0^\kappa \psi(2\kappa^{-1}s)e^{-s}s^{k-3}ds$$

$$\le Cc_1^{-1}(\kappa) \int_0^\infty \psi(2\kappa^{-1}s)e^{-s}s^{k-3}ds, \quad (13.19)$$

where $c_1(\kappa) = 1 - e^{-\kappa}\sum_{r=0}^{k-3}\kappa^r/r!$. Because f is uniformly continuous on the compact metric space (Σ_2^k, d_E), ψ is bounded and $\lim_{d\to 0}\psi(d) = 0$. Also, it is easy to check that $\lim_{\kappa\to\infty} c_1(\kappa) = 1$. Because $e^{-s}s^{k-3}$ is integrable on $(0, \infty)$, using the Lebesgue DCT on the integral in inequality (13.19), we conclude that

$$\lim_{\kappa\to\infty} \sup_{m\in\Sigma_2^k} |I(m)| = 0.$$

Hence assumption A3 is also satisfied. This completes the proof. $\square$

Proof of Theorem 13.10 Given $\kappa \ge 0$, define

$$\phi(t) = \exp\left(\frac{-\kappa}{2}t^2\right), \quad t \in [0, \sqrt{2}]$$

and then $K(m; \mu, \kappa) = c_1^{-1}(\kappa)\phi(d_E^2(m, \mu))$, with c_1 defined in Lemma 13.11. Then $|\phi'(t)| \le \sqrt{2}\kappa$, so that

$$|\phi(t) - \phi(s)| \le \sqrt{2}\kappa |s - t|, \quad s, t \in [0, \sqrt{2}],$$

which implies that

$$|K(m; \mu, \kappa) - K(m; \nu, \kappa)| \le Cc_1^{-1}(\kappa)\kappa |d_E(m, \mu) - d_E(m, \nu)|$$
$$\le C\kappa c_1^{-1}(\kappa)d_E(\mu, \nu). \tag{13.20}$$

For $\kappa \le \mathcal{K}$, Lemma 13.11 implies that

$$\kappa c_1^{-1}(\kappa) \le \mathcal{K}c_1^{-1}(\mathcal{K}) = C\mathcal{K}^{k-1}c_2^{-1}(\mathcal{K})$$
$$\le C\mathcal{K}^{k-1}c_2^{-1}(1)$$

provided $\mathcal{K} \ge 1$. Hence, for any $\mathcal{K} \ge 1$,

$$\sup_{\kappa \in [0, \mathcal{K}]} \kappa c_1^{-1}(\kappa) \le C\mathcal{K}^{k-1}$$

and from inequality (13.20), $a_1 = k - 1$ follows.

By direct computation, one can show that

$$\frac{\partial}{\partial \kappa} K(m; \mu, \kappa) = C \exp\left\{ -\frac{1}{2}\kappa d_E^2(m, \mu) - \kappa \right\}$$
$$\times c_1^{-2}(\kappa)\kappa^{2-k}\Big[\sum_{r=k-1}^{\infty} \frac{\kappa^{r-1}}{r!}\{k - 2 - \frac{r}{2}d_E^2(m, \mu)\}\Big]. \tag{13.21}$$

Denote by S the sum in the second line of equation (13.21). Because $d_E^2(m, \mu) \le 2$, it can be shown that

$$|k - 2 - \frac{r}{2}d_E^2(m, \mu)| \le \begin{cases} k - 2 & \text{if } k - 1 \le r \le 2k - 4, \\ r - k + 2 & \text{if } 2k - 3 \le r, \end{cases}$$

so that

$$|S| \le (k - 2) \sum_{r=k-1}^{2k-4} \frac{\kappa^{r-1}}{r!} + \sum_{r=2k-3}^{\infty} \frac{\kappa^{r-1}}{r!}(r - k + 2)$$
$$= (k - 2)\kappa^{k-2} \sum_{r=0}^{k-3} \frac{\kappa^r}{(r + k - 1)!} + \kappa^{2k-4} \sum_{r=0}^{\infty} \frac{\kappa^r}{(r + 2k - 3)!}(r + k - 1)$$
$$\le C\kappa^{k-2}e^{\kappa} + \kappa^{2k-4}e^{\kappa}.$$

Plug the above inequality into equation (13.21) to get

$$\left|\frac{\partial}{\partial \kappa} K(m; \mu, \kappa)\right| \le Cc_1^{-2}(\kappa)\kappa^{2-k} \exp\left\{ -\frac{1}{2}\kappa d_E^2(m, \mu)\right\}(C\kappa^{k-2} + \kappa^{2k-4})$$
$$\le Cc_1^{-2}(\kappa)(C + \kappa^{k-2}). \tag{13.22}$$

For $\kappa \leq \mathcal{K}$ and $\mathcal{K} \geq 1$, using Lemma 13.11, we bound the expression in inequality (13.22) by

$$Cc_1^{-2}(\mathcal{K})(C + \mathcal{K}^{k-2}) = C\mathcal{K}^{2k-6}c_2^{-2}(\mathcal{K})(C + \mathcal{K}^{k-2})$$

$$\leq C\mathcal{K}^{2k-6}c_2^{-2}(1)(C + \mathcal{K}^{k-2}) \leq C\mathcal{K}^{3k-8} \qquad (13.23)$$

for $\mathcal{K}$ sufficiently large. Because K is a continuously differentiable in κ, from equation (13.23) it follows that there exists $\mathcal{K}_1 > 0$ such that, for all $\mathcal{K} \geq \mathcal{K}_1, \kappa_1, \kappa_2 \leq \mathcal{K}$,

$$\sup_{m,\mu} |K(m; \mu, \kappa_1) - K(m; \mu, \kappa_2)| \leq \sup_{m,\mu \in \Sigma_2^k, \kappa \in [0,\mathcal{K}]} \left| \frac{\partial}{\partial \kappa} K(m; \mu, \kappa) \right| |\kappa_1 - \kappa_2|$$

$$\leq C\mathcal{K}^{3k-8} |\kappa_1 - \kappa_2|.$$

This proves assumption A7 with $a_2 = 3k - 8$ and completes the proof. $\square$

Proof of Theorem 13.12 From Theorem 13.7, given any $\delta > 0$, δ covering number of the complex sphere $\mathbb{C}S^{k-2}$ is bounded by $C\delta^{-(2k-3)} + C$. The extrinsic distance d_E on Σ_2^k can be bounded by the chord distance on $\mathbb{C}S^{k-2}$ as follows. For $u, v \in \mathbb{C}S^{k-2}$,

$$\|u - v\|_2^2 = 2 - 2\text{Re}(u^*v) \geq 2 - 2|u^*v| = 2(1 - |u^*v|)$$

$$\geq (1 + |u^*v|)(1 - |u^*v|) = 1/2d_E^2([u], [v]).$$

Hence $d_E([u], [v]) \leq \sqrt{2} \|u - v\|_2$, so that given any $\epsilon > 0$, the shape image of a δ-cover for $\mathbb{C}S^{k-2}$ with $\delta = \epsilon/\sqrt{2}$ provides an ϵ-cover for Σ_2^k. Hence the ϵ-covering size for Σ_2^k can be bounded by $C\epsilon^{-(2k-3)} + C$. $\square$

13.9 References

Ghosh and Ramamoorthi (2003) may be used as a general reference to non-parametric Bayes theory. Hjort et al. (2010) contains surveys of various topics in the field, including recent work. The stick-breaking representation of Sethuraman (1994) has proved useful both for theory and for posterior computations. The chapter is based on Bhattacharya and Dunson (2010a) and Bhattacharya and Dunson (2011).

14

Nonparametric Bayes regression, classification and hypothesis testing on manifolds

This chapter develops nonparametric Bayes procedures for classification, hypothesis testing and regression. The classification of a random observation to one of several groups is an important problem in statistics. This is the objective in medical diagnostics, the classification of subspecies, and, more generally, the target of most problems in image analysis. Equally important is the estimation of the regression function of Y given X and the prediction of Y given a random observation X. Here Y and X are, in general, manifold-valued, and we use nonparametric Bayes procedures to estimate the regression function.

14.1 Introduction

Consider the general problem of predicting a response $Y \in \mathbb{Y}$ based on predictors $X \in \mathbb{X}$, where $\mathbb{Y}$ and $\mathbb{X}$ are initially considered to be arbitrary metric spaces. The spaces can be discrete, Euclidean, or even non-Euclidean manifolds. In the context of this book, such data arise in many chapters. For example, for each study subject, we may obtain information on an unordered categorical response variable such as the presence/absence of a particular feature as well as predictors having different supports including categorical, Euclidean, spherical, or on a shape space. In this chapter we extend the methods of Chapter 13 to define a very general nonparametric Bayes modeling framework for the conditional distribution of Y given $X = x$ through joint modeling of $Z = (X, Y)$. The flexibility of our modelling approach will be justified theoretically through Theorems, Propositions, and Corollaries 14.1, 14.2, 14.3, 14.4, and 14.5. For example, using results from Sections 14.2 and 14.3.2, we will show that the joint model can approximate any continuous positive density, to any level of accuracy. In other words, our model has full support. Under some additional conditions on prior and model choice, we prove consistency in estimating the true data-generating distribution, given a random sample, in both the weak and strong senses. This in turn implies consistency of the estimated regression or classification function, that is, the conditional distribution of Y as a function of $X = x$.

Apart from establishing flexibility, we will also present efficient algorithms for getting random draws from the posterior of the regression function. Several applications will be presented at the end, which apply our methods and compare them with other standard estimates.

A problem closely related to classification is testing of differences in the distribution of features across various groups. In this setting, the non-parametric Bayes literature is surprisingly limited, perhaps due to the computational challenges that arise in calculating Bayes factors. Here we modify the methodology developed for the classification problem to obtain an easy-to-implement approach for nonparametric Bayes testing of differences between groups, with the data within each group constrained to lie on a compact metric space, and prove consistency of this testing procedure. We also present a novel algorithm to estimate the Bayes factor. The method is applied to hypothesis testing problems on spheres and shape spaces.

As in the last chapter, all proofs will be presented at the end, in Section 14.6.

The material of this chapter is based on Bhattacharya and Dunson (2010b) and Dunson and Bhattacharya (2010).

14.2 Regression using mixtures of product kernels

Suppose that $Y \in \mathbb{Y}$ and $X = \{X_1, \ldots, X_p\} \in \mathbb{X} = \prod_{j=1}^p \mathbb{X}_j$ with $X_j \in \mathbb{X}_j$, for $j = 1, \ldots, p$. We let the sample spaces $\mathbb{X}_j$ and $\mathbb{Y}$ be very general topological spaces ranging from subsets of $\mathbb{R}$ or $\{1, 2, \ldots, \infty\}$ to arbitrary manifolds. We assume that the pair (X, Y) has a joint density f with respect to some fixed product base measure on the product space. We model f as

$$f(x, y) = \int \left\{ \prod_{j=1}^p K^{(x_j)}(x_j; \theta^{(x_j)}) \right\} K^{(y)}(y; \theta^{(y)}) P(d\theta), \quad \theta = \{\theta^{(x_1)}, \ldots, \theta^{(x_p)}, \theta^{(y)}\},$$

$$(14.1)$$

where $K^{(x_j)}$ and $K^{(y)}$ are some parametric densities on $\mathbb{X}_j$, $j = 1, \ldots, p$, and $\mathbb{Y}$, respectively, with known expressions. The parameter P is a mixing distribution on $\mathbb{X} \times \mathbb{Y}$; it is assigned a prior Π_P. In particular, we assume Π_P is chosen so that

$$P = \sum_{h=1}^\infty w_h \delta_{\theta_h}, \quad \theta_h = \{\theta_h^{(x_1)}, \ldots, \theta_h^{(x_p)}, \theta_h^{(y)}\} \overset{\text{iid}}{\sim} P_0 = (\prod_{j=1}^p P_0^{(x_j)}) P_0^{(y)}, \quad (14.2)$$

where P_0 is a base measure constructed as a product. The prior (14.2) encompasses a broad class of sampling priors, with the Dirichlet process $DP(w_0 P_0)$ arising as a special case by letting $w_h = V_h \prod_{l<h}(1 - V_l)$ with $V_h \overset{\text{iid}}{\sim} \text{Beta}(1; w_0)$, for $h = 1, \ldots, \infty$. Besides P, there can be other parameters in the model, such as scale parameters, shape parameters, and so on, which can easily be taken into account. Model (14.1)–(14.2) implies the following model for the conditional density $f(y|x)$:

$$f(y|x) = \sum_{h=1}^{\infty} \left\{ \frac{w_h \prod_{j=1}^{p} K^{(x_j)}(x_j; \theta_h^{(x_j)})}{\sum_{l=1}^{\infty} w_l \prod_{j=1}^{p} K^{(x_j)}(x_j; \theta_l^{(x_j)})} \right\} K^{(y)}(y; \theta_h^{(y)})$$

$$= \sum_{h=1}^{\infty} w_h(x) K^{(y)}(y; \theta_h^{(y)}),$$

which expresses the conditional density as a predictor-dependent mixture of kernels that do not depend on x.

Given a training sample of size n, let y_i and $x_i = \{x_{i1}, \ldots, x_{ip}\}$ denote the response and predictor values for subject i and assume $(x_i, y_i) \overset{\text{iid}}{\sim} f$, for $i = 1, \ldots, n$. To generate $\{w_h, \theta_h\}_{h=1}^{\infty}$ from their joint posterior and hence $f(x, y)$ or $f(y|x)$ from their respective posteriors given the training sample, as in Chapter 13, we introduce latent class variables $S_1, \ldots S_n$, and express model (14.1)–(14.2) in the following way. The full posterior is

$$(x_i, y_i, S_i|-) \sim w_{S_i} \left\{ \prod_{j=1}^{p} K^{(x_j)}(x_{ij}; \theta_{S_i}^{(x_j)}) \right\} K^{(y)}(y_i; \theta_{S_i}^{(y)}), \quad i = 1, \ldots, n.$$

Then, conditionally on the latent class status for the different subjects, the response and different predictors are independent, with the parameters in the different likelihoods assigned independent priors. The dependence comes in through sharing a common cluster allocation across the different data types. This conditional independence greatly facilitates posterior computation in very general problems involving mixtures of different complicated and high-dimensional data types. The method is illustrated in detail in the context of classification in Section 14.3.1.

If the product kernel and the prior on the parameters satisfy the assumptions in Chapter 13, the induced prior on the space of all densities on the product space has full support in the KL sense, as shown in Theorem 13.1. Then, using the Schwartz theorem, we have weak posterior consistency for the joint density of X and Y. This in turn implies consistency in estimating the conditional probability $\Pr(Y \in B|X \in A)$, $A \subset \mathbb{X}$ and $B \subset \mathbb{Y}$, provided the true joint distribution gives positive mass to $A \times \mathbb{Y}$ and zero probability to the boundaries of $A \times B$ and $A \times \mathbb{Y}$.

To prove consistency in estimating the conditional density function $f(\cdot \,|x)$, we need to show strong consistency of the joint density $f(\cdot\,,\,\cdot)$. This follows from Corollary 13.5 under necessary assumptions. Then L_1 consistency for the conditional density function follows from Proposition 14.1. In its proof, $\lambda = \lambda_1 \times \lambda_2$ denotes the base measure on $\mathbb{X} \times \mathbb{Y}$ with respect to which all densities are defined.

Proposition 14.1 *Let $(x_1, y_1), \ldots, (x_n, y_n)$ be i.i.d. f_t. Let g_t and $f_t(\cdot\,|x)$ be the X-marginal density and the Y-conditional density function given $X = x$, respectively, under f_t. Let f be a joint density model for (X, Y). Strong consistency for the posterior of f implies that the posterior probability of*

$$\left\{ f : \mathrm{E}_{f_t} \left| \frac{f(Y|X)}{f_t(Y|X)} - 1 \right| < \epsilon \right\}$$

converges to 1 as $n \to \infty$ a.s. for any $\epsilon > 0$.

Hence, for example, if we have scalar response Y such as image size and predictor X on a manifold such as image shape, then we may use a discrete mixture of products of log-normal and Watson kernels as the joint model, and then we can consistently estimate the conditional density of size given shape, in the sense of Proposition 14.1, under mild assumptions.

It is interesting that such a rich model can be induced through the very simple structure on the joint density through expressions (14.1)–(14.2), which do not directly model dependence between Y and X or between the different elements of X. In fact, it can be shown that the dependence only comes in through the sharing of a common cluster allocation latent class variable across the different data types. Such shared latent class models are useful not only in modeling of conditional distributions in regression and classification but also in data fusion and combining information from disparate data sources.

14.3 Classification

In this section we focus on the special case of classification where Y takes finitely many values, say, $Y \in \mathbb{Y} = \{1, \ldots, c\}$. The goal is to model the classification function $p(y, x) \equiv \Pr(Y = y|X = x)$ flexibly as a function of $x \in \mathbb{X}$ for each $y \in \mathbb{Y}$. To do so, we use the approach in Section 14.2 and model the joint of (X, Y) via a joint density f as in expression (14.1). The base measure on $\mathbb{Y}$ is the counting measure, $\lambda_2 = \sum_{j=1}^{c} \delta_j$. In expression (14.1), we let the Y-kernel $K^{(y)}$ be a c-dimensional probability vector v taking values from the simplex $S_{c-1} = \{v \in [0, 1]^c : \sum v_j = 1\}$. Hence the joint density model simplifies to

$$f(x, y; P, \phi) = \int v_y K(x; \theta, \phi) P(d\theta \, dv), \quad (x, y) \in \mathbb{X} \times \mathbb{Y}, \quad (14.3)$$

with $K(\cdot; \theta, \phi)$ being some density with respect to λ_1 on $\mathbb{X}$ with parameters θ and ϕ. While defining f, we have integrated out θ using the mixing distribution P that takes a form as in expression (14.2). Hence the parameters used in defining the joint density are the random distribution P and the scale/shape parameters ϕ. By setting appropriate priors on them, we induce a prior on the joint of X and Y and hence on the probability functions $p(j, \cdot)$, $j = 1, \ldots, c$. By sampling from their posteriors given a training sample, we estimate these functions and classify Y based on X.

This joint model can be interpreted in the following hierarchical way. First draw parameters (P, ϕ) from their prior, denoted by Π_1. Then draw (θ, v) from P. Given (θ, v, ϕ), X and Y are conditionally independent with X having the conditional density $K(\cdot; \theta, \phi)$ with respect to λ_1 and Y follows a multinomial with

$$\Pr(Y = j \mid \theta, v, \phi) = v_j, \quad 1 \le j \le c.$$

In the next section we present an algorithm to get draws from the posterior. In Section 14.3.2 we will provide sufficient conditions for the model to have full support in the uniform distance and in the KL sense. We also theoretically prove our estimated classification functions to be consistent, without any parametric assumptions on the truth. This is not just of theoretical interest; it is important to verify that the model is sufficiently flexible to approximate any classification function, with the accuracy of the estimate improving as the amount of training data grows. This is not automatic for nonparametric models in which there is often concern about overfitting.

14.3.1 Posterior computation

Given a training sample $(\mathbf{x}_n, \mathbf{y}_n)$, we classify a new subject based on the predictive probability of allocating it to category j, which is expressed as

$$\Pr(y_{n+1} = j \mid x_{n+1}, \mathbf{x}_n, \mathbf{y}_n), \quad j \in \mathbb{Y}, \quad (14.4)$$

where x_{n+1} denotes the feature for the new subject and y_{n+1} is its unknown class label. It follows from Theorem 14.3 and Corollary 14.4 that the classification rule is consistent if the kernel and prior are chosen correctly. For the prior we let $P \sim \Pi_P = \mathrm{DP}(w_0 P_0)$ independently of $\phi \sim \pi$, with $P_0 = P_{01} \times P_{02}$, P_{01} a distribution on the θ space, P_{02} a Dirichlet distribution $\mathrm{Diri}(\mathbf{a})$ ($\mathbf{a} = (a_1, \ldots, a_c)$) on S_{c-1}, and π a base distribution on the

ϕ space. With such a choice for the base P_{02}, we achieve conjugacy as is illustrated below. Because it is not possible to get a closed form expression for the predictive probability posterior distribution, we need an MCMC algorithm to approximate it.

Using the stick-breaking representation (14.2) for P and introducing cluster allocation indices $S = (S_1, \ldots, S_n)$, the generative model (14.3) can be expressed in hierarchical form as

$$x_i \sim K(\cdot\,; \theta_{S_i}, \phi), \quad y_i \sim \text{Multi}(1, \ldots, c; \nu_{S_i}), \quad S_i \sim \sum_{j=1}^{\infty} w_j \delta_j,$$

where $w_j = V_j \prod_{h<j}(1 - V_h)$ is the probability that subject i is allocated to cluster $S_i = j$ $(j = 1, \ldots, \infty)$ and $\phi \sim \pi$, $V_j \sim \text{Beta}(1, w_0)$, $\theta_j \sim P_{01}$ and $\nu_j \sim \text{Diri}(\mathbf{a})$, $j = 1, \ldots, \infty$, are mutually independent.

We apply the exact block Gibbs sampler (Yau et al., 2011) for posterior computation. The joint posterior density of $\{V_j, \theta_j, \nu_j\}_{j=1}^{\infty}$, S, and ϕ given the training data is proportional to

$$\left\{ \prod_{i=1}^{n} K(x_i; \theta_{S_i}, \phi) \nu_{S_i y_i} w_{S_i} \right\} \left\{ \prod_{j=1}^{\infty} \text{Beta}(V_j; 1, w_0) P_{01}(d\theta_j) \text{Diri}(\nu_j; \mathbf{a}) \right\} \pi(\phi).$$

To avoid the need for posterior computation for infinitely many unknowns, we introduce slice sampling latent variables $u = \{u_i\}_{i=1}^{n}$ drawn i.i.d from Unif(0,1) such that the augmented posterior density becomes

$$\pi(u, V, \theta, \nu, S, \phi \mid \mathbf{x}_n, \mathbf{y}_n) \propto \left\{ \prod_{i=1}^{n} K(x_i; \theta_{S_i}, \phi) \nu_{S_i y_i} I(u_i < w_{S_i}) \right\}$$

$$\times \left\{ \prod_{j=1}^{\infty} \text{Beta}(V_j; 1, w_0) P_{01}(d\theta_j) \text{Diri}(\nu_j; \mathbf{a}) \right\} \pi(\phi).$$

Letting $\max(S)$ denote the largest of labels $\{S_i\}$, the conditional posterior distribution of $\{(V_j, \theta_j, \nu_j), j > \max(S)\}$ is the same as the prior, and we can use this to bypass the need for updating infinitely many unknowns in the Gibbs sampler. After choosing initial values, the sampler iterates through the following steps.

1. Update S_i, $i = 1, \ldots, n$, independently by sampling from multinomial distributions with

   $$\Pr(S_i = h) \propto K(x_i; \theta_h, \phi) \nu_{h y_i} \quad \text{for } h \in \{h : 1 \leq h \leq H, w_h > u_i\},$$

   with H being the smallest index satisfying $1 - \min(u) < \sum_{h=1}^{H} w_h$. In implementing this step, draw $V_h \sim \text{Beta}(1, w_0)$ and $(\theta_h, \nu_h) \sim P_0$ for $h > \max(S)$ as needed.

2. Update ϕ by sampling from the full conditional posterior, which is proportional to

$$\pi(\phi) \prod_{i=1}^{n} K(x_i; \theta_{S_i}, \phi).$$

If direct sampling is not possible, rejection sampling or Metropolis–Hastings (MH) sampling can be used.

3. Update the atoms (θ_j, ν_j), $j = 1, \ldots, \max(S)$ from the full conditional posterior distribution, which is equivalent to independently sampling from

$$\pi(\theta_j \mid -) \propto P_{01}(d\theta_j) \prod_{i:S_i=j} K(x_i; \theta_j, \phi)$$

$$(\nu_j \mid -) \sim \text{Diri}\left(a_1 + \sum_{i:S_i=j} I(y_i = 1), \ldots, a_c + \sum_{i:S_i=j} I(y_i = c)\right).$$

Hence the Dirichlet choice for P_{02} yields conjugacy for ν. In most applications, the first component P_{01} can also be chosen conjugate or approximately conjugate.

4. Update the stick-breaking random variables V_j, for $j = 1, \ldots, \max(S)$, from their conditional posterior distributions given the cluster allocation S but marginalizing out the slice sampling latent variables u. In particular, they are independent, with

$$V_j \sim \text{Beta}\left(1 + \sum_i I(S_i = j), w_0 + \sum_i I(S_i > j)\right).$$

5. Update the slice sampling latent variables from their conditional posterior by letting

$$u_i \sim \text{Unif}(0, w_{S_i}), \quad i = 1, \ldots, n.$$

These steps are repeated a large number of iterations, with a burn-in discarded to allow convergence. Given a draw from the posterior, the predictive probability of allocating a new observation to category l, $l = 1, \ldots, c$, as defined through expression (14.4), is proportional to

$$\sum_{j=1}^{\max(S)} w_j \nu_{jl} K(x_{n+1}; \theta_j, \phi) + \left(1 - \sum_{j=1}^{\max(S)} w_j\right) \int \nu_l K(x_{n+1}; \theta, \phi) P_0(d\theta \, d\nu).$$

We can average these conditional predictive probabilities across the MCMC iterations after burn-in to estimate predictive probabilities. For a moderate to large training sample size n, $\sum_{j=1}^{\max(S)} w_j$ is nearly 1 with high probability, so that an accurate approximation can be obtained by setting the final term equal to zero and hence bypassing the need to calculate the integral.

14.3.2 Support of the prior and consistency

In this section we theoretically justify the classification model specification (14.3) by showing flexibility in approximating any joint density and consistency of the posterior estimate of the classification functions. Because the non-Euclidean predictor spaces of interest in this book are mostly compact, we assume compact support in all our theorems.

In this context the results of Chapter 13 cannot be applied directly because the Y-kernel has only one parameter that is not from $\mathbb{Y}$.

We assume that X has a marginal density g_t on $\mathbb{X}$ and $\Pr(Y = y|X = x) = p_t(y, x)$, $y \in \mathbb{Y} = \{1, \ldots, c\}$, and $x \in \mathbb{X}$. Hence the joint distribution of (X, Y) has a density $f_t(x, y) = g_t(x)p_t(y, x)$, which is modeled by a specification $f(x, y; P, \phi)$ as in equation (14.3). Denote by Π_1 the chosen prior on parameters (P, ϕ), such as $\mathrm{DP}(w_0 P_0) \times \pi$. Let Π denote the prior induced on the space $\mathcal{D}(\mathbb{X} \times \mathbb{Y})$ of all joint densities through Π_1 and equation (14.3). Under minor assumptions on Π_1 and hence Π, Theorem 14.2 shows that the prior probability of any L^∞ and KL neighborhood of any continuous density is positive. For the sake of illustration, in this theorem and the subsequent ones we choose the X kernel $K(\cdot\,; \theta, \phi)$ to be a location–scale kernel with $\theta \in \mathbb{X}$ being the location parameter while $\phi \in \mathbb{R}^+$ is the (inverse) scale parameter. When $\mathbb{X}$ is a Riemannian manifold, K may be chosen to be a parametric density with respect to the invariant volume form, such as Gaussian on $\mathbb{R}^d$, a Fisher distribution on a sphere, and a complex Watson on the planar shape space. The theorems can easily be extended to more general predictor spaces, involving combinations of discrete, categorical, and continuous predictors. That is left to the reader.

Theorem 14.2 *Under the assumptions*

A1 $(\mathbb{X}, \rho)$ *is a compact metric space.*

A2 K *is continuous in its arguments.*

A3 *For any continuous function g from $\mathbb{X}$ to $\mathbb{R}$,*

$$\lim_{\phi \to \infty} \sup_{x \in \mathbb{X}} \left| g(x) - \int_{\mathbb{X}} K(x; \theta, \phi)g(\theta)\lambda_1(d\theta) \right| = 0.$$

A4 *For any $\phi > 0$, there exists $\tilde{\phi} \geq \phi$ such that $(P_t, \tilde{\phi}) \in \mathrm{supp}(\Pi_1)$, where $P_t \in \mathcal{M}(\mathbb{X} \times S_{c-1})$ is defined as*

$$P_t(d\theta \, dv) = \sum_{j \in \mathbb{Y}} f_t(\theta, j) \delta_{e_j}(dv) \lambda_1(d\theta),$$

with $\mathcal{M}(.)$ *denoting the space of all probability distributions and* $e_j \in \mathbb{R}^c$ *a zero vector with a single one in position* j.

A5 $f_t(\cdot, j)$ *is continuous for all* $j \in \mathbb{Y}$*, given any* $\epsilon > 0$*:*

$$\Pi\Big(\{f \in \mathcal{D}(\mathbb{X} \times \mathbb{Y}) : \sup_{x \in \mathbb{X}, y \in \mathbb{Y}} |f(x, y) - f_t(x, y)| < \epsilon\}\Big) > 0.$$

This in turn implies that f_t *is in the KL support of* Π.

Assumptions A2 and A3 place minor regularity conditions on the X-kernel K. If $K(x; \theta, \phi)$ is symmetric in x and θ, as will be the case in most examples, A3 implies that $K(.; \theta, \phi)$ converges to δ_θ in the weak sense uniformly in θ as $\phi \to \infty$. This justifies the names "location" and "inverse scale" for the parameters. Assumption A4 provides a minimal condition on the support of the prior on (P, ϕ). We may take the prior to have full support and the assumption will be automatically satisfied.

Suppose we have an i.i.d. sample $(\mathbf{x}_n, \mathbf{y}_n) \equiv (x_i, y_i)_{i=1}^n$ from f_t. Because f_t is unobserved, we take the likelihood function to be

$$\prod_{i=1}^n f(x_i, y_i; P, \phi).$$

Using the prior Π on f and the observed sample, we find the posterior distribution of f as in Section 14.3.1. Using the Schwartz (1965) theorem (Proposition 13.2), Theorem 14.2 implies weak posterior consistency for estimating the joint distribution of (x, y). This in turn implies that for any subset A of $\mathbb{X}$, with $\lambda_1(A) > 0$, $\lambda_1(\partial A) = 0$, and $y \in \mathbb{Y}$, the posterior conditional probability of $Y = y$ given $X \in A$ converges to the true conditional probability almost surely. Here ∂A denotes the boundary of A.

Under stronger assumptions on the kernel and the prior, we prove strong posterior consistency for the joint model.

Theorem 14.3 *In addition to assumptions A1–A5, assume*

A6 *There exist positive constants* Φ_0, a_1, A_1 *such that for all* $\Phi \geq \Phi_0$*,* $\theta_1, \theta_2 \in \mathbb{X}$*,*

$$\sup_{x \in \mathbb{X}, \phi \in [0, \Phi]} |K(x; \theta_1, \phi) - K(x; \theta_2, \phi)| \leq A_1 \Phi^{a_1} \rho(\theta_1, \theta_2).$$

A7 *There exist positive constants* a_2, A_2 *such that for all* $\phi_1, \phi_2 \in [0, \Phi]$*,* $\Phi \geq \Phi_0$*,*

$$\sup_{x, \theta \in \mathbb{X}} |K(x; \theta, \phi_1) - K(x; \theta, \phi_2)| \leq A_2 \Phi^{a_2} |\phi_1 - \phi_2|.$$

A8 *There exist positive constants a_3, A_3, A_4 such that given any $\epsilon > 0$, $\mathbb{X}$ can be covered by at most $A_3 \epsilon^{-a_3} + A_4$ many subsets of diameter at most ϵ; that is, its ϵ-covering number is of order at most ϵ^{-a_3}.*

A9 *$\Pi_1(\mathcal{M}(\mathbb{X}) \times (n^a, \infty))$ is exponentially small for some $a < (a_1 a_3)^{-1}$.*

Then the posterior probability of any total variation neighborhood of f_t converges to 1 almost surely.

Given the training data, we can classify a new feature based on a draw from the posterior of the predictive probability function p. As a corollary to Theorem 14.3, we show that it converges to the truth p_t in the L^1 sense as the training sample size increases.

Corollary 14.4 *(a) Strong consistency for the posterior of f implies that, for any $\epsilon > 0$,*

$$\Pi\left\{f : \max_{y \in \mathbb{Y}} \int_{\mathbb{X}} |p(y, x) - p_t(y, x)| g_t(x) \lambda_1(dx) < \epsilon \Big| \mathbf{x_n}, \mathbf{y_n}\right\}$$

converges to 1 as $n \to \infty$, almost surely.
(b) With $\mathbb{X}$ being compact and f_t continuous and strictly positive, part (a) implies that

$$\Pi\left\{f : \max_{y \in \mathbb{Y}} \int_{\mathbb{X}} |p(y, x) - p_t(y, x)| w(x) \lambda_1(dx) < \epsilon \Big| \mathbf{x_n}, \mathbf{y_n}\right\}$$

converges to 1 a.s. for any nonnegative function w with $\sup_x w(x) < \infty$.

Part (a) of Corollary 14.4 holds even when $\mathbb{X}$ is noncompact. It just needs strong posterior consistency for the joint model. From part (b), it would seem intuitive that pointwise posterior consistency can be obtained for the predictive probability function. However, this is not immediate because the convergence rate may depend on the choice of the weight function w.

Assumption A9 is hard to satisfy, especially when the feature space is high dimensional. Then a_1 and a_3 turn out to be very big, so that the prior is required to have very light tails and place small mass at high precisions. This is undesirable in applications and instead we can let Π_1 depend on the sample size n and obtain weak and strong consistency under weaker assumptions.

Theorem 14.5 *Let $\Pi_1 = \Pi_{11} \otimes \pi_n$, where π_n is a sequence of densities on $\mathbb{R}^+$. Assume the following.*

A10 *The prior Π_{11} has full support.*
A11 *For any $\beta > 0$, there exists a $\phi_0 \geq 0$, such that for all $\phi \geq \phi_0$,*

$$\liminf_{n \to \infty} \exp(n\beta) \pi_n(\phi) = \infty.$$

A12 *For some $\beta_0 > 0$ and $a < (a_1 a_3)^{-1}$,*

$$\lim_{n\to\infty} \exp(n\beta_0)\pi_n\{(n^a, \infty)\} = 0.$$

(a) Under assumptions A1–A3, A5, and A10–A11, the posterior probability of any weak neighborhood of f_t converges to 1 almost surely (b) Under assumptions A1–A3, A5–A8, and A10–A12, the posterior probability of any total variation neighborhood of f_t converges to 1 almost surely.

We do not present the proof. It is similar to the proofs of theorems 2.7 and 2.10 in Bhattacharya and Dunson (2011). With $\Pi_{11} = DP(w_0 P_0)$ and $\pi_n = \text{Gamma}(a, b_n)$, the conditions in Theorem 14.5 are satisfied (for example) when P_0 has full support and $b_n = b_1 n / \{\log(n)\}^{b_2}$ for any $b_1, b_2 > 0$. Then, from Corollary 14.4, we have L^1 consistency for the classification function estimate.

14.4 Nonparametric Bayes testing

14.4.1 Hypotheses and Bayes factor

A problem analogous to classification is testing of differences between groups. In particular, instead of wanting to predict the class label y_{n+1} for a new subject based on training data $(\mathbf{x}_n, \mathbf{y}_n)$, the goal is to test whether the distribution of the features differs across the classes. Although our methods can allow testing of pairwise differences between groups, we focus for simplicity in exposition on the case in which the null hypothesis corresponds to *homogeneity across the groups*. Formally, the alternative hypothesis H_1 corresponds to any joint density in $\mathcal{D}(\mathbb{X}\times\mathbb{Y})$ excluding densities of the form

$$H_0 : f(x, y) = g(x)p(y) \tag{14.5}$$

for all (x, y) outside of a λ-null set. Note that the prior on f induced through model (14.3) will in general assign zero probability to H_0 and hence is an appropriate model for the joint density under H_1.

As a model for the joint density under the null hypothesis H_0 in equation (14.5), we replace $P(d\theta\,dv)$ in equation (14.3) by $P_1(d\theta)P_2(dv)$ so that the joint density becomes

$$f(x, y; P_1, P_2, \phi) = g(x; P_1, \phi)p(y; P_2),$$

where

$$g(x; P_1, \phi) = \int_{\mathbb{X}} K(x; \theta, \phi)P_1(d\theta), \quad p(y; P_2) = \int_{S_{c-1}} v_y P_2(dv). \tag{14.6}$$

We set priors Π_1 and Π_0 for the parameters in the models under H_1 and H_0, respectively. The *Bayes factor* in favor of H_1 over H_0 is then the ratio of the marginal likelihoods under H_1 and H_0:

$$\text{BF}(H_1 : H_0) = \frac{\int \prod_{i=1}^n f(x_i, y_i; P, \phi)\Pi_1(dP \, d\phi)}{\int \prod_{i=1}^n g(x_i; P_1, \phi)p(y_i; P_2)\Pi_0(dP_1 \, dP_2 \, d\phi)}.$$

The priors should be suitably constructed so that we get consistency of the Bayes factor and computation is straightforward and efficient. The prior Π_1 on (P, ϕ) under H_1 can be constructed as in Section 14.3. To choose a prior Π_0 for (P_1, P_2, ϕ) under H_0, we take (P_1, ϕ) to be independent of P_2 so that the marginal likelihood becomes a product of the X and Y marginals if H_0 is true. Dependence in the priors for the mixing measures would induce dependence between the X and Y densities, and it is important to maintain independence under H_0.

Expression (14.6) suggests that under H_0 the density of Y depends on P_2 only through the c-dimensional vector

$$p = (p(1; P_2), p(2; P_2), \ldots, p(c; P_2))' \in S_{c-1}.$$

Hence, it is sufficient to choose a prior for p, such as Diri($\mathbf{b}$) with $\mathbf{b} = (b_1, \ldots, b_c)'$, instead of specifying a full prior for P_2. To independently choose a prior for (P_1, ϕ), we recommend the marginal induced from the prior Π_1 on (P, ϕ) under H_1. Under this choice, the marginal likelihood under H_0 is

$$\int \prod_{i=1}^n g(x_i; P_1, \phi)\Pi_1(dPd\phi) \int_{S_{c-1}} \prod_{j=1}^c p_j^{\sum_{i=1}^n I(y_i=j)} \text{Diri}(dp; \mathbf{b})$$

$$= \frac{D(\mathbf{b}_n)}{D(\mathbf{b})} \int \prod_{i=1}^n g(x_i; P_1, \phi)\Pi_1(dP \, d\phi), \quad (14.7)$$

with $\mathbf{b}_n$ being the c-dimensional vector with jth coordinate $b_j + \sum_{i=1}^n I(y_i = j)$, $1 \le j \le c$, D being the normalizing constant for Dirichlet distribution given by $D(\mathbf{a}) = \frac{\prod_{j=1}^c \Gamma(a_j)}{\Gamma(\sum_{j=1}^c a_j)}$, and with Γ denoting the Gamma function. The marginal likelihood under H_1 is

$$\int \prod_{i=1}^n f(x_i, y_i; P, \phi)\Pi_1(dP \, d\phi). \quad (14.8)$$

The Bayes factor in favor of H_1 against H_0 is the ratio of the marginal (14.8) to equation (14.7).

14.4.2 Consistency of the Bayes factor

Let Π be the prior induced on the space of all densities $\mathcal{D}(\mathbb{X} \times \mathbb{Y})$ through Π_1. For any density $f(x,y)$, let $g(x) = \sum_j f(x,j)$ denote the marginal density of X while $p(y) = \int_\mathbb{X} f(x,y)\lambda_1(dx)$ denotes the marginal probability vector of Y. Let f_t, g_t and p_t be the corresponding values for the true distribution of (X, Y). The Bayes factor in favor of the alternative, as obtained in the last section, can be expressed as

$$\mathrm{BF} = \frac{D(\mathbf{b})}{D(\mathbf{b}_n)} \frac{\int \prod_i f(x_i, y_i)\Pi(df)}{\int \prod_i g(x_i)\Pi(df)}. \tag{14.9}$$

Theorem 14.6 proves consistency of the Bayes factor at an exponential rate if the alternative hypothesis of dependence holds.

Theorem 14.6 *If X and Y are not independent under the true density f_t and if the prior Π satisfies the KL condition at f_t, then there exists a $\beta_0 > 0$ for which $\lim \inf_{n \to \infty} \exp(-n\beta_0)\mathrm{BF} = \infty$ a.s. f_t^∞.*

14.4.3 Computation

We introduce a latent variable $z = I(H_1 \text{ is true})$ that takes value 1 if H_1 is true and 0 if H_0 is true. Assuming equal prior probabilities for H_0 and H_1, the conditional likelihood of $(\mathbf{x}_n, \mathbf{y}_n)$ given z is

$$\Pi(\mathbf{x}_n, \mathbf{y}_n | z = 0) = \frac{D(\mathbf{b}_n)}{D(\mathbf{b})} \int \prod_{i=1}^n g(x_i; P_1, \phi)\Pi_1(dP\, d\phi) \text{ and}$$

$$\Pi(\mathbf{x}_n, \mathbf{y}_n | z = 1) = \int \prod_{i=1}^n f(x_i, y_i; P, \phi)\Pi_1(dP\, d\phi).$$

In addition, the Bayes factor can be expressed as

$$\mathrm{BF} = \frac{\Pr(z = 1 | \mathbf{x}_n, \mathbf{y}_n)}{\Pr(z = 0 | \mathbf{x}_n, \mathbf{y}_n)}. \tag{14.10}$$

Next introduce latent parameters θ, v, V, S, ϕ as in Section 14.3.1 such that the joint of the parameters and data become

$$\Pi(\mathbf{x}_n, \mathbf{y}_n, \theta, V, S, \phi, z = 0) = \frac{D(\mathbf{b}_n)}{D(\mathbf{b})}\pi(\phi) \prod_{i=1}^n \{w_{S_i} K(x_i; \theta_{S_i}, \phi)\} \tag{14.11}$$

$$\times \prod_{j=1}^\infty \{\mathrm{Beta}(V_j; 1, w_0)P_{01}(d\theta_j)\},$$

$$\Pi(\mathbf{x}_n, \mathbf{y}_n, \theta, \nu, V, S, \phi, z = 1) = \pi(\phi) \prod_{i=1}^{n} \{w_{S_i} \nu_{S_i y_i} K(x_i; \theta_{S_i}, \phi)\}$$

$$\times \prod_{j=1}^{\infty} \{\text{Beta}(V_j; 1, w_0) P_0(d\theta_j \, d\nu_j)\}. \tag{14.12}$$

Marginalize out ν from equation (14.12) to get

$$\Pi(\mathbf{x}_n, \mathbf{y}_n, \theta, V, S, \phi, z = 1) = \pi(\phi) \prod_{j=1}^{\infty} \frac{D(\mathbf{a} + \tilde{a}_j(S))}{D(\mathbf{a})}$$

$$\times \prod_{i=1}^{n} \{w_{S_i} K(x_i; \theta_{S_i}, \phi)\} \prod_{j=1}^{\infty} \text{Beta}(V_j; 1, w_0) P_{01}(d\theta_j)\}, \tag{14.13}$$

with $\tilde{a}_j(S)$, $1 \le j < \infty$ being c-dimensional vectors with lth coordinate $\sum_{i:S_i=j} I(y_i = l)$, $l \in \mathbb{Y}$. Integrate out z by adding equations (14.11) and (14.13) and the joint posterior of (θ, V, S, ϕ) given the data becomes

$$\Pi(\theta, V, S, \phi | \mathbf{x}_n, \mathbf{y}_n) \propto \{C_0 + C_1(S)\} \pi(\phi) \prod_{i=1}^{n} \{w_{S_i} K(x_i; \theta_{S_i}, \phi)\}$$

$$\times \prod_{j=1}^{\infty} \{\text{Beta}(V_j; 1, w_0) P_{01}(d\theta_j)\} \tag{14.14}$$

$$\text{with } C_0 = \frac{D(\mathbf{b}_n)}{D(\mathbf{b})} \text{ and } C_1(S) = \prod_{j=1}^{\infty} \frac{D(\mathbf{a} + \tilde{a}_j(S))}{D(\mathbf{a})}.$$

To estimate the Bayes factor, first make repeated draws from the posterior in equation (14.14). For each draw, compute the posterior probability distribution of z from equations (14.11) and (14.13) and take their average after discarding a suitable burn-in. The averages estimate the posterior distribution of z given the data, from which we can get an estimate for BF from equation (14.10). The sampling steps are accomplished as follows:

1. Update the cluster labels S given (θ, V, ϕ) and the data from their joint posterior, which is proportional to

$$\{C_0 + C_1(S)\} \prod_{i=1}^{n} \{w_{S_i} K(x_i; \theta_{S_i}, \phi)\}. \tag{14.15}$$

Introduce slice sampling latent variables u as in Section 14.3.1 and replace w_{S_i} by $I(u_i < w_{S_i})$ to make the total number of possible states finite. However, unlike in Section 14.3.1, the S_i are no longer conditionally independent. We propose using a Metropolis–Hastings block update step in which a candidate for $(S_1, \ldots, S_n)$, or some subset of this vector if n is large, is sampled independently from multinomials with $\Pr(S_i = h) \propto K(x_i; \theta_h, \phi)$, for $h \in A_i$, where $A_i = \{h : 1 \leq h \leq H, w_h > u_i\}$ and H is the smallest index satisfying $1 - \min(u) < \sum_{h=1}^{H} w_h$. In implementing this step, draw $V_j \sim \text{Beta}(1, w_0)$ and $\theta_j \sim P_{01}$ for $j > \max(S)$ as needed. The acceptance probability is simply the ratio of $C_0 + C_1(S)$ calculated for the candidate value and the current value of S.

2. Update $\phi, \{\theta_j, V_j\}_{j=1}^{\max(S)}, \{u_i\}_{i=1}^{n}$ as in steps 2–5 of the algorithm in Section 14.3.1.

3. Compute the full conditional posterior distribution of z, which is given by

$$
\Pr(z|\theta, S, \mathbf{x}_n, \mathbf{y}_n) \propto
\begin{cases}
\frac{D(\mathbf{b}_n)}{D(\mathbf{b})} & \text{if } z = 0, \\
\prod_{j=1}^{\max(S)} \frac{D(\mathbf{a} + \bar{a}_j(S))}{D(\mathbf{a})} & \text{if } z = 1.
\end{cases}
$$

14.5 Examples

In this section we present some data examples of classification and testing of hypothesis problems, where we apply the methods of the earlier sections and compare with other inference methods introduced in earlier chapters or from other sources. The first three examples consist of data simulated from known distributions on various dimensional spheres while the last two constitute real data on directions and shapes.

14.5.1 Classification

We draw i.i.d. samples on $S^9 \times \mathbb{Y}$, $\mathbb{Y} = \{1, 2, 3\}$ from

$$
f_t(x, y) = \frac{1}{3} \sum_{l=1}^{3} I(y = l) \text{vMF}(x; \mu_l, 200),
$$

where $\mu_1 = (1, 0, \ldots)'$, $\mu_j = \cos(0.2)\mu_1 + \sin(0.2)v_j$, $j = 2, 3$, $v_2 = (0, 1, \ldots)'$ and $v_3 = (0, 0.5, \sqrt{0.75}, 0, \ldots)'$. Hence, the three response classes $y \in \{1, 2, 3\}$ are equally likely and the distribution of the features within each class is a vMF on S^9 with distinct location parameters. We purposely choose the separation between the kernel locations to be small, so that the classification task is challenging.

The approach described in Section 14.3.1 is implemented to perform nonparametric Bayes classification using a vMF kernel. The hyperparameters are chosen to be $w_0 = 1$, DP base $P_0 = \text{vMF}(\mu_n, 10) \otimes \text{Diri}(1, 1, 1)$, with μ_n being the feature sample extrinsic mean, and the prior π on ϕ

as in equation (13.5) with $a = 1$, $b = 0.1$. Cross-validation is used to assess classification performance, with posterior computation applied to data from a training sample of size 200, and the results used to predict y given the x values for subjects in a test sample of size 100. The MCMC algorithm was run for 5×10^4 iterations after a 10^4 iteration burn-in. Based on examination of trace plots for the predictive probabilities of y for representative test subjects, the proposed algorithm exhibits good rates of convergence and mixing. The out-of-sample misclassification rates for categories $y = 1$, 2, and 3 were 18.9%, 9.7%, and 12.5%, respectively, with the overall rate being 14%.

As an alternative method for flexible model-based classification, a discriminant analysis approach is considered that models the conditional density of x given y as a finite mixture of 10-dimensional Gaussians. In the literature it is very common to treat data lying on a hypersphere as if the data had support in a Euclidean space to simplify the analysis. Using the EM algorithm to fit the finite mixture model, we encountered singularity problems when allowing more than two Gaussian components per response class. Hence, we present the results only for mixtures of one or two multivariate Gaussian components. In the one-component case, we obtained class-specific misclassification rates of 27%, 12.9%, and 18.8%, with the overall rate being 20%. The corresponding results for the two-component mixture were 21.6%, 16.1%, and 28.1%, with an overall misclassification rate of 22%.

Hence, the results from a parametric Gaussian discriminant analysis and a mixture of Gaussians classifier were much worse than those for our proposed Bayesian nonparametric approach. There are several possible factors contributing to the improvement in performance. First, the discriminant analysis approach requires separate fittings of different mixture models to each of the response categories. When the amount of data in each category is small, it is difficult to reliably estimate all these parameters, leading to high variance and unstable estimates. In contrast, the approach of joint modeling of f_t using a Dirichlet process mixture favors a more parsimonious representation. Second, inappropriately modeling the data as having support on a Euclidean space has some clear drawbacks. The size of the space over which the densities are estimated is increased from a compact subset S^9 to an unbounded space $\mathbb{R}^{10}$. This can lead to an inflated variance and difficulties with convergence of the EM and MCMC algorithms. In addition, the properties of the approach are expected to be poor even in larger samples. As Gaussian mixtures give zero probability to the embedded hypersphere, one cannot expect consistency.

14.5.2 Hypothesis testing

An i.i.d sample of size 100 on $S^9 \times \mathbb{Y}$, $\mathbb{Y} = \{1, 2, 3\}$, is drawn from the distribution

$$f_t(x, y) = \frac{1}{3} \sum_{l=1}^{3} I(y = l) \sum_{j=1}^{3} w_{lj} \text{vMF}(x; \mu_j, 200),$$

where μ_j, $j = 1, 2, 3$, are as in the earlier example and the weights $\{w_{lj}\}$ are chosen so that $w_{11} = 1$ and $w_{lj} = 0.5$ for $l = 2, 3$ and $j = 2, 3$. Hence, in group $y = 1$, the features are drawn from a single von Mises–Fisher (vMF) density, while in groups $y = 2$ and 3 the feature distributions are equally weighted mixtures of the same two vMFs.

Letting f_j denote the conditional density of X given $Y = j$ for $j = 1, 2, 3$, respectively, the global null hypothesis of no difference in the three groups is $H_0 : f_1 = f_2 = f_3$, while the alternative H_1 is that they are not all the same. We set the hyperparameters as $w_0 = 1$, $P_0 = \text{vMF}(\mu_n, 10) \otimes \text{Diri}(\mathbf{a})$, with μ_n being the X-sample extrinsic mean, $\mathbf{b} = \mathbf{a} = \hat{p} = (0.28, 0.36, 0.36)$ – the sample proportions of observations from the three groups, and a prior π on ϕ as in the earlier example. We run the proposed MCMC algorithm for calculating the Bayes factor (BF) in favor of H_1 over H_0 for 6×10^4 iterations, updating cluster labels S in four blocks of 25 each in every iteration. The trace plots exhibit good rates of convergence of the algorithm. After discarding a burn-in of 4×10^4 iterations, the estimated BF was 2.23×10^{15}, suggesting strong evidence in the data in favor of H_1. We tried multiple starting points and different hyperparameter choices and found the conclusions to be robust, with the estimated BFs not exactly the same but within an order of magnitude. Similar estimates were also obtained using substantially shorter and longer chains.

One can also use the proposed methodology for pairwise hypothesis testing of $H_{0,ll'} : f_l = f_{l'}$ against the alternative $H_{1,ll'} : f_l \neq f_{l'}$ for any pair, l, l', with $l \neq l'$. The analysis is otherwise implemented exactly as in the global hypothesis testing case. The resulting BFs in favor of $H_{1,ll'}$ over $H_{0,ll'}$ for the different possible choices of (l, l') are shown in Table 14.1. We obtain very large BFs in testing differences between groups 1 and 2 and 1 and 3, but a moderately small BF for testing a difference between groups 2 and 3, yielding no evidence of significant difference between groups 2 and 3. These conclusions are all consistent with the truth. We have noted a general tendency for the BF in favor of the alternative to be large when the alternative is true even in modest sample sizes, suggesting a rapid rate of convergence under the alternative, in agreement with our

Table 14.1 *Nonparametric Bayes and frequentist test results for data simulated for three groups with the second and third groups identical.*

Groups	BF	p-Value
(1,2,3)	2.3×10^{15}	2×10^{-6}
(1,2)	2.4×10^{4}	1.8×10^{-4}
(1,3)	1.7×10^{6}	1.5×10^{-5}
(2,3)	0.235	0.43

theoretical results. When the null is true, the BF appears to converge to zero based on empirical results in our simulations, but at a slow rate.

For comparison, consider the frequentist nonparametric test, introduced in Section 4.5, for detecting differences in groups based on extrinsic means of the f_l. The test statistic used has an asymptotic $\mathcal{X}^2_{d(L-1)}$ distribution, where $d = 9$ is the feature space dimension and L is the number of groups that we are comparing. The corresponding p-values are shown in Table 14.1. The conclusions are all consistent with those from the nonparametric Bayes approach.

14.5.3 Testing with no differences in means

In this example, i.i.d. samples are drawn on $S^2 \times \mathbb{Y}$, $\mathbb{Y} = \{1, 2\}$ from the distribution

$$f_t(x, y) = \frac{1}{2} \sum_{l=1}^{2} I(y = l) \sum_{j=1}^{3} w_{lj} \text{vMF}(x; \mu_j, 200),$$

where $w = \begin{bmatrix} 1 & 0 & 0 \\ 0 & 0.5 & 0.5 \end{bmatrix}$, $\mu_1 = (1, 0, 0)^T$, $\mu_j = \cos(0.2)\mu_1 + \sin(0.2)v_j$ ($j = 2, 3$), and $v_2 = -v_3 = (0, 1, 0)^T$. In this case the features are drawn from two equally likely groups, one of which is a vMF and the other is an equally weighted mixture of two different vMFs. The locations μ_j are chosen such that both the groups have the same extrinsic mean μ_1.

We draw 10 samples of 50 observations each from the model f_t and carry out hypothesis testing to test for association between X and Y via the Bayesian method and the nonparametric (frequentist) asymptotic chi-squared test. The prior, hyperparameters, and the algorithm for the Bayes factor (BF) computation are as in the earlier example. In each case the p-values are insignificant, often over 0.5, but with very high BFs, often exceeding 10^6. The values are listed in Table 14.2.

Table 14.2 *Nonparametric Bayes and frequentist test results for 10 simulations of 50 observations each for two groups with same population means.*

BF	6.1e9	6.4e8	1.3e9	4.3e8	703.1
p-value	1.00	0.48	0.31	0.89	0.89
BF	4.4e7	42.6	4.7e6	1.9e6	379.1
p-value	0.49	0.71	0.53	0.56	0.60

The reason for the failure of the frequentist test is because it relies on comparing the group-specific sample extrinsic means, and in this example the difference between them is little (by choice). The other method, on the other hand, compares the full conditionals and hence can detect differences that are not in the means.

14.5.4 Magnetization direction data

In this example from Embleton and McDonnell (1980), measurements of remanent magnetization in red silts and claystones were made at four locations. This results in samples from four groups of directions on the sphere S^2; the sample sizes are 36, 39, 16, and 16. The goal is to compare the magnetization direction distributions across the groups and test for any significant difference. Figure 4.1, which shows the three-dimensional plot of the sample clouds, suggests no major differences. To test that statistically, we calculate the BF in favor of the alternative, as in Section 14.5.2. The estimated BF was nearly 1, suggesting no evidence in favor of the alternative hypothesis that the distribution of magnetization directions vary across locations.

To assess sensitivity to the prior specification, the analysis was repeated with different hyperparameter values of **a, b** equal to the proportions of samples within each group and P_{01} corresponding to a uniform on the sphere. In addition, different starting clusterings in the data were tried, with a default choice obtained by implementing k-means with 10 clusters. In each case, we obtain BFs approximately 1, so the results were robust.

In Example 7.7 of Fisher et al. (1987), a coordinate-based parametric test was conducted to compare mean direction in these data, producing a p-value of $1 - 1.4205 \times 10^{-5}$ based on a $\mathcal{X}_6^2$ statistic. These authors also compared the mean directions for the first two groups and obtained a nonsignificant p-value. Repeating this two-sample test using our Bayesian nonparametric method, we obtained a Bayes factor of 1.00. The nonparametric frequentist test from Section 4.9.1 yields p-values of 0.06 and 0.38 for the two tests.

14.5.5 Volcano location data

Consider the data analyzed in Section 4.9.2. We are interested in testing if there is any association between the location and type of the volcano. Considered are the three most common types: strato, shield, and submarine volcanoes, with data available for 999 volcanoes of these types worldwide. Their location coordinates are shown in Figure 4.3. Denoting by X the volcano location that lies on S^2 and by Y its type, which takes values from $\mathbb{Y} = \{1, 2, 3\}$, we compute the BF for testing if X and Y are independent.

As should be apparent from Figures 4.2 and 4.3, the volcano data are particularly challenging in terms of density estimation because the locations tend to be concentrated along fault lines. Potentially, data on distance to the closest fault could be utilized to improve performance, but we do not have access to such data. Without such information, the data present a challenging test case for the methodology in that it is clear that one may need to utilize very many vMF kernels to accurately characterize the density of volcano locations across the globe, with the use of moderate to large numbers of kernels leading to challenging mixing issues. Indeed, we did encounter a sensitivity to the starting cluster configuration in our initial analysis.

We found that one of the issues that exacerbated the problem with mixing of the cluster allocation was the ordering in the weights on the stick-breaking representation utilized by the exact block Gibbs sampler. Hence, the computational algorithm was slightly modified to use, instead, the finite Dirichlet approximation to the Dirichlet process proposed in Ishwaran and Zarepour (2002). The finite Dirichlet treats the components as exchangeable and, therefore, eliminates sensitivity to the indices on the starting clusters, which were obtained using k-means for 50 clusters. We used $K = 50$ as the dimension of the finite Dirichlet and hence the upper bound on the number of occupied clusters. Another issue that led to mixing problems was the use of a hyperprior on ϕ. In particular, when the initial clusters were not well chosen, the kernel precision would tend to drift toward smaller than optimal values, and as a result too few clusters would be occupied to adequately fit the data. To address this issue, we chose and fixed the kernel precision parameter ϕ by cross-validation. In particular, the sample was split into training and test sets, and then the Bayesian nonparametric analysis on the training data was run separately for a wide variety of ϕ values between 0 and 1000. We chose the value that produced the highest expected posterior log likelihood in the test data, leading to $\hat{\phi} = 80$. In this analysis and the subsequent analyses for estimating the BF, the prior on the mixture weights was chosen to be $\text{Diri}(w_0/K\mathbf{1}_K)$ ($K = 50$). The other hyperparameters were chosen to be $w_0 = 1$, $\mathbf{a} = \mathbf{b} = (0.71, 0.17, 0.11) =$ the sample proportions of different volcano types, $P_{01} = \text{vMF}(\mu_0, 10)$, and μ_0 the X-sample extrinsic mean. We collected 5×10^4 MCMC iterations after discarding a burn-in of 10^4. Using a fixed bandwidth considerably improved the algorithm convergence rate.

Based on the complete data set of 999 volcanoes, the resulting BF in favor of the alternative was estimated to be more than 10^{100}, providing conclusive evidence that the different types of volcanos have a different spatial distribution across the globe. For the same fixed $\hat{\phi}$ value, the analysis was repeated for a variety of alternative hyperparameter values and different starting points, obtaining similar BF estimates and the same conclusion. We also repeated the analysis for a randomly selected subsample of 300 observations, obtaining BF $= 5.4 \times 10^{11}$. The testing is repeated for other subsamples, each resulting in a very high BF.

For comparison, we perform the asymptotic $\mathcal{X}^2$ test as described in Section 4.9.2, obtaining a p-value of 3.6×10^{-7}, which again favors H_1. The large sample sizes for the three types (713, 172, 114) justify the use of asymptotic theory.

We also perform a coordinate-based test by comparing the means of the latitude–longitude coordinates of the three subsamples using a $\mathcal{X}^2$ statistic. The three coordinate means are $(12.6, 27.9)$, $(21.5, 9.2)$, and $(9.97, 21.5)$ (latitude, longitude). The value of the statistic is 17.07 and the asymptotic p-value equals 1.9×10^{-3}, which is larger by orders of magnitude than its coordinate-free counterpart, but still significant. Coordinate-based methods, however, can be very misleading because of the discontinuity at the boundaries. They heavily distort the geometry of the sphere, which is evident from the figures.

14.6 Proofs of Proposition 14.1; Theorems 14.2, 14.3, and 14.6; and Corollary 14.4

Proof of Proposition 14.1 Denote by g the X-marginal density corresponding to f. Express $\mathrm{E}_{f_t}\left|\dfrac{f(Y|X)}{f_t(Y|X)} - 1\right|$ as

$$\int_{\mathbb{X}\times\mathbb{Y}} \left| \frac{f(y|x)}{f_t(y|x)} - 1 \right| f_t(x,y)\lambda(dx\,dy)$$

$$= \int_{\mathbb{X}\times\mathbb{Y}} |f_t(x,y) - f(y|x)g_t(x)|\lambda(dx\,dy)$$

$$= \int_{\mathbb{X}\times\mathbb{Y}} |f_t(x,y) - f(x,y) + f(y|x)g(x) - f(y|x)g_t(x)|\lambda_1(dx)\lambda_2(dy)$$

$$\leq \|f_t - f\|_{L^1} + \|g_t - g\|_{L^1} \leq 2\|f_t - f\|_{L^1}.$$

Hence any neighborhood of $f_t(.|.)$ of the form $\left\{ \mathrm{E}_{f_t} \left| \frac{f(Y|X)}{f_t(Y|X)} - 1 \right| < \epsilon \right\}$ contains an L^1 neighborhood of f_t. Therefore, strong consistency of the posterior distribution of f implies the desired result. □

Before proving Theorem 14.2, we prove the following lemma.

Lemma 14.7 *Under assumptions A3 and A5,*

$$\lim_{\phi\to\infty} \sup\{|f(x,y;P_t,\phi) - f_t(x,y)| : (x,y) \in \mathbb{X}\times\mathbb{Y}\} = 0,$$

with $f(x,y,P,\phi)$ defined in equation (14.3).

Proof From the definition of P_t we can write

$$f(x,y;P_t,\phi) = \int_{\mathbb{X}} K(x;\mu,\phi)g_y(\mu)\lambda_1(d\mu)$$

for $g_y(\mu) = f_t(\mu,y)$. Then from assumption A5 it follows that g_y is continuous for all $y \in \mathbb{Y}$. Hence, from assumption A3 it follows that

$$\lim_{\phi\to\infty} \sup_{x\in\mathbb{X}} \left| f_t(x,y) - \int_{\mathbb{X}} K(x;\mu,\phi)f_t(\mu,y)\lambda_1(d\mu) \right| = 0$$

for any $y \in \mathbb{Y}$. Since $\mathbb{Y}$ is finite, the proof is complete. □

Proof of Theorem 14.2. Throughout this proof we will view $\mathcal{M}(\mathbb{X}\times S_{c-1})$ as a topological space under the weak topology. From Lemma 14.7 it follows that there exists a $\phi_t \equiv \phi_t(\epsilon) > 0$ such that

$$\sup_{x,y} |f(x,y;P_t,\phi) - f_t(x,y)| < \frac{\epsilon}{3} \qquad (14.16)$$

for all $\phi \geq \phi_t$. From assumption A4, it follows that by choosing ϕ_t sufficiently large, we can ensure that $(P_t,\phi_t) \in \mathrm{supp}(\Pi_1)$. From assumptions A1 and A2 it follows that K is uniformly continuous at ϕ_t, that is, there exists an open set $W(\epsilon) \subseteq \mathbb{R}^+$ containing ϕ_t such that

$$\sup_{x,\mu\in\mathbb{X}} |K(x;\mu,\phi) - K(x;\mu,\phi_t)| < \frac{\epsilon}{3}, \qquad \forall\phi \in W(\epsilon).$$

This in turn implies that, for all $\phi \in W(\epsilon)$, $P \in \mathcal{M}(\mathbb{X} \times S_{c-1})$,

$$\sup_{x,y} |f(x, y; P, \phi) - f(x, y; P, \phi_t)| < \frac{\epsilon}{3} \tag{14.17}$$

because the left expression in equation (14.17) is

$$\sup_{x,y} \left| \int v_y \{K(x; \mu, \phi) - K(x; \mu, \phi_t)\} P(d\mu \, dv) \right| \leq \sup_{x,\mu \in \mathbb{X}} |K(x; \mu, \phi) - K(x; \mu, \phi_t)|.$$

Because $\mathbb{X}$ is compact and $K(.; ., \phi_t)$ is uniformly continuous on $\mathbb{X} \times \mathbb{X}$, we can cover $\mathbb{X}$ by finitely many open sets $U_1, \ldots U_K$ such that

$$\sup_{\mu \in \mathbb{X}, x, \tilde{x} \in U_i} |K(x; \mu, \phi_t) - K(\tilde{x}; \mu, \phi_t)| < \frac{\epsilon}{12} \tag{14.18}$$

for each $i \leq K$. For fixed x, y, ϕ, $f(x, y; P, \phi)$ is a continuous function of P. Hence, for $x_i \in U_i$, $y = j \in \mathbb{Y}$,

$$\mathcal{W}_{ij}(\epsilon) = \{P \in \mathcal{M}(\mathbb{X} \times S_{c-1}) : |f(x_i, j; P, \phi_t) - f(x_i, j; P_t, \phi_t)| < \frac{\epsilon}{6}\},$$

$1 \leq i \leq K$, $1 \leq j \leq c$, define open neighborhoods of P_t. Let $\mathcal{W}(\epsilon) = \bigcap_{i,j} \mathcal{W}_{ij}(\epsilon)$, which is also an open neighborhood of P_t. For a general $x \in \mathbb{X}$, $y \equiv j \in \mathbb{Y}$, find a U_i containing x. Then, for any $P \in \mathcal{W}(\epsilon)$,

$$|f(x, y; P, \phi_t) - f(x, y; P_t, \phi_t)|$$
$$\leq |f(x, j; P, \phi_t) - f(x_i, j; P, \phi_t)| + |f(x_i, j; P, \phi_t) - f(x_i, j; P_t, \phi_t)|$$
$$+ |f(x_i, j; P_t, \phi_t) - f(x, j; P_t, \phi_t)|. \tag{14.19}$$

Denote the three terms to the right in equation (14.19) as T_1, T_2, and T_3. Because $x \in U_i$, it follows from equation (14.18) that $T_1, T_3 < \frac{\epsilon}{12}$. Because $P \in \mathcal{W}_{ij}(\epsilon)$, $T_2 < \frac{\epsilon}{6}$ by definition of $\mathcal{W}_{ij}(\epsilon)$. Hence $\sup_{x,y} |f(x, y; P, \phi_t) - f(x, y; P_t, \phi_t)| < \frac{\epsilon}{3}$. Therefore,

$$\mathcal{W}_2(\epsilon) \equiv \{P : \sup_{x,y} |f(x, y; P, \phi_t) - f(x, y; P_t, \phi_t)| < \frac{\epsilon}{3}\}$$

contains $\mathcal{W}(\epsilon)$. Because $(P_t, \phi_t) \in \text{supp}(\Pi_1)$ and $\mathcal{W}_2(\epsilon) \times W(\epsilon)$ contains an open neighborhood of (P_t, ϕ_t), therefore

$$\Pi_1(\mathcal{W}_2(\epsilon) \times W(\epsilon)) > 0.$$

Let $(P, \phi) \in \mathcal{W}_2(\epsilon) \times W(\epsilon)$. Then, for $(x, y) \in \mathbb{X} \times \mathbb{Y}$,

$$|f(x, y; P, \phi) - f_t(x, y)|$$
$$\leq |f(x, y; P, \phi) - f(x, y; P, \phi_t)| + |f(x, y; P, \phi_t) - f(x, y; P_t, \phi_t)|$$
$$+ |f(x, y; P_t, \phi_t) - f_t(x, y)|. \tag{14.20}$$

The first term to the right in equation (14.20) is $< \frac{\epsilon}{3}$ since $\phi \in W(\epsilon)$. The second one is $< \frac{\epsilon}{3}$ because $P \in W_2(\epsilon)$. The third one is also $< \frac{\epsilon}{3}$, which follows from equation (14.16). Therefore,

$$\Pi_1\Big(\{(P, \phi): \sup_{x,y} |f(x, y; P, \phi) - f_i(x, y)| < \epsilon\}\Big) > 0.$$

This completes the proof. $\qquad\qquad\qquad\qquad\qquad\qquad\qquad\qquad\qquad\qquad\qquad$ $\square$

Proof of Theorem 14.3 For a density $f \in \mathcal{D}(\mathbb{X} \times \mathbb{Y})$, let $p(y)$ be the marginal probability of Y being y and $f(x|y)$ be the conditional density of X at x given $Y = y$. Then p can be viewed as a vector in S_{c-1} while $f(\cdot|y) \in \mathcal{D}(\mathbb{X})$ $\forall y \in \mathbb{Y}$. Endow the density spaces $\mathcal{D}(\mathbb{X} \times \mathbb{Y})$ and $\mathcal{D}(\mathbb{X})$ with the respective total variation distances $\| \cdot \|$. Similarly equip S_{c-1} with the L_1 distance. For $f_1, f_2 \in \mathcal{D}(\mathbb{X} \times \mathbb{Y})$, $f_i(x, y) = p_i(y)f_i(x|y)$, $i = 1, 2$,

$$\|f_1 - f_2\| = \int |f_1(x, y) - f_2(x, y)|\lambda(dx\, dy)$$

$$= \sum_{j=1}^{c} \int_{\mathbb{X}} |p_1(j)f_1(x|j) - p_2(j)f_2(x|j)|\lambda_1(dx)$$

$$\leq \max_j \|f_1(.|j) - f_2(.|j)\| + \sum_j |p_1(j) - p_2(j)|. \qquad (14.21)$$

Hence an ϵ diameter ball in $\mathcal{D}(\mathbb{X} \times \mathbb{Y})$ contains the intersection of c many $\frac{\epsilon}{2}$ diameter balls from $\mathcal{D}(\mathbb{X})$ with a $\frac{\epsilon}{2}$ diameter subset of S_{c-1}.

Represent the class of joint densities of the form (14.3) by $\mathcal{D}_0$, that is,

$$\mathcal{D}_0 = \{f(\cdot\,; P, \phi) \in \mathcal{D}(\mathbb{X} \times \mathbb{Y}) : P \in \mathcal{M}(\mathbb{X} \times S_{c-1}),\ \phi \in \mathbb{R}^+\},$$

and define

$$\tilde{\mathcal{D}}_n = \bigcup_{j \in \mathbb{Y}} \{f(.|j) \in \mathcal{D}(\mathbb{X}) : f \in \mathcal{D}_0, \phi \in [0, n^a]\}.$$

Any element of $\tilde{\mathcal{D}}_n$ can be expressed as

$$f(x|j) = \frac{\int_{\mathbb{X} \times S_{c-1}} v_j K(x; \mu, \phi) P(d\mu\, dv)}{\int_{\mathbb{X} \times S_{c-1}} v_j P(d\mu\, dv)} = \int_{\mathbb{X}} K(x; \mu, \phi) P_j(d\mu)$$

$$\text{with } P_j(d\mu) = \frac{\int_{S_{c-1}} v_j P(d\mu\, dv)}{\int_{\mathbb{X} \times S_{c-1}} v_j P(d\mu\, dv)}.$$

Hence $f(\cdot|j)$ is as in equation (13.1) with $M = \mathbb{X}$. Therefore, from Theorem 13.4, under assumptions A6–A8, the ϵ L_1-metric entropy $N(\epsilon, \tilde{\mathcal{D}}_n)$ is of order at most $n^{aa_1 a_3}$, which is $o(n)$. Next define

$$\mathcal{D}_n = \{f \in \mathcal{D}_0 : \phi \in [0, n^a]\}.$$

By definition of $\tilde{\mathcal{D}}_n$,

$$\mathcal{D}_n = \{f \in \mathcal{D}(\mathbb{X} \times \mathbb{Y}) : f(\cdot|j) \in \tilde{\mathcal{D}}_n \quad \forall j \in \mathbb{Y}\}. \tag{14.22}$$

Hence from equations (14.21) and (14.22), $N(\epsilon, \mathcal{D}_n)$ is also $o(n)$. Therefore, Proposition 13.3 implies strong posterior consistency under assumptions A1–A9.   □

Proof of Corollary 14.4   (a) Note that

$$\int_{\mathbb{X}} |p(y,x) - p_t(y,x)| g_t(x) \lambda_1(dx)$$

$$= \int_{\mathbb{X}} |f_t(x,y) - f(x,y) + p(y,x)g(x) - p(y,x)g_t(x)| \lambda_1(dx)$$

$$\leq \int_{\mathbb{X}} |f_t(x,y) - f(x,y)| \lambda_1(dx) + \int_{\mathbb{X}} |g_t(x) - g(x)| \lambda_1(dx)$$

$$\leq 2 \int_{\mathbb{X}} |f(x,y) - f_t(x,y)| \lambda_1(dx),$$

and hence any neighborhood of $p_t(y, .)$ of the form $\{\int_{\mathbb{X}} |p(y,x) - p_t(y,x)| g_t(x)$ $\lambda_1(dx) < \epsilon\}$ contains an L^1 neighborhood of f_t. Now part (a) follows from strong consistency of the posterior distribution of f.

(b) Because $\mathbb{X}$ is compact, f_t being continuous and positive implies that $c = \inf_{x \in \mathbb{X}} g_t(x) > 0$. Hence

$$\int_{\mathbb{X}} |p(y,x) - p_t(y,x)| w(x) \lambda_1(dx)$$

$$\leq c^{-1} \sup(w(x)) \int_{\mathbb{X}} g_t(x) |p(y,x) - p_t(y,x)| \lambda_1(dx).$$

Now the result follows from part (a).   □

The proof of Theorem 14.6 uses Lemma 14.8. This lemma is fundamental to proving weak posterior consistency using the Schwartz theorem and its proof follows from the discussion in Section 13.3.1.

Lemma 14.8   *(a) If Π includes f_t in its KL support, then*

$$\lim \inf_{n \to \infty} \exp(n\beta) \int \prod_i \frac{f(x_i, y_i)}{f_t(x_i, y_i)} \Pi(df) = \infty$$

a.s. f_t^∞ for any $\beta > 0$. (b) If U is a weak open neighborhood of f_t and Π_0 is a prior on $\mathcal{D}(\mathbb{X} \times \mathbb{Y})$ with support in U^c, then there exists a $\beta_0 > 0$ for which

$$\lim_{n\to\infty} \exp(n\beta_0) \int \prod_i \frac{f(x_i, y_i)}{f_t(x_i, y_i)} \Pi_0(df) = 0$$

a.s. f_t^∞.

Proof of Theorem 14.6   Express BF as

$$\text{BF} = \left\{ \prod_i p_t(y_i) \right\} \frac{D(\mathbf{b})}{D(\mathbf{b}_n)} \frac{\int \prod_i \frac{f(x_i,y_i)}{f_t(x_i,y_i)} \Pi(df)}{\int \prod_i \frac{g(x_i)p_t(y_i)}{f_t(x_i,y_i)} \Pi(df)} = \frac{T_1 T_2}{T_3}$$

with $T_1 = \{\prod_i p_t(y_i)\} \frac{D(\mathbf{b})}{D(\mathbf{b}_n)}$, $T_2 = \int \prod_i \frac{f(x_i,y_i)}{f_t(x_i,y_i)} \Pi(df)$, and $T_3 = \int \prod_i \frac{g(x_i)p_t(y_i)}{f_t(x_i,y_i)} \Pi(df)$. Because Π satisfies the KL condition, Lemma 14.8(a) implies that $\liminf_{n\to\infty} \exp(n\beta)T_2 = \infty$ a.s. for any $\beta > 0$.

Let U be the space of all dependent densities, that is,

$$U^c = \{f \in \mathcal{D}(\mathbb{X} \times \mathbb{Y}) : f(x, y) = g(x)p(y) \text{ a.s. } \lambda(dx\,dy)\}.$$

The prior Π induces a prior Π_0 on U^c via $f \mapsto \{\sum_j f(\cdot, j)\}p_t$ and T_3 can be expressed as $\int \prod_i \frac{f(x_i,y_i)}{f_t(x_i,y_i)} \Pi_0(df)$. It is easy to show that U is open under the weak topology and hence under H_1 is a weak open neighborhood of f_t. Then, using Lemma 14.8(b), it follows that $\lim_{n\to\infty} \exp(n\beta_0)T_3 = 0$ a.s. for some $\beta_0 > 0$.

The proof is complete if we can show that $\liminf_{n\to\infty} \exp(n\beta)T_1 = \infty$ a.s. for any $\beta > 0$ or $\log(T_1) = o(n)$ a.s. For a positive sequence a_n diverging to ∞, Stirling's formula implies that $\log\Gamma(a_n) = a_n \log(a_n) - a_n + o(a_n)$. Express $\log(T_1)$ as

$$\sum_i \log(p_t(y_i)) - \log(D(\mathbf{b}_n)) + o(n). \tag{14.23}$$

Because $p_t(j) > 0 \; \forall j \le c$, by the SLLN,

$$\sum_i \log(p_t(y_i)) = n \sum_j p_t(j) \log(p_t(j)) + o(n) \text{ a.s.} \tag{14.24}$$

Let $b_{nj} = b_j + \sum_i I(y_i = j)$ be the jth component of $\mathbf{b}_n$. Then $\lim_{n\to\infty} b_{nj}/n = p_t(j)$, that is, $b_{nj} = np_t(j) + o(n)$ a.s., and hence Stirling's formula implies that

$$\log(\Gamma(b_{nj})) = b_{nj} \log(b_{nj}) - b_{nj} + o(n)$$
$$= np_t(j) \log(p_t(j)) - np_t(j) + \log(n)b_{nj} + o(n) \text{ a.s.},$$

which implies that

$$\log(D(\mathbf{b}_n)) = \sum_{j=1}^{L} \log(\Gamma(b_{nj})) - \log \Gamma\left(\sum_j b_j + n \right)$$

$$= n \sum_j p_t(j) \log(p_t(j)) + o(n) \text{ a.s.} \qquad (14.25)$$

From equations (14.23), (14.24), and (14.25), $\log(T_1) = o(n)$ a.s. follows and this completes the proof. $\square$

Appendix

A Differentiable manifolds

A *d-dimensional differentiable manifold* M is a separable metric space with the following properties:

(i) Every $p \in M$ has an open neighborhood U_p and a homeomorphism $\psi_p : U_p \to B_p$, where B_p is an open subset of $\mathbb{R}^d$.

(ii) The maps ψ_p are (smoothly) compatible; that is, if $U_p \cap U_q \neq \phi$, then $\psi_p \circ \psi_q^{-1}$ is a C^∞ (infinitely differentiable) diffeomorphism on $\psi_q(U_p \cap U_q)$ ($\subseteq B_q$) onto $\psi_p(U_p \cap U_q)$ ($\subseteq B_p$).

The pair (U_p, ψ_p) in (ii) is called a *coordinate neighborhood* of p, and $\psi_p(p') = (x_1(p'), \ldots, x_d(p'))$, $p' \in U_p$, are sometimes referred to as *local coordinates* of p'. The collection $\{(U_p, \psi_p) : p \in M\}$ is called an *atlas* or a *differential structure* of M. In general, there are many atlases that are compatible with a given atlas or differential structure. One therefore defines a differentiable manifold as given by a maximal atlas, that is, the collection of all coordinate neighborhoods compatible with a given one of interest. Property (ii) of a differentiable manifold M allows one to extend differential calculus on a Euclidean space to M, as we shall see next.

Unless otherwise specified, we assume the manifold M is connected.

Given two differentiable manifolds M, N of dimensions d and k, and atlases $\{(U_p, \psi_p) : p \in M\}$, $\{(V_q, \phi_q) : q \in N\}$, respectively, a function $g : M \to N$ is said to be *r-times continuously differentiable*, in symbols $g \in C^r(M \to N)$, if, for each $p \in M$ and $q = g(p)$, g is *r*-times continuously differentiable when expressed in local coordinates $\psi_p(p') = (x_1(p'), \ldots, x_d(p'))$ for $p' \in U_p$ and $\phi_q(q') = (y_1(q'), \ldots, y_k(q'))$ for $q' \in V_q$. That is, assuming without loss of generality that $g(U_p) \subset V_q$, the function $h(x_1, \ldots, x_d) \equiv \phi_q \circ g \circ \psi_p^{-1}(x_1, \ldots, x_d)$ is *r*-times continuously differentiable on $\psi_p(U_p) \subset \mathbb{R}^d$ into $\phi_q(V_q) \subset \mathbb{R}^k$. If this holds for all positive integers r, then g is infinitely differentiable: $g \in C^\infty(M \to N)$. If $N = \mathbb{R}$, one simply writes g is C^r or $g \in C^r(M)$, g is C^∞ or $g \in C^\infty(M)$, and so on. The set of all real-valued C^∞ functions on M is denoted by $C^\infty(M)$. In view of (ii), this definition of differentiability does not depend on the particular coordinate neighborhoods chosen for p and q.

For the extension to a manifold of the notion of derivatives of a function f on $\mathbb{R}^d$ as providing local linear approximations, and for various other purposes to be encountered, one needs to introduce the notion of tangent vectors and tangent spaces. One way to introduce them is to consider a C^1-function γ on an interval $(-a, a)$, $a > 0$, taking values in a manifold M. Then let $\gamma: (-a, a) \to M$ be a continuously differentiable function (curve), with $\gamma(0) = p$. Expressing γ in local coordinates, $x(t) \equiv \psi_p \circ \gamma(t) = (x_1(t), \ldots, x_d(t))$, say, is a differentiable curve in $\mathbb{R}^d$, with a tangent vector at $\psi_p(p)$ given by $x'(0) = ((d/dt)x_1(t), \ldots, (d/dt)x_d(t))_{t=0} = \lim_{t\downarrow 0}\{x(t) - x(0)\}/t$. For $f \in C^1(M)$, $f \circ \gamma$ is a real-valued C^1 function on $(-a, a)$, whose derivative at 0 is well defined and is given by

$$\tau_p(f) = (d/dt)\{f \circ \gamma(t)\}_{t=0}$$
$$= (d/dt)\{f \circ \psi_p^{-1} x(t)\}_{t=0} = \langle x'(0), \text{grad}(f \circ \psi_p^{-1})\{x(0)\}\rangle. \quad \text{(A.1)}$$

Here $\text{grad}(g)\{x(0)\} = (\partial g(x)/\partial x_1, \ldots, \partial g(x)/\partial x_d)_{x=x(0)}$ and $\langle \cdot, \cdot \rangle$ denotes the Euclidean inner product on the appropriate tangent space $T_{x(0)}\mathbb{R}^d$ (which may be identified with $\mathbb{R}^d$ in the present case). Note that τ_p is linear in f on the vector space $C^1(M)$; it depends only on the derivative $x'(0)$ of the curve $x(t)$ at $t = 0$, and it is determined by it, although there are infinitely many C^1-curves γ with the same derivative (of $x(t)$) at 0. The linear function τ_p is called a *tangent vector* at p. In local coordinates, it is the directional derivative at p in the direction $x'(0)$. The set of all such vectors is a d-dimensional vector space, called the *tangent space* at p, denoted as T_pM or $T_p(M)$, or simply T_p when the manifold M is clearly specified from the context. Given a coordinate neighborhood (U_p, ψ_p) of p, in local coordinates, T_p is spanned by the basis $\{\partial/\partial x_1, \ldots, \partial/\partial x_d\}_{x=x(0)}$, that is, by the basis of derivatives in the directions $\{e_i : i = 1, \ldots, d\}$, where e_i has 1 as its ith coordinate and 0s as the remaining $d - 1$ coordinates.

The linear functional τ_p on the vector space $C^1(M)$, defined by equation (A.1), clearly satisfies the Leibnitz rule for differentiation of products of functions: $\tau_p(fg) = \tau_p(f)g + f\tau_p(g)$ on $C^1(M)$. This is easily checked by observing that $(fg) \circ \psi_p^{-1} = (f \circ \psi_p^{-1})(g \circ \psi_p^{-1})$ and applying the usual Leibnitz rule in the last equality in equation (A.1). Conversely, one can show that if a linear functional on $C^1(M)$ satisfies the Leibnitz rule, then it is a tangent vector at p, in the sense defined by equation (A.1) (see, e.g., Boothby, 1986, chapter IV).

The definition of a tangent vector as given by the first relation in equation (A.1) does not depend on the coordinate system chosen, but its representation in terms of $x'(0)$ does. As we shall see, one can relate representations such as those given in equation (A.1) in two different coordinate systems by a linear map, or a Jacobian, on the tangent space T_p.

Example A.1 Some common examples of manifolds are the *regular submanifolds of a Euclidean space* $\mathbb{R}^n$, defined as the set $M = \{h(x) = 0 : x \in \mathbb{R}^n\}$, where $h(x) = (h^1(x), \ldots, h^{n-d}(x))$ is an infinitely differentiable map on an open subset V of $\mathbb{R}^n$ into $\mathbb{R}^{n-d}$ ($1 \le d < n$), and $\text{Grad } h(x)$ is of full rank $n-d$. Here $\text{Grad } h(x)$ is the $(n - d) \times n$ matrix $\{(\partial h^i(x)/\partial x_j)\}_{1 \le i \le n-d, 1 \le j \le n}$, whose rows are $(\text{grad}(h^i(x)))_{1 \le i \le n-d}$. It follows from the implicit function theorem that, with the relative topology of $\mathbb{R}^n$, M is a d-dimensional differentiable manifold, that is, it satisfies both the defining properties (i) and (ii) stated at the outset, if one chooses an atlas $\{(U_x, \psi_x) : x \in M\}$ where $U_x = O_x \cap M$, with O_x a sufficiently small open ball in $\mathbb{R}^n$ centered at x, and ψ_x is the restriction to U_x of a C^∞ diffeomorphism θ_x of O_x onto an open set $\theta_x(O_x) \subset \mathbb{R}^n$, such that $\theta_x(U_x)$ is diffeomorphic to an open subset B_x, say, of $\mathbb{R}^d$.

For submanifolds, to find/represent the tangent space T_x at $x \in M = \{h(x) = 0 : x \in \mathbb{R}^n\}$, one may proceed directly. Let $\gamma : (-a, a) \to \mathbb{R}^n$ be a differentiable curve in $\mathbb{R}^n$ with $\gamma(t) = (x_1(t), \dots, x_n(t)) = x(t) \in M$, $x(0) = x$. That is, γ is also a differentiable curve in M, with $\gamma(0) = x$. Then the relations $h^i(x(t)) = 0$, $1 \leq i \leq n - d$, yield, on differentiation, $\langle \text{grad } h^i(x(0)), x'(0) \rangle = 0$, $i = 1, \dots, n - d$. Thus the tangent vector at $x = x(0)$ (represented by a vector in the tangent space of $\mathbb{R}^n$ at x) is orthogonal to the $n - d$ linearly independent vectors grad $h^i(x)$, $1 \leq i \leq n - d$. Thus the d-dimensional tangent space T_x of M at x is represented by the d-dimensional subspace of the tangent space of $\mathbb{R}^n$ at x orthogonal to grad $h^i(x)$, $1 \leq i \leq n - d$.

A special submanifold of interest is the d-dimensional sphere $S^d = \{x \in \mathbb{R}^{d+1} : x_1^2 + \cdots + x_{d+1}^2 = 1\}$. It follows from the above that the tangent space $T_x S^d$ may be represented as the d-dimensional linear subspace of $T_x \mathbb{R}^{d+1} \equiv \mathbb{R}^{d+1}$ spanned by vectors (in $\mathbb{R}^{d+1}$) orthogonal to x, since Grad $h(x)$ here equals $2x$.

We now turn to the notion of the *differential* of a map $h \in C^1(M \to N)$, where M, N are differentiable manifolds of dimensions n and k, respectively. First consider the case $M = \mathbb{R}^n$ and $N = \mathbb{R}^k$. The local linear approximation of h in a neighborhood of a point x_0 in $\mathbb{R}^n$ is given by the linear map represented by the *Jacobian matrix* $J(x) = [((\partial h^i(x)/\partial x_j))_{1 \leq i \leq k, 1 \leq j \leq n}]_{x=x_0}$, writing $h(x) = (h^1(x), \dots, h^k(x))'$. Given a vector $v \in \mathbb{R}^n$, one has the approximation $h(x_0 + v) \approx h(x_0) + J(x_0)v$ (treating $h(x)$ and v as $(k \times 1)$ and $(n \times 1)$ column vectors). One should think of v, $J(x_0)v$ as tangent vectors: $v \in T_{x_0}\mathbb{R}^n$ and $J(x_0)v \in T_{h(x_0)}\mathbb{R}^k$. The transformation $v \to J(x_0)v$ defines a linear map: $T_{x_0}\mathbb{R}^n \to T_{h(x_0)}\mathbb{R}^k$, called the *differential of h at x_0*, denoted as $d_{x_0}h$.

For general differentiable manifolds M, N, let $h \in C^1(M \to N)$. If $f \in C^1(N)$, a continuously differentiable real-valued function on N, then $f \circ h \in C^1(M)$, and one may use a tangent space approximation of $f \circ h$ near a point $p \in M$, using the differential $d_p h : T_p M \to T_{h(p)} N$ formally defined as the linear map $d_p h(\tau) = \eta$ ($\tau \in T_p M$, $\eta \in T_{h(p)} N$), where

$$\eta(f) = \tau(f \circ h), \quad \forall f \in C^1(N). \tag{A.2}$$

Note that $f \circ h$ is linear in f and τ is linear, and hence the left side is a linear function of $f \in C^1(N)$ that obeys the Leibnitz rule and, therefore, defines a tangent vector $\eta \in T_{h(p)} N$ equation [see (A.1)]. In terms of our more explicit definition of tangent vectors, consider a tangent vector $\tau \in T_p M$ defined by a C^1 curve γ passing through $p = \gamma(0)$. Then $\tilde{\gamma} \equiv h \circ \gamma$ is a C^1 curve passing through $q = h(p) = \tilde{\gamma}(0)$. Let (U, ψ) and (V, ϕ) be coordinate neighborhoods of p and $q = h(p)$, respectively. Writing $x(t) = \psi \circ \gamma(t)$ and $y(t) = \phi \circ h \circ \gamma(t) = \phi \circ h \circ \psi^{-1} \circ x(t)$, the tangent vector $\eta \in T_{h(p)} N$ is given in local coordinates by $y'(0)$, namely,

$$y'(0) = (d/dt)(\phi \circ h \circ \psi^{-1} \circ x(t))_{t=0} = J(x(0))x'(0), \tag{A.3}$$

where $J(x(0))$ is the Jacobian at $x(0)$ of the transformation $\tilde{h} \equiv \phi \circ h \circ \psi^{-1}$ on $\psi(U) \subset \mathbb{R}^d$ into $\phi(V) \subset \mathbb{R}^k$, given by $[((\partial(\tilde{h})^i(x)/\partial x_j))_{1 \le i \le k, 1 \le j \le d}]_{x=x(0)}$. Thus, in local coordinates, the differential of h is given by the linear map $J(x(0))$ on $T_{x(0)}U$ (identified with $\mathbb{R}^d$) into $T_{h(x(0))}V$ (identified with $\mathbb{R}^k$). For $f \in C^1(N)$, and with $\tilde{\gamma} = h \circ \gamma$ in place of γ in equation (A.1), one obtains

$$
\begin{aligned}
\eta(f) &= (d/dt)\{f \circ \tilde{\gamma}(t)\}_{t=0} = (d/dt)\{f \circ \phi^{-1}(y(t))\}_{t=0} \\
&= \langle y'(0), \mathrm{Grad}(f \circ \phi^{-1})(y(0)) \rangle \\
&= \langle J(x(0))x'(0), \mathrm{Grad}(f \circ \phi^{-1})(y(0)) \rangle \\
&= \langle x'(0), J(x(0))^t \mathrm{Grad}(f \circ \phi^{-1})(y(0)) \rangle \\
&= \langle x'(0), J(x(0))^t \mathrm{Grad}(f \circ \phi^{-1})(\phi \circ h \circ \psi^{-1}(x(0))) \rangle \\
&= \langle x'(0), \mathrm{Grad}(f \circ h \circ \psi^{-1})(x(0)) \rangle, \tag{A.4}
\end{aligned}
$$

where A^t denotes the transpose of a matrix A. Writing $\tilde{h} = \varphi \circ h \circ \psi^{-1}$, the last equality follows from the rule for differentiating the composite function:

$$
g(x) = (f \circ h \circ \psi^{-1})(x) = (f \circ \phi^{-1}) \circ (\phi \circ h \circ \psi^{-1})(x) \equiv (f \circ \phi^{-1}) \circ \tilde{h}(x),
$$

$$
\begin{aligned}
\partial g(x)/\partial x_i &= \sum_{j=1}^{k} \{\partial(f \circ \phi^{-1})(y)/\partial y_j\}_{y=\tilde{h}(x)} \partial \tilde{h}^j(x)/\partial x_i \\
&= \sum_{j=1}^{k} \{\partial(f \circ \phi^{-1})(y)/\partial y_j\}_{y=\tilde{h}(x)} (J(x))_{ji}.
\end{aligned}
$$

The last expression of equation (A.4) equals $\tau(f \circ h)$ (see equation (A.1)), establishing equation (A.2).

A differentiable manifold M is said to be *orientable* if it has an atlas $\{(U_p, \psi_p); p \in M\}$ such that the determinant of the map $\psi_p \circ \psi_q^{-1}$, defined at the beginning, has a positive determinant for all p, q such that $U_p \cap U_q$ is not empty. For such a manifold, one can also easily find an atlas such that the above maps have negative determinants. These positive and negative orientations are the only possibilities on an orientable manifold. There are many examples of nonorientable manifolds (see Do Carmo, 1992).

Next, consider the notion of a *vector field* on a manifold M, which is a smooth assignment $p \mapsto \tau_p$ of tangent vectors, or velocities, on M. On the Euclidean space $E^d \approx \mathbb{R}^d$, such an assignment is determined by a smooth vector-valued function $u(x) = (u_1(x), \dots, u_d(x))$, $x \in \mathbb{R}^d$, with $\tau_p(x) = \sum u_i(x)\partial/\partial x_i$. Given such a vector (or velocity) field, the path $x(t)$ of a particle starting at a given point x_0 is determined, at least in a neighborhood of $t = 0$, governed by the ordinary differential equation $dx(t)/dt = u(x(t))$, $x(0) = x_0$. Note that for smooth functions f on $\mathbb{R}^d$, one has $df(x(t))/dt = \sum u_i(x(t))(\partial f/\partial x_i)(x(t))$, that is, $\tau_{x(t)}(f) = \sum u_i(x(t))(\partial f/\partial x_i)(x(t))$. This is possible because one has a well-defined field of basis vectors $\{\partial/\partial x_i, i = 1, \dots, d\}$, or a *tangent frame*, on all of $\mathbb{R}^d$. Because tangent vectors at different points are not naturally related to each other on a general manifold M, to define the smoothness of such an assignment $q \mapsto \tau_q$, one needs to introduce a differential structure on the *tangent bundle* $TM = \{(q, \tau_q) : q \in M, \tau_q \in T_q M\}$. This is determined by the coordinate maps $\Psi_p : \{(q, \tau_q) : q \in U_p, \tau_q \in T_q M\} \equiv TU_p \mapsto B_p \times R_d$, defined by $\Psi_p(q, \tau_q) = (x, u)$, where $x = \psi_p(q)$ and $u = u(q) = (u_1(q), \ldots, u_d(q))$ is determined by $d\psi_p(\tau_q) = \sum u_i(q) \partial/\partial x_i$. Here $\{\partial/\partial x_i, i = 1, \ldots, d\}$ is the Euclidean tangent frame on B_p. It is easy to check that this defines a differential structure on TM, satisfying conditions (i) and (ii), making it a $2d$-dimensional differentiable manifold. This also defines a tangent frame $\{E_{i,p} : i = 1, \ldots, d\}$ on U_p given by $E_{i,p} = d\psi_p - 1(\partial/\partial x_i), i = 1, \ldots, d$, corresponding to the frame $\{\partial/\partial x_i, i = 1, \ldots, d\}$ on B_p. We will refer to $\{E_{i,p} : i = 1, \ldots, d\}$ as the *coordinate frame* on U_p. A *vector field* W on M is now defined as a C^∞ map on M into TM of the form $q \mapsto (q, \tau_q)$, that is, a smooth section of TM. That is, in local coordinates, for each p, the vector field $\sum u_i(\psi_p^{-1}(x))\partial/\partial x_i$ on B_p is smooth: $x \mapsto u_i \circ \psi_p^{-1}(x)$ is C^∞ for each $i = 1, \ldots, d$.

For the final notion of this section, consider differentiable manifolds M and N of dimensions d and k, respectively, $k \geq d$ (usually, $k > d$). One defines $j \in C^\infty(M \to N)$ to be an *embedding* of M into N, if j is a homeomorphism onto $j(M)$ with its relative topology in N, and its differential $d_p j$ is injective (i.e., one-to-one) on $T_p M$ into $T_{j(p)} N$ for every $p \in M$. We will be mostly interested in embeddings of M into a Euclidean space N. Simple examples of such embeddings are provided by regular submanifolds as considered under Example 1, with j as the inclusion map.

Because most of our manifolds in this book are compact, the following simple lemma is useful.

Lemma A.2 *Let M be a compact differentiable manifold, and let $F \in C^\infty(M \to N)$ be a one-to-one map whose differential $d_p F$ is injective at every $p \in M$. Then F is an embedding.*

Proof Because F is continuous and one-to-one, to establish that F is a homeomorphism on M onto $F(M)$, it is enough to show that F^{-1} is continuous. The continuous image of a compact set under a continuous map is compact. Hence $F(M)$ is compact and, therefore, so is every closed subset of $F(M)$. The inverse image under F^{-1} of a closed and, therefore, compact set C of M is $F(C)$, a compact and, therefore, closed subset of $F(M)$. This proves F^{-1} is continuous. □

It can be shown that if $F \in C^\infty(M \to \mathbb{R}^k)$ is an embedding, then $F(M)$ is a regular submanifold of $\mathbb{R}^k$ (see, e.g., Boothby, 1986, p. 68). We will, however, directly establish this submanifold property for our special manifolds.

The shape spaces of special interest in this book are not regular submanifolds, defined directly by an inclusion map in a Euclidean space $\mathbb{R}^k$. Instead, they are often quotients of a high-dimensional sphere S^d under the action of a (Lie) group G acting on it. In general, a *Lie group G* is a group that is also a manifold such that the group operation of multiplication $(g_1, g_2) \to g_1 g_2$ is $C^\infty(G \times G \to G)$ and the inverse operation $g \to g^{-1}$ is $C^\infty(G \to G)$. We also allow the group G to be a *discrete group*, that is, G is countable and has the discrete topology, which is thought of as a manifold of dimension zero. The groups G are *groups of transformations*, that is, maps g on M, with $g_1 g_2$ as the *composition* $g_1 \circ g_2$ of the maps g_1 and g_2. That is, each g in G is a one-to-one map $g: p \to gp$ on M onto M. One requires that $(g, p) \to gp$ is $C^\infty(G \times M \to M)$. If G is discrete, this simply means that each map $g: p \to gp$ is $C^\infty(M \to M)$. The *quotient space* M/G is the space whose elements are the *orbits* $O_p = \{gp : g \in G\}$, $p \in M$. Equivalently, M/G is the space of *equivalence classes* of elements of M, where the equivalence relation $\sim$ is given by $p \sim q$ if $q = gp$ for some $g \in G$, that is, if p and q belong to the same orbit. For the *quotient topology* of M/G, a set $V \subset M/G$ is defined to be *open* if the set of orbits in V comprises an open subset of M. We will generally assume that the map $p \to O_p$ $(M \to M/G)$ is an *open map*. That is, if $U \subset M$ is open, then the set $\{O_p, p \in U\}$ is an open subset of M/G, that is, the union of the orbits $\{O_p, p \in U\}$ is open as a subset of M. The following lemma indicates the possibility of M/G being a manifold. Its proof may be found in Boothby (1986).

Lemma A.3 *If G is a compact Lie group acting freely on M, then M/G is a differentiable manifold.*

For each specific case of interest in this monograph, the manifold structure of M/G is explicitly constructed.

B Riemannian manifolds

In this section we sketch briefly some of the basic concepts of Riemannian geometry and refer the reader to Do Carmo (1992), Boothby (1986), and Gallot et al. (1990) for details and many additional properties. Our main objective here is to describe and explain those notions that appear in the statements and proofs of the main results of intrinsic analysis in this monograph.

On a differentiable manifold M one may define a distance metrizing its topology in many ways. For extrinsic analysis, this distance is chosen to be that induced by an appropriate embedding in a Euclidean space. This is, however, not enough to measure lengths of curves, accelerations of a particle moving along a curve, or quantizing the shape of M by notions of curvature. For these, one first needs a *metric tensor* – a smooth symmetric positive definite bilinear form on the tangent bundle TM. This form provides an inner product $\langle \cdot, \cdot \rangle_p$ on each tangent space $T_p M$. In local coordinates, for the coordinate frame $\{E_i = d\psi_p^{-1}(\partial/\partial x_i) : i = 1, \ldots, d\}$ on U_p, one can define on $B_p = \psi_p(U_p)$ the functions $g_{ij}(x) = \langle E_i, E_j \rangle(q)$ $(i, j = 1, \ldots, d)$, where $x = \psi_p(q)$. The smoothness of the bilinear form means that these functions are C^∞ on B_p for all i, j. It is simple to check that this property does not depend on the particular coordinate system chosen. A differentiable manifold M endowed with a metric tensor is called a *Riemannian manifold* and is denoted by (M, g), where g is the metric tensor as expressed above. There are again many ways to construct such a metric tensor on M. For regular submanifolds M of a Euclidean space E^k, a natural metric is a smooth restriction of the Euclidean metric on E^k to TM, regarding the tangent space T_pM as a subset of the tangent space $T_p(E^k) \approx \mathbb{R}^k$. This, for example, is the metric tensor generally used on S^d ($k = d + 1$). The shape manifolds considered in this monograph are not directly regular Euclidean submanifolds. But they are often quotients of S^d under a Lie group G of isometries of S^d. Here an *isometry* g on M is a diffeomorphism of M onto itself, such that the metric tensor is invariant under the differential dg_p: $\langle dg_p(u), dg_p(v) \rangle_{g(p)} = \langle u, v \rangle_p$ for all $u, v \in T_pM$ and all $p \in M$. On S^d, the isometries are given by orthogonal transformations, that is, by $gp = Op$, where O is a $(d + 1) \times (d + 1)$ orthogonal matrix operating on vectors in $\mathbb{R}^{d+1}$. The group G is then a closed subgroup of the group $O(d + 1)$ of all orthogonal transformations.

Let (M, g) be a Riemannian manifold in the ensuing discussion. We will write $|u| = \langle u, u \rangle_p^{1/2}$ for the length, or norm, of a vector $u \in T_p(M)$. The length of a smooth curve $\gamma(t)$, $a \leq t \leq b$, is given by $s(t) = \int_{[a,b]} \|d\gamma(t)/dt\| dt$. This definition does not depend on the parametrization of the curve, as is easily checked. Consider now the class of all smooth curves from $p = \gamma(a)$ to $q = \gamma(b)$ for a given pair of points p, q in M. The minimum length of these curves is taken to be the distance between p and q, denoted $d_g(p, q)$ also called the *geodesic distance* between p and q. In this book it is also called the *intrinsic distance*. The variational equation for this minimization turns out to be that of "zero acceleration." From now on, unless stated otherwise, we will parametrize a curve by its length s, or a (positive) constant multiple of it. In this case, $\|\dot{\gamma}(t)\| = 1$ for all t. To understand the appropriate notion of a second derivative of $\gamma(t)$, that is, the first derivative of $\dot{\gamma} = (d\gamma(t)/dt)$ on M, consider the case of a regular submanifold M of E^k ($k > d$). The velocity $\dot{\gamma}(t)$ lies in $T_{\gamma(t)}M$ contained in $T_{\gamma(t)}E^k$, by the definition of tangent vectors. However, the derivative of $\dot{\gamma}(t)$ does not, in general, lie in $T_{\gamma(t)}M$. Intuitively, for the particle living in M and moving with velocity $\dot{\gamma}(t)$ along $\gamma(t)$, the "space outside of M" does not exist. Its acceleration is then the component of $d\dot{\gamma}(t)/dt$ that lies in $T_{\gamma(t)}M$, denoted $D\dot{\gamma}(t)/dt$. Thus one removes from $d\dot{\gamma}(t)/dt$ the component orthogonal to $T_{\gamma(t)}M$. The quantity $Du(t)/dt$ so defined is called the *covariant derivative* of a (tangent) curve $u(t)$. The equation $D\dot{\gamma}(t)/dt = D[d\gamma(t)/dt]/dt = 0$ is the equation of zero acceleration. The solution (for all t) is called a *geodesic* (curve), and on it lies the segment joining p and q that minimizes the length between p and q. On Euclidean spaces with a Euclidean metric, the geodesics are the straight lines. On S^d, the geodesics are given by great circles. Between any two given points p, q on S^d, there are two segments of the geodesic joining p and q. The segment with the smaller distance is the length minimizing curve between p and q. In the case $q = -p$ (antipodal point of p), the two segments have the same length. The geodesic curve continuously goes around this great circle as time progresses.

On a general manifold M, it does not make sense to speak about the "space outside of M," and hence one cannot define the covariant derivative, or the derivative of a tangent curve, in the above manner. An important milestone in differential geometry was reached when Levi-Civita (1917) developed the notion of an "affine connection" on a general Riemannian manifold, which provided for differentiation of a vector field along a curve. For details, see Boothby (1986) and Do Carmo (1992).

Certain properties of geodesics are fairly obvious. First, a geodesic $\gamma(t)$ is determined by its value $\gamma(t_0)$ at any given point $t = t_0$ and by its derivative $\dot{\gamma}(t_0)$ at that point. This is a consequence of a basic theorem in ordinary differential equations, because $\gamma(t)$ is governed by a second-order equation, namely, its acceleration is zero. Second, if $\gamma(t)$ is a geodesic, so is $\gamma_a(t) = \gamma(at)$, because $D[d\gamma_a(t)/dt]/dt = D[a\dot{\gamma}(at)]/dt = a^2(D\dot{\gamma})_{at} = 0$. The geodesic γ_a has the initial value $\gamma_a(0) = \gamma(0)$ at time $t = 0$, the same as that of γ, and an initial velocity $a\dot{\gamma}(0)$. One may thus write a geodesic γ in terms of its two parameters, namely, $\gamma(0)$ and $\dot{\gamma}(0)$, say, q and v: $\gamma(t) = \gamma(t; q, v)$, and note that $\gamma(at; q, v) = \gamma(t; q, av)$.

It is known that on a connected Riemannian manifold that is complete in the geodesic distance, every geodesic γ can be continued indefinitely, that is, $\gamma(t)$ is defined for all $t \in (-\infty, \infty)$. Indeed, the Hopf–Rinow theorem says that the properties of (1) completeness of M under the geodesic distance, (2) the indefinite continuity of geodesics, and (3) the compactness of all closed bounded subsets are equivalent (Do Carmo, 1992, p. 147). Unless stated otherwise, we will assume that the Riemannian manifolds under consideration are connected and complete.

Given an initial value p and an initial velocity v of a geodesic γ, let $r(v)$ be the supremum of all t such that the geodesic γ is distance minimizing between p and $\gamma(t; p, v)$. If $r(v)$ is finite, $\gamma(r(v); p, v)$ is called the *cut-point* of p along γ. Otherwise, one says that the cut-point does not exist. The *cut-locus* $C(p)$ of p is the set of all cut-points of geodesics starting at p. For $q \in M \setminus C(p)$, there is a unique v such that $\gamma(t; p, v) = q$ for some $t < r(v)$. Note that on S^d, $C(p) = \{-p\}$, and cut-points and cut-loci do not exist on Euclidean spaces. Define the *exponential map* $\exp_p$ on T_pM as $\exp_p(v) = \gamma(1; p, v)$, where, as above, $\gamma(t; p, v)$ is the position at time t of the geodesic starting at p with initial velocity v ($v \in T_pM$). It is known that $\exp_p$ is a diffeomorphism on an open subset of T_pM onto $M \setminus C(p)$ (see Gallot et al., 1990, pp. 95, 133). The inverse $\exp_p^{-1}$ of the exponential map on $M \setminus C(p)$ is sometimes also called the *log map*. An open neighborhood U of a point $q \in M$ is said to be a *normal neighborhood* of p if there exists a star-shaped open subset V of T_pM (i.e., $v \in V$ implies $cv \in V$ for all $0 \leq c \leq 1$) such that $\exp_p$ is a diffeomorphism on V onto U. If $\exp_p(\sum x^i v_i) = q \in U$, for a chosen orthonormal basis $\{v_1, \ldots, v_d\}$ of T_pM, then $(x^1, \ldots, x^d)$ are called the *normal coordinates* of q in U. Thus $M \setminus C(p)$ is a normal neighborhood of p.

A subset C of M is said to be *convex* if, for every pair, $p, q \in C$, there is a unique distance minimizing geodesic joining p and q lying entirely in C.

The metric tensor g on an orientable Riemannian manifold (M, g) may be used to define a volume measure (or volume form) on it. If $\{(U_p, \psi_p) : p \in M\}$ is an atlas with a positive orientation, define the *volume measure* on U_p, denoted *vol*, by assigning the volume for all subsets $\psi_p^{-1}(B)$ of U_p (B, a Borel subset of B_p) by

$$\text{vol}(\psi_p^{-1}(B)) = \int_B \sqrt{(\det g(x))} dx_1 \cdots dx_d. \tag{B.1}$$

Here $\det g(x)$ is the determinant of the matrix $((g_{ij}(x)))$, $g_{ij}(x) = \langle E_i, E_j \rangle (\psi_p^{-1}(x))$ and $\{E_i = d\psi_p^{-1}(\partial/\partial x_i), i = 1, \ldots, d\}$ is the coordinate frame on U_p. It is not difficult to check that this definition of volume measure does not depend on the coordinate system. To define the measure on all of M, one pieces this construction together by means of a "partition of unity" (see, e.g., Do Carmo, 1992, pp. 44–45). To motivate equation (B.1), thinking infinitesimally, the "unit cube,"

$$C = \{t_1 F_1 + \cdots + t_d F_d : 0 \leq t_i \leq 1 \text{ for } i = 1, \ldots, d\},$$

formed on $T_p M$ with respect to an orthonormal basis $\{F_1, \ldots, F_d\}$ of $T_p M$, has volume one. Now

$$D = \{t_1 E_1 + \cdots + t_d E_d : 0 \leq t_i \leq 1 \text{ for } i = 1, \ldots, d\}$$

is the parallelepiped formed by the coordinate vectors $\{E_i = d\psi_p^{-1}(\partial/\partial x_i) : i = 1, \ldots, d\}$ at p. Expressing $E_i = \sum a_{ij} F_j$, one has $g_{ij} = \langle E_i, E_j \rangle (p) = \sum_k a_{ik} a_{jk}$, which is the (i, j) element of the matrix AA', where $A = ((a_{ij}))$. Hence $\det g = (\det A)^2$, so that $\det A = \sqrt{\det g}$, which is the Jacobian determinant of the linear transformation $E = AF$. Thus multiplication of the infinitesimal volume $dx_1 \cdots dx_d$ in B_p by $\sqrt{\det g(x)}$ yields the corresponding infinitesimal volume on U_p.

The volume measure (form) defined by equation (B.1) is clearly invariant under isometries. In the case (M, g) is compact, one may normalize it to get a probability measure, which is often called the *uniform distribution* on M. The volume measure may sometimes be determined (up to a scalar multiple) by the requirement of invariance under isometries.

It is a useful fact that $\text{vol}(C(p)) = 0$ for all $p \in M$ (Gallot et al., 1990, p. 141).

Next we describe the splitting of the tangent space $T_q N$ of a Riemannian manifold N of dimension n ($q \in N$) into vertical and horizontal subspaces under a differentiable map (submersion) F on N onto a Riemannian manifold M of dimension $d < n$, such that $d_q F$ is of rank d. In this case, for every $p \in M$, $F^{-1}(p)$ is a submanifold (*fiber*) of N. For $q \in F^{-1}(p)$, tangent vectors $v \in T_q N$ are said to be *vertical* if $v \in T_q F^{-1}(p) = V_q$, say. The vector subspace of $T_q N$ orthogonal to the vertical tangent space V_q is called the *horizontal subspace* of $T_q N$ and is denoted as H_q. If, in addition, $d_q F|_{H_q}: H_q \to T_{F(q)} M$ is an isometry (i.e., $\langle q, h \rangle_q = \langle d_q F(g), d_q F(h) \rangle \forall g, h \in H_q$), then F is called a *Riemannian submersion*. Note that H_q has dimension d and V_q has dimension $n - d$. An important class of examples is provided by the projection $F: N \to M = N/G$, where G is a compact Lie group of certain isometries of N.

Example We now consider the construction of geodesics on $SO(m)$. Let $M(m)$ be the space of all $m \times m$ matrices with real coordinates. It is an m^2-dimensional vector space with respect to coordinatewise addition and may be identified with $\mathbb{R}^{m^2}$. A Riemannian structure on $M(m)$ is then provided by the usual Riemannian structure on $\mathbb{R}^{m^2}$. Thus its tangent bundle is isomorphic to $\mathbb{R}^{m^2}$ and the metric tensor is provided by that on $\mathbb{R}^{m^2}$, which may be expressed by the inner product $\langle A, B \rangle =$ Trace $(A'B)(A, B \in M(m))$. The squared Euclidean distance between A and B is Trace $((A - B)(A - B)')$. Consider now $SO(m)$ as a submanifold of $M(m)$: $SO(m) =$ $\{A \in M(m) : A'A = I_m, \text{Det } A = +1\}$. Let us determine its tangent space at the identity $I = I_m$. Consider a smooth function $t \to \gamma(t) : \mathbb{R} \to SO(m)$, with $\gamma(0) = I$. Because $\gamma(t)'\gamma(t) = I$, differentiation yields $\dot{\gamma}(t)'\gamma(t) + \gamma(t)'\dot{\gamma}(t) = 0$. In particular, at $t = 0$, $\dot{\gamma}(0)' + \dot{\gamma}(0) = 0$. In other words, $\dot{\gamma}(0)$ is an $m \times m$ skew-symmetric matrix. In particular, one may consider the map $\gamma(t) = e^{tA}$ for an arbitrary skew-symmetric matrix A, noting that $\gamma(t)'\gamma(t) = e^{t(A+A')} = e^0 = I$. The curve $\gamma(t) = e^{tA}$ satisfies $\gamma(0) = I$, $\dot{\gamma}(0) = A$. Hence $T_I SO(m)$ is the vector space of all skew-symmetric matrices. One may now easily argue that the tangent space of $SO(m)$ at B is given by $T_B SO(m) = \{BA : A \text{ skew-symmetric}\}$. We now show that, for any given skew-symmetric A, the curve $\gamma(t) = e^{tA}$ is the geodesic satisfying $\gamma(0) = I$, $\dot{\gamma}(0) = A$. That is, we wish to prove that the covariant derivative of $\dot{\gamma}(t)$ vanishes. Now it is simple to check that $\dot{\gamma}(t) = d\gamma(t)/dt = e^{tA}A$ and $d^2\gamma(t)/dt^2 = e^{tA}A^2$. One then needs to prove that $e^{tA}A^2$ is orthogonal to the tangent space of $SO(m)$ at e^{tA}, namely, $T_{e^{tA}}SO(m) = \{e^{tA}D : D \text{ skew symmetric}\}$. Now

$$\text{Trace}((e^{tA}A^2)'e^{tA}D) = \text{Trace}((A')^2 e^{tA'} e^{tA}D) = \text{Trace}((A')^2 D)$$
$$= \text{Trace}(A^2 D) = \Sigma_{i,j}c_{ij}d_{ij}, \tag{B.2}$$

where $A^2 = ((c_{ij}))$ is symmetric. Because the sum over $i < j$ cancels the sum over $i > j$, the sum in equation (B.2) vanishes. Thus $\gamma(t) = e^{tA}$ is the geodesic satisfying $\gamma(0) = I$, $\dot{\gamma}(0) = A$. It also follows that the curve $\gamma(t) = Be^{tA}$ is the geodesic satisfying $\gamma(0) = B$, $\dot{\gamma}(0) = BA$ for $B \in SO(m)$ and A skew-symmetric. Recall that the *exponential map* $\exp_B$ on $T_B SO(m)$ is defined by $\exp_B(BA) = \gamma(1) \equiv \gamma(1; B, BA)$ where γ is the geodesic starting at $\gamma(0) = B$ with velocity $\dot{\gamma}(0) = BA$. Hence $\exp_B(BA) = Be^A$. In particular, for $B = I$, $\exp_I(A) = e^A$. The name "exponential map" on a general Riemannian manifold comes from such a representation on spaces of matrices.

C Dirichlet processes

Nonparametric inference from the Bayesian perspective requires putting a prior distribution on the space of all probability measures on the measurable space $(X, \mathcal{B})$ of observations.

C.1 Finite X

We first consider a finite X with k elements $a_1, \ldots, a_k$, say (and $\mathcal{B}$ is the class of all subsets). Then the unknown probability $P \in \mathcal{P}$ on $(X, \mathcal{B})$, which is the object of inference, is determined by $\theta_i = P(\{a_i\})$, $1 \le i \le k$, and this is a finite-dimensional (i.e., parametric) problem, and a convenient conjugate prior for P is the multivariate

Dirichlet, or Beta, distribution $D_\alpha = D(\alpha_1, \dots, \alpha_k)$ for $(\theta_1, \dots, \theta_k)$, with $\theta_k = 1 - \theta_1 \cdots - \theta_{k-1}$. First, consider the case $\alpha_i > 0$ for all i. Then $\theta = (\theta_1, \dots, \theta_{k-1})$ has the density on the set $\{(\theta_1, \dots, \theta_{k-1}) : \theta_i > 0 \text{ for all } i, \sum_{1 \le \theta \le k-1} \theta_i \le 1\}$, given by

$$\pi(\theta_1, \dots, \theta_k; \alpha_1, \dots, \alpha_k) = c(\alpha_1, \dots, \alpha_k) \theta_1^{\alpha_1 - 1} \cdots \theta_{k-1}^{\alpha_{k-1}-1} \theta_k^{\alpha_k - 1}, \qquad \text{(C.1)}$$

where $\theta_k = 1 - \theta_1 - \cdots - \theta_{k-1}$.

One may also define $D_\alpha = D(\alpha_1, \dots, \alpha_k)$, where some α_i are zero. If $\alpha_i = 0$, then the Dirichlet assigns probability one to $\{\theta_i = 0\}$ and a distribution such as given by equation (C.1) in the variables θ_j for which $\alpha_j > 0$. This defines, for arbitrary nonnegative $\alpha_1, \dots, \alpha_k$, the distribution $D_\alpha = D(\alpha_1, \dots, \alpha_k)$ on the *simplex* $= \Delta_k = \{(\theta_1, \dots, \theta_k) : \theta_i \ge 0 \text{ for all } i, \sum_{1 \le i \le k} \theta_i = 1\}$. Note that, under $D(\alpha_1, \dots, \alpha_k)$, the distribution of θ_i is Beta$(\alpha_i, \alpha(\mathcal{X}) - \alpha_i)$, where

$$\alpha(\mathcal{X}) = \alpha_1 + \cdots + \alpha_k.$$

Before proceeding further, we recall a fruitful representation of a random P with distribution $D(\alpha_1, \dots, \alpha_k)$. For $c > 0$, a *Gamma(c) distribution* is defined by its density $\Gamma(c)^{-1} \exp\{-z\} z^{c-1}$ $(z > 0)$. If $c = 0$, define Gamma(0) to be the distribution degenerate at 0. Suppose Z_i, $1 \le i \le k$, are independent random variables, with Z_i having the distribution Gamma(α_i), and let $S_k = Z_1 + \cdots + Z_k$. If $\alpha_i > 0$ for all $i = 1, \dots, k$. Then the usual transformation rule yields that Z_i / S_k, $1 \le i \le k - 1$, have the joint density (C.1), and they are independent of S_k, which is Gamma$(\alpha_1 + \cdots + \alpha_k)$. In particular, $(Z_1 / S_k, \dots Z_k / S_k)$ has the Dirichlet distribution $D(\alpha_1, \dots, \alpha_k)$ and it is independent of S_k. If a subset of the α are zero, then the corresponding relationship holds among the remaining variables. Now, inserting the degenerate variables (with values 0) also, the representation holds for the general case. The following lemma is proved using this representation.

Lemma C.1 *Suppose U_1 and U_2 are independent random vectors with Dirichlet distributions $D_\alpha = D(\alpha_1, \dots, \alpha_k)$ and $D_\beta = D(\beta_1, \dots, \beta_k)$, respectively, on Δ_k, and let Y be independent of $\{U_1, U_2\}$ and have the Beta distribution Beta$(\alpha_1 + \cdots + \alpha_k, \beta_1 + \cdots + \beta_k)$. Then $Y U_1 + (1 - Y) U_2$ has the distribution $D(\alpha_1 + \beta_1, \dots, \alpha_k + \beta_k) = D_{\alpha+\beta}$.*

Proof Assume Z_i $(i = 1, \dots, k)$ and Z_i' $(i = 1, \dots, k)$ are $2k$ independent random variables, with Z_i being Gamma(α_i) $(i = 1, \dots, k)$ and Z_i' Gamma(β_i) $(i = 1, \dots, k)$. Write $S_j = \sum_{1 \le i \le j} Z_i$ and $S_j' = \sum_{1 \le i \le j} Z_i'$. Then $Y U_1 + (1 - Y) U_2$ has the same distribution as

$$[S_k / (S_k + S_k')](Z_1 / S_k, \dots, Z_k / S_k)$$
$$+ [S_k' / (S_k + S_k')](Z_1' / S_k', \dots, Z_k' / S_k'), \qquad \text{(C.2)}$$

because $(Z_1 / S_k, \dots, Z_k / S_k)$ is $D(\alpha_1, \dots, \alpha_k)$ and $(Z_1' / S_k', \dots, Z_k' / S_k')$ is $D(\beta_1, \dots, \beta_k)$, independent of each other and of $V = S_k / (S_k + S_k')$ and $(1 - V) = S_k' / (S_k + S_k')$, with V distributed as Beta$(\alpha_1 + \cdots + \alpha_k, \beta_1 + \cdots + \beta_k)$. But equation (C.2) equals

$$((Z_1 + Z'_1)/[S_k + S'_k], \ldots, (Z_k + Z'_k)/[S_k + S'_k]),$$

which has the desired distribution $D(\alpha_1 + \beta_1, \ldots, \alpha_k + \beta_k) = D_{\alpha+\beta}$, because $Z_i + Z'_i$ are independent Gamma$(\alpha_i + \beta_i)$, $i = 1, \ldots, k$.   □

If the random distribution P on Δ_k has the Dirichlet distribution $D_\alpha = D(\alpha_1, \ldots, \alpha_k)$, and if $X_1, \ldots, X_n$ are i.i.d. observations from P, conditional on P (i.e., given $(\theta_1, \ldots, \theta_k)$), then the likelihood function is proportional to

$$\theta_1^{\alpha_1 - 1 + n_1} \cdots \theta_{k-1}^{\alpha_{k-1} - 1 + n_{k-1}} \theta_k^{\alpha_k - 1 + n_k},$$

where $n_i = \sum \delta_{X_j}(\{a_i\})$ is the number of observations having value $a_i \in X$. Here δ_x is the point mass at x, that is, $\delta_x(\{x\}) = 1$, $\delta_x(X \backslash \{x\}) = 0$. Hence the posterior distribution of P (or of $(\theta_1, \ldots, \theta_k)$) is $D(\alpha_1 + n_1, \ldots, \alpha_k + n_k)$. If $\alpha_i + n_i = 0$, this is interpreted, as before, as $\theta_i = 0$ with posterior probability one.

When $\alpha = (\alpha_1, \ldots, \alpha_k)$ is viewed as a measure on X : $\alpha(\{a_i\}) = \alpha_i$ $(1 \leq i \leq k)$, then the posterior may be expressed as the Dirichlet distribution $D_{\alpha + \sum_{1 \leq j \leq n} \delta_{X_j}}$ with measure $\alpha + \sum_{1 \leq j \leq n} \delta_{X_j}$. We define the Dirichlet distribution D_{δ_x} to be the distribution degenerate at $x \in X$. That is, $D_{\delta_x}(\{\theta_i = 1\}) = 1$ if $x = a_i$ and this probability is zero if $x \neq a_i$. We will make use of the fact that if Y is Beta$(1, \alpha(X))$ independent of a P that is Dirichlet D_α, then

$$Y\delta_x + (1 - Y)P \text{ has the distribution } D_{\alpha + \delta_x}. \tag{C.3}$$

One may derive this from Lemma C.1 by taking $\alpha = \delta_x$ and $\beta = \alpha$ in the lemma. Next note that, conditionally given P (i.e., given $(\theta_1, \ldots, \theta_k)$), a single observation X from P has the marginal distribution

$$\Pr(X = a_i) = c(\alpha_i, \alpha(X) - \alpha_i) \int_0^1 \theta_i \theta_i^{\alpha_i - 1} (1 - \theta_i)^{\alpha(X) - \alpha_i - 1} d\theta_i = \alpha_i / \alpha(X)$$

$$(i = 1, \ldots, k). \tag{C.4}$$

Here $c(a, b) = \Gamma(a + b)/\Gamma(a)\Gamma(b)$ is the normalizing constant of the Beta(a, b) distribution. Thinking of the problem of a single observation X from P, conditionally given P, and using the fact that the conditional distribution of P, given X is $D_{\alpha + \delta_X}$, it follows that the marginal distribution of P, namely, the prior $D_\alpha = D(\alpha_1, \ldots, \alpha_k)$, satisfies the identity

$$D_\alpha(B) = \sum_{1 \leq i \leq k} D_{\alpha + \delta_{a_i}}(B) \alpha_i / \alpha(X) \quad (B, \text{ a Borel subset of } \Delta_k). \tag{C.5}$$

Lemma C.2 *Suppose (i) P is Dirichlet $D_\alpha = D(\alpha_1, \ldots, \alpha_k)$, (ii) X is independent of P having the distribution $\overline{\alpha} = \alpha/\alpha(X)$ on X, and (iii). Y is independent of P and X and has the Beta distribution Beta$(1, \alpha(X))$ on $[0, 1]$. Then $Y\delta_X + (1 - Y)P$ has the same distribution as P, namely, $D_\alpha = D(\alpha_1, \ldots, \alpha_k)$.*

Proof Conditionally given $X = a_i$, the distribution of the random measure $Y\delta_X + (1 - Y)P = Q_X$, say, is $D_{\alpha + \delta_{a_i}}$, by equation (C.3). Now apply equation (C.5) to see that the (marginal) distribution of Q_X is D_α.   □

C.2 General X

We now turn to the general case of a Polish space X, with its Borel sigma-field. Recall that a Polish space is a topological space that is homeomorphic to a complete separable metric space. In this case, the set $\mathcal{P}$ of all probability measures on $(X, \mathcal{B})$ is also Polish under the weak topology (see, e.g., Parthasarathy, 1967, theorem 6.5, p. 46; Bhattacharya and Waymire, 2007, pp. 68, 69). Let $\mathcal{B}(\mathcal{P})$ denote the Borel sigma-field of $\mathcal{P}$. If X is a compact metric space, so is $\mathcal{P}$, under the weak topology (Bhattacharya and Waymire, 2007, proposition 5.5, p. 66).

Let α be a nonzero finite measure on $(X, \mathcal{B})$. We will construct the *Dirichlet distribution* D_α on $\mathcal{P}$ [i.e., on $\mathcal{B}(\mathcal{P})$] having the following finite-dimensional distributions: Let $\{B_1, \ldots, B_k\}$ be an arbitrary partition of X, $k > 1$. That is, B_i are measurable, nonempty, pairwise disjoint, and $\cup B_i = X$. Write $\theta_i = P(B_i)$, $1 \le i \le k$, $P \in \mathcal{P}$. Then the distribution of $(\theta_1, \ldots, \theta_k)$ is k-dimensional Dirichlet $D(\alpha(B_1), \ldots, \alpha(B_k))$. In other words, under D_α, the set $\{P \in \mathcal{P} : ((\theta_1, \ldots, \theta_k) \in C\}$ has the probability $D(\alpha(B_1), \ldots, \alpha(B_k))(C)$ for every Borel subset C of the simplex $\Delta_k = \{(\theta_1, \ldots, \theta_k) : \theta_k \ge 0 \text{ for all } i, \sum_{1 \le i \le k} \theta_i = 1\}$. One can show that this assignment of finite-dimensional distributions satisfies the Kolmogorov consistency theorem and, hence, defines a unique probability measure on the product sigma-field generated by the individual maps into $[0,1]$ defined by $P \mapsto P(B)$, $B \in \mathcal{B}$. Although this sigma-field suffices when X is countable, it is quite inadequate for most purposes when X is uncountable. For example, when X is uncountable, singletons $\{Q\}$ ($Q \in \mathcal{P}$) do not belong to this sigma-field, and non-constant continuous functions on $\mathcal{P}$ are not measurable with respect to it. Ferguson (1973), who is the founder of the Dirichlet distribution on $\mathcal{P}$ and thus of nonparametric Bayes theory, provided a construction of this measure on $\mathcal{B}(\mathcal{P})$. We will, however, present a more convenient construction due to Sethuraman (1994), which immediately yields some important information about the distribution and which is very useful for purposes of simulation. A random probability P, defined on some probability space $(\Omega, \mathcal{F}, \Gamma)$ with values in $\mathcal{P}$ and measurable with respect to $\mathcal{B}(\mathcal{P})$, is called a *Dirichlet process* with α as its *base measure* if it has the Dirichlet distribution D_α on $(\mathcal{P}, \mathcal{B}(\mathcal{P}))$. The proof of Sethuraman's result given below is adapted from Ghosh and Ramamoorthi (2003, pp. 103, 104).

Theorem C.3 *Let α be a finite nonzero measure on $(X, \mathcal{B})$. Suppose two independent i.i.d. sequences θ_n ($n = 1, 2, \ldots$) and Y_n ($n = 1, 2, \ldots$) are defined on a probability space $(\Omega, \mathcal{F}, \mu)$, with θ_n distributed as* $\text{Beta}(1, \alpha(X))$ *on $[0, 1]$, and Y_n having the distribution $\overline{\alpha} = \alpha / \alpha(X)$ on X. Let*

$$p_1 = \theta_1, \ p_n = \theta_n \prod_{1 \le i \le n-1} (1 - \theta_i) \ (n = 2, \ldots). \tag{C.6}$$

Then the random probability measure Q defined by

$$Q(\omega, B) = \sum_{1 \le n < \infty} p_n(\omega) \delta_{Y_n(\omega)}(B), \quad B \in \mathcal{B}(X) \tag{C.7}$$

has the Dirichlet distribution D_α.

Proof First note that $\omega \mapsto Q(\omega, .)$ is a measurable map on Ω into $\mathcal{P}$, with respect to the sigma-field $\mathcal{F}$ on Ω and the Borel sigma-field on $\mathcal{P}$, because each term in the summation in equation (C.7) is. Thus one only needs to show that, for every finite partition $\{B_1, \ldots, B_k\}$ of $\mathcal{X}$, the distribution of $(Q(\cdot, B_1), \ldots, Q(\cdot, B_k))$ has the Dirichlet distribution $D(\alpha(B_1), \ldots, \alpha(B_k))$. For this, write $\delta_{Y_{i:k}}$ as the restriction of δ_{Y_i} to the partition, that is, $\delta_{Y_{i:k}}$ assigns its entire mass 1 to the set of the partition to which Y_i belongs. Also, let P_k be Dirichlet $D(\alpha(B_1), \ldots, \alpha(B_k))$, independent of the two sequences θ_n $(n = 1, 2, \ldots)$ and Y_n $(n = 1, 2, \ldots)$. By Lemma C.2, $Q_1 \equiv p_1 \delta_{Y_{1:k}} + (1 - p_1)P_k$ has the Dirichlet distribution $D(\alpha(B_1), \ldots, \alpha(B_k))$. For the induction argument, we will make use of the identity $\prod_{1 \leq i \leq n}(1 - \theta_i) = 1 - \sum_{1 \leq i \leq n} p_i$. Suppose that

$$Q_n \equiv \sum_{1 \leq i \leq n} p_i \delta_{Y_{i:k}} + \left(\prod_{1 \leq i \leq n} (1 - \theta_i) \right)P_k \equiv \sum_{1 \leq i \leq n} p_i \delta_{Y_{i:k}} + \left(1 - \sum_{1 \leq i \leq n} p_i\right)P_k \quad (C.8)$$

has the Dirichlet distribution $D(\alpha(B_1), \ldots, \alpha(B_k))$. Now

$$Q_{n+1} = \sum_{1 \leq i \leq n+1} p_i \delta_{Y_{i:k}} + \left(\prod_{1 \leq i \leq n+1} (1 - \theta_i) \right)P_k$$

$$= \sum_{1 \leq i \leq n} p_i \delta_{Y_{i:k}} + p_{n+1}\delta_{Y_{n+1:k}} + (1 - \theta_{n+1})\left(1 - \sum_{1 \leq i \leq n} p_i\right)P_k$$

$$= \sum_{1 \leq i \leq n} p_i \delta_{Y_{i:k}} + \left(1 - \sum_{1 \leq i \leq n} p_i\right)(\theta_{n+1}\delta_{Y_{n+1:k}} + (1 - \theta_{n+1})P_k).$$

By Lemma C.2, the distribution of $\theta_{n+1}\delta_{Y_{n+1:k}} + (1 - \theta_{n+1})P_k$ is that of P_k, namely, $D(\alpha(B_1), \ldots, \alpha(B_k))$, and it is independent of $\{\theta_i, Y_{i:k} : i = 1, \ldots, n\}$. Hence Q_{n+1} has the same distribution as $\sum_{1 \leq i \leq n} p_i \delta_{Y_{i:k}} + (1 - \sum_{1 \leq i \leq n} p_i)P_k = Q_n$. This completes the induction argument proving that Q_n has the Dirichlet distribution $D(\alpha(B_1), \ldots, \alpha(B_k))$ for all $n = 1, 2, \ldots$. Letting $n \to \infty$ in equation (C.8), and noting that $\prod_{1 \leq i \leq n}(1 - \theta_i) \to 0$ almost surely as $n \to \infty$ (by the strong law of large numbers applied to the i.i.d. sequence $\log(1 - \theta_i)$), it follows that the distribution of $(Q(\cdot, B_1), \ldots, Q(\cdot, B_k))$ is $D(\alpha(B_1), \ldots, \alpha(B_k))$, where Q is the random probability defined by equation (C.7). □

As an immediate consequence of Theorem C.3, we have the following result. We refer to Ghosh and Ramamoorthi (2003, proposition 2.2.4), for the fact that the set of all discrete distributions on $(\mathcal{X}, \mathcal{B})$ belongs to the Borel sigma-field of $\mathcal{P}$.

Corollary C.4 *The Dirichlet distribution D_α assigns probability one to the set of discrete distributions on $(\mathcal{X}, \mathcal{B})$.*

Proof The Dirichlet process Q in equation (C.7) assigns, for every ω, its entire mass on the countable set $\{Y_n(\omega) : n = 1, 2, \ldots\}$. □

We now state for general $\mathcal{X}$ the obvious analog of the posterior distribution derived in Section C.1 for finite $\mathcal{X}$ (see equation (C.1)).

Theorem C.5   *The posterior distribution of the Dirichlet process P with base measure α, given (conditionally i.i.d.) observations $X_1, \ldots, X_n$ from it, is Dirichlet with base measure $\alpha + \sum_{1 \le j \le n} \delta_{X_j}$.*

Let $\{B_1, \ldots, B_k\}$ be a given partition of $\mathcal{X}$, and let $\alpha{:}k$, $\delta_{X_{j;k}}$ be the restrictions, respectively, of α and δ_{X_j} to this partition; that is, $\delta_{X_{j;k}}$ is the probability measure that assigns mass 1 to the set of the partition to which X_j belongs and zero to others. From the argument in the case of finite $\mathcal{X}$, it is clear that, given only the information about the sets of partition to which $X_1, \ldots, X_n$ belong, the posterior distribution of $(P(B_1), \ldots, P(B_k))$ is Dirichlet $D_{\alpha{:}k + \sum_{1 \le j \le n} \delta_{X_{j;k}}}$. One may intuitively argue that as the partition gets finer and finer, in the limit the distribution of P, given $X_1, \ldots, X_n$, is obtained as Dirichlet with base measure $\alpha + \sum_{1 \le j \le n} \delta_{X_j}$. For a complete argument we refer to Sethuraman (1994) and Ghosh and Ramamoorthi (2003).

C.3 Dirichlet process mixture density

In case the sample $X = (X_1, \ldots, X_n)$ is drawn from a continuous distribution, absolutely continuous with respect to some base measure on $\mathcal{X}$, which is typically set to be the volume form if $\mathcal{X}$ is a Riemannian manifold, the Dirichlet process is no longer appropriate as a prior on the unknown distribution. An alternative approach is to model the unknown density as mixture of some known family of densities $K(\cdot; \theta)$, mixed across the parameter $\theta \in \Theta$ with respect to some unknown probability distribution P on Θ. In particular, we set X_i i.i.d. $f \in \mathcal{D}$, the space of all densities on $\mathcal{X}$, with

$$f(x; P) = \int_{\Theta} K(x; \theta) P(d\theta). \tag{C.9}$$

Then, by setting a DP prior on parameter P, we induce a corresponding prior on $\mathcal{D}$ through model (C.9). To justify the use of this model, we need to verify that it is flexible enough, that is, the prior probability of any neighborhood of any density is positive. Further when estimating the unknown density using draws from its posterior distribution, we require posterior consistency without requiring any parametric assumptions to be made on the true density. The posterior of P given the data and model (C.9) becomes proportional to the prior times the likelihood given f, which is

$$\prod_{i=1}^{n} \left(\int K(X_i; \theta) P(d\theta) \right). \tag{C.10}$$

It is difficult to perform exact sampling of P from this posterior. To simplify computations, we introduce latent parameters $\theta = (\theta_1, \ldots, \theta_n)$ and express the joint of the sample X and θ as follows. Given P, θ_i are i.i.d. P and given θ, $X_i \sim K(X_i; \theta_i)$. Hence, given P, (θ, X) has the likelihood

$$\prod_{i=1}^{n}(P(d\theta_i)K(X_i;\theta_i)).$$

One may integrate out θ and arrive at the marginal likelihood of sample X as in equation (C.10). Because the θ_i, $i = 1,\ldots,n$, are drawn from P, which is discrete, there therefore must be many ties in their values. We use the stick-breaking representation for P as in equation (C.7) and express it as

$$P = \sum_{j=1}^{\infty} p_j \delta_{\theta'_j}.$$

As a result, given $p = \{p_j\}_1^{\infty}$ and $\theta' = \{\theta'_j\}_1^{\infty}$, $\theta_i = \theta'_j$ with probability p_j, $i = 1,\ldots,n$. Introduce another set of latent variables $S = (S_1,\ldots,S_n)$, called *latent class variables*, such that $\theta_i = \theta'_{S_i}$, where the S_i's are i.i.d. multinomials taking value j with probability p_j, $j = 1,\ldots,\infty$. Then the likelihood of (X,S) given (p,θ') becomes

$$\prod_{i=1}^{n}(p_{S_i}K(X_i;\theta'_{S_i})).$$

If one denotes the prior on (p,θ') as π, the joint posterior of (p,θ',S) given X is proportional to

$$\pi(p,\theta')\prod_{i=1}^{n}(p_{S_i}K(X_i;\theta'_{S_i})). \tag{C.11}$$

We use a full conditional Gibbs sampling to get repeated draws of (p,θ',S) from the above posterior and hence of P. This suggests that a DP mixture can be used for clustering of data when the total number of clusters is not prefixed. In any iteration, observation i is allotted to the cluster j for which $S_i = j$. To make the total number of clusters finite in any iteration, one can introduce a set of latent slice variables $u = (u_1,\ldots,u_n)$ drawn from uniform on $[0,1]$ and replace p_{S_i} by $I(u_i < p_{S_i})$. Here I denotes the indicator function. As a result, the marginal of (p,θ',S) after integrating out u is as in equation (C.11). Then, due to the stick-breaking nature of p, the maximum number of clusters occupied in any iteration is bounded by $\min\{j : \sum_{l=1}^{j} p_l > 1 - \min(u)\}$.

We conclude this appendix by recalling that for a locally compact metric space X, such as a d-dimensional compact manifold, and a measure μ on $(X,\mathcal{B})$ that is finite on compact subsets of X, the space $L^1(X,\mathcal{B},\mu)$ of (equivalence classes of) μ-integrable functions is a separable Banach space under the L^1-norm (see, e.g., Dieudonné, 1970, p. 155). In particular, the space of probability measures that are absolutely continuous with respect to μ is a complete separable metric space in the L^1-norm and, therefore, in the total variation distance. One may consider an even stronger distance on the space of probabilities with continuous densities with respect to a finite μ on a compact metric space X, namely, the supremum distance. Because the space $C(X)$ of continuous functions on a compact metric is a complete separable metric space under the supremum distance (see, e.g.,

Bhattacharya and Waymire, 2007, p. 189), the set of continuous densities, which is a closed subset of $C(X)$, is a separable complete metric space in this distance.

D Parametric models on S^d and Σ_2^k

One of the early parametric models on the circle S^1 is due to von Mises (1918)with a density (with respect to Lebesgue measure for arc length) given by

$$g(\theta; \mu, \kappa) = c(\kappa)e^{\kappa \cos(\theta - \mu)}, \quad 0 \le \theta < 2\pi \quad (\kappa \ge 0, \ 0 \le \mu < 2\pi). \tag{D.1}$$

Here $c(\kappa) = \left(\int_0^{2\pi} \exp\{\kappa \cos \theta\} d\theta \right)^{-1}$ is the normalizing constant. If $\kappa = 0$, then the distribution is the uniform distribution. Suppose $\kappa > 0$. Then the distribution is symmetric about μ, and μ is the extrinsic as well as the intrinsic mean, and it is also the mode of the distribution.

One may also consider the one-parameter family with density

$$g(\theta; \kappa) = C(\kappa)e^{\kappa \cos \theta}, \quad 0 \le \theta < 2\pi, \quad (\kappa \ge 0). \tag{D.2}$$

To test the hypothesis of uniformity of the distribution of "fractional parts" $\theta = x - [x]$ ($[x]$ = integer part of x) of atomic weights x of elements, von Mises used the fractional parts of 24 elements, deemed as a random sample of all elements. A test for $\kappa = 0$ with this model yields a p-value of the order 10^{-7}, leading to the rejection of the hypothesis (Mardia and Jupp, 2000, p. 99).

The *von Mises–Fisher distribution* on S^d ($d > 1$) has the following density with respect to the uniform distribution on the sphere (Fisher, 1953; also see Mardia and Jupp, 2000, p. 168):

$$f(x; \mu, \kappa) = c_d(\kappa) \exp\{\kappa \langle x, \mu \rangle\}, \quad x \in S^d \ (\kappa \ge 0, \mu \in S^d). \tag{D.3}$$

Here $\langle, \rangle$ denotes the inner product in $\mathbb{R}^{d+1}$. The case $\kappa = 0$ corresponds to the uniform distribution on S^d. Assume $\kappa > 0$, unless otherwise specified. Note that this distribution is invariant under all rotations around the axis defined by μ. Indeed, if O is an orthogonal $(d + 1) \times (d + 1)$ matrix for which $O\mu = \mu$, then $f(x; \mu, \kappa) = f(Ox; \mu, \kappa)$. In particular, this means that the mean of this distribution, considered as a probability measure on $\mathbb{R}^{d+1}$, is invariant under all such transformations. Hence this mean is of the form $a\mu$, $a > 0$. Therefore, the extrinsic mean of equation (D.3) on the sphere S^d, which is given by the projection of $a\mu$ on S^d, is μ. That the scaler a is positive follows from the fact that f attains its maximum at $x = \mu$ (and minimum at $x = -\mu$). Another way of viewing this is to take the average of x over the small $(d - 1)$-dimensional sphere (small circle, in the case $d = 2$) $\{x \in S^d : \langle x, \mu \rangle = r\} = S_r$, say, $-1 \le r \le 1$. This average is the center s_r of the disc whose boundary is S_r. Note that $s_1 = \mu$, $s_{-1} = -\mu$ and, in general, $s_r = b(r)\mu$, where b is odd: $b(-r) = -b(r)$. The (overall) mean on $\mathbb{R}^{d+1}$ is $a\mu$, where a is the weighted average of $b(r)$, with weights proportional to $\exp\{\kappa r\}v(r)$, with $v(r)$ being the $((d - 1)$-dimensional) "volume" (surface area) of S_r. Since v is symmetric: $v(-r) = v(r)$, it follows that $a > 0$.

One may find the normalizing constant $c_d(\kappa)$ by a similar argument. Writing $r = \cos\theta$, where θ is the angle between x and μ, the radius of S_r is $\sin\theta = (1 - r^2)^{1/2}$ and $v(r) = \left(2\pi^{d/2}/\Gamma(d/2)\right)\left(1 - r^2\right)^{\frac{d-1}{2}}$. Therefore,

$$
\begin{aligned}
c_d(\kappa)^{-1} &= \int_{-1}^{1} e^{\kappa r} \frac{2\pi^{d/2}}{\Gamma(d/2)} \left(1 - r^2\right)^{\frac{d-1}{2}} \frac{dr}{\sqrt{1 - r^2}} \\
&= \frac{2\pi^{d/2}}{\Gamma(d/2)} \int_{-1}^{1} e^{\kappa r} (1 - r^2)^{\frac{d-2}{2}} dr.
\end{aligned}
\tag{D.4}
$$

It follows from proposition 2.2 in Bhattacharya and Patrangenaru (2003) that the intrinsic mean of equation (D.3) is also μ.

To find the MLE of μ based on i.i.d. observations $X_1, \ldots, X_n$ from equation (D.3), one may write the likelihood function as

$$
l(\mu, \kappa : X_1, \ldots, X_n) = C_d(\kappa)^n \exp\left\{n\kappa|\overline{X}|\langle \overline{X}/|\overline{X}|, \mu\rangle\right\}.
\tag{D.5}
$$

For each $\kappa > 0$, the maximum of l is attained at $\mu = \overline{X}/|\overline{X}|$ ($\overline{X} \neq 0$, with probability one). Hence, the MLE of μ is the extrinsic sample mean. The MLE of κ is not explicitly computable (see Fisher, 1953).

It is an interesting (and simple to check) fact that the von Mises–Fisher distribution (D.3) is the conditional distribution, given $|X| = 1$, of a Normal random vector X on $\mathbb{R}^{d+1}$ with mean μ and covariance matrix $\kappa^{-1}I_{d+1}$. A more general family of distributions on S^d may be obtained as the conditional distribution, given $|X| = 1$, of a Normal X on $\mathbb{R}^{d+1}$ with mean γ and covariance matrix Γ. Its density with respect to the uniform distribution on S^d may be expressed as

$$
\begin{aligned}
f_1(x; \gamma, \Gamma) &= c_1(\Gamma) \exp\{-\frac{1}{2}\langle x - \gamma, \Gamma^{-1}(x - \gamma)\rangle\} \\
&= c_2(\gamma, \Gamma) \exp\{\langle x, \Gamma^{-1}\gamma\rangle - \frac{1}{2}\langle x, \Gamma^{-1}x\rangle\}, \quad (x \in S^d).
\end{aligned}
\tag{D.6}
$$

Letting $\kappa = |\Gamma^{-1}\gamma|$, one may write $\Gamma^{-1}\gamma = \kappa\mu$ ($\mu \in S^d$). Also write $A = -\frac{1}{2}\Gamma^{-1}$. One then obtains the *Fisher–Bingham distribution* (Bingham, 1974) with density (with respect to the uniform distribution on S^d)

$$
f(x; \kappa, \mu, A) = c(\kappa, A) \exp\{\kappa\langle x, \mu\rangle + \langle x, Ax\rangle\}, \quad x \in S^d
$$

$$
(\kappa \geq 0, \mu \in S^d, A \text{ a } (d+1) \times (d+1) \text{ symmetric matrix}).
\tag{D.7}
$$

Observe that on replacing A by $A + cI_{d+1}$ for some scalar c, one does not change the above distribution. Hence, for the purpose of identifiability, we let

$$
\text{Trace } A = 0.
\tag{D.8}
$$

One may also take $\kappa = -|\Gamma^{-1}\gamma|$ and replace μ by $-\mu$ without changing the distribution. Hence we choose $\kappa \geq 0$.

Turning to *axial distributions*, consider a random vector $X \in S^d$ that has the same distribution as $-X$. This defines a distribution on the real projective space $\mathbb{R}P^d$ of $[X] = \{X, -X\}$. Recall that $\mathbb{R}P^d$ is the quotient of S^d under the two-element group $G = \{e, x \mapsto -x\}$. One may get a density of f on $\mathbb{R}P^d$ also by changing $\langle x, \mu \rangle$ to $\langle x, \mu \rangle^2$ in equation (D.3) and, more generally, in equation (D.7):

$$f([x]; \kappa, \mu, A) = C_3(\kappa, A) \exp\{\kappa \langle x, \mu \rangle^2 + \langle x, Ax \rangle\}. \tag{D.9}$$

This is a density with respect to the uniform distribution on $\mathbb{R}P^d$ induced from that on S^d by the quotient map. In the special case $A = 0$ (the null matrix), one has the *Dimroth–Watson distribution* (Dimroth, 1963; Watson, 1965) with density

$$f([x]; \kappa, \mu) = c_4(\kappa) \exp\{\kappa \langle x, \mu \rangle^2\}. \tag{D.10}$$

We next turn to the so-called complex Bingham distribution introduced by Kent (1994) on the planar shape space Σ_2^k. Let a point $m = [z]$ in Σ_2^k be expressed by a representative point $\{(z_1, \ldots, z_{k-1})' : \sum_{i=1}^{k-1} |z_i|^2 = 1\}$. A very useful system of coordinates for a complex projective space $\Sigma_2^k = \mathbb{C}P^{k-2}$ was given by Kent (1994) as follows. Let $z_j = r_j^{1/2} \exp\{i\theta_j\}$, where $r_j = |z_j|^2$, $\theta_j \in (-\pi, \pi]$ $(1 \le j \le k-1)$. Because $\sum_{j=1}^{k-1} r_j = 1$, $r_{k-1} = 1 - \sum_{i=1}^{k-2} r_j$ and $r = (r_1, \ldots, r_{k-2})$ belongs to the simplex

$$S_{k-2} = \{r = (r_1, \ldots, r_{k-2})\} : \sum_{i=1}^{k-2} r_j \le 1, r_j \ge 0 \; \forall j = 1, \ldots, k-2\}. \tag{D.11}$$

A point z in $\mathbb{C}S^{k-2} \sim S^{2k-3}$ is then represented by the coordinates $(r_1, \ldots, r_{k-2}, \theta_1, \ldots, \theta_{k-1})$. Consider the distribution on $\mathbb{C}S^{k-2}$ having the constant density $(1/(k-2)!)(2\pi)^{-(k-1)}$ with respect to the $(2k-3)$-dimensional Lebesgue measure on $S_{k-2} \times (-\pi, \pi]^{k-1}$. In these coordinates $\theta_1, \ldots, \theta_{k-1}$ are i.i.d. uniform on $(-\pi, \pi]$, $r = (r_1, \ldots, r_{k-2})$ has the uniform distribution on S_{k-2}, and $\theta = (\theta_1, \ldots, \theta_{k-1})$ and $r = (r_1, \ldots, r_{k-2})$ are independent. To derive the corresponding distribution on $\mathbb{C}P^{k-2}$, consider $\theta_1, \ldots, \theta_{k-1}$ defined up to rotation around z_{k-1}; that is, let $\varphi_j = \theta_j - \theta_{k-1}$ $(1 \le j \le k-1)$, identified so as to belong to $(-\pi, \pi]$. Then $\varphi_{k-1} = 0$, and, conditionally given θ_{k-1}, the free coordinates $\varphi_i, \ldots, \varphi_{k-2}$ are again i.i.d. uniform on $(-\pi, \pi]^{k-2}$. The resulting distribution on $\mathbb{C}P^{k-2}$, represented as $S_{k-2} \times (-\pi, \pi]^{k-2}$, has the density $(1/(k-2)!)(2\pi)^{-(k-2)}$, with r uniformly distributed on S_{k-2}, $\varphi = (\varphi_1, \ldots, \varphi_{k-2})$ uniformly distributed on $(-\pi, \pi]^{k-2}$, and r and φ independent. Let us denote this distribution by ν. The *complex Bingham distribution* CB(A) has the density (with respect to the distribution ν on $\mathbb{C}P^{k-2}$)

$$C(A) \exp\{z^* A z\} \qquad (z \in \mathbb{C}S^{k-2}), \tag{D.12}$$

where $[z] \in \mathbb{C}P^{k-2}$ may be thought of as the orbit of z under all rotations in the plane, or $[z] = \{e^{i\theta} z : -\pi < \theta \le \pi\}$, and A is a $(k-1) \times (k-1)$ Hermitian matrix, $A^* = A$. Note that if one replaces A by $cI_{k-1} + A$ for some $c \in \mathbb{R}$, the distribution does not change. Hence, without loss of generality, we assume that the eigenvalues $\lambda_1, \ldots, \lambda_{k-1}$ of A satisfy $\lambda_1 \le \lambda_2 \le \ldots \le \lambda_{k-1} = 0$. There exists a special unitary matrix U (i.e., $UU^* = I_{k-1}$, det $U = 1$) such that $A = U\Lambda U^*$, where $\Lambda = \text{Diag}(\lambda_1, \ldots, \lambda_{k-1})$ and the exponent in equation (D.12) may be expressed as $\sum_{j=1}^{k-1} \lambda_j |w_j|^2$. Here the jth column of U, say U_j, is a unit eigenvector of A with eigenvalue λ_j, and $w_j = U_j^* z$ $(1 \leq j \leq k - 1)$. One may more simply take $A =$ $\text{Diag}(\lambda_1, \ldots, \lambda_{k-1})$ with $\lambda_1 \leq \cdots \leq \lambda_{k-1} = 0$ and consider the complex Bingham distribution with density

$$C(A) \exp\left\{ \sum_{j=1}^{k-2} \lambda_j r_j \right\}. \tag{D.13}$$

An important special case of equation (D.12) is the *complex Watson distribution* with density (with respect to ν):

$$f([z]; \mu, \sigma) = c(\sigma) \exp\{|z^* \mu|^2 / \sigma^2\},$$

$$z \in \mathbb{C}S^{k-2}, \quad [z] \in \mathbb{C}P^{k-2}, \tag{D.14}$$

with parameter $\mu \in \mathbb{C}S^{k-2}$ and $\sigma > 0$. In this case, $A = -\mu\mu^* = ((-\mu_j\overline{\mu_{j'}}))$ has rank one, with all columns being scalar multiples of μ. Arguing as in the case of the von Mises–Fisher distribution in equation (D.3), one shows that $[\mu]$ is the extrinsic mean.

For drawing samples from parametric models on manifolds, which are useful in simulation studies, we refer to Diaconis *et al.* (2012).

A more general class of models on the planar shape space than considered here may be found in Micheas *et al.* (2006).

References

Amit, Y. 2002. *2D Object Detection and Recognition: Models, Algorithms and Networks.* Lecture Notes in Biomathematic, MIT Press, Cambridge, MA.

Anderson, C.R. 1997. *Object recognition using statistical shape analysis.* Ph.D. Thesis, University of Leeds.

Babu, G. J. and Singh, K. 1984. On one term Edgeworth correction by Efron's bootstrap. *Sankhya Ser. A,* **46**, 219–232.

Bandulasiri, A. and Patrangenaru, V. 2005. Algorithms for nonparametric inference on shape manifolds. *Proc. JSM 2005,* Minneapolis, pp. 1617–1622.

Bandulasiri, A., Bhattacharya, R. N., and Patrangenaru, V. 2009. Nonparametric inference on shape manifolds with applications in medical imaging. *J. Multivariate Analysis,* **100**, 1867–1882.

Barron, A. R., Schervish, M., and Wasserman, L. 1999. The consistency of posterior distribution in nonparametric problems. *Ann. Statist.,* **27**, 536–561.

Barron, A. R. 1989. Uniformly powerful goodness of fit tests. *Ann. Statist.,* **17**, 107–124.

Beran, R. J. 1968. Testing for uniformity on a compact homogeneous space. *J. Appl. Probability,* **5**, 177–195.

Beran, R. J. 1987. Prepivoting to reduce level error of confidence sets. *Biometrica,* **74**, 457–468.

Beran, R. J. and Fisher, N. I. 1998. Nonparametric comparison of mean axes. *Ann. Statist.,* **26**, 472–493.

Berthilsson, R. and Astrom, K. 1999. Extension of affine shape. *J. Math. Imaging Vision,* **11**, 119–136.

Berthilsson, R. and Heyden, A. 1999. Recognition of planar objects using the density of affine shape. *Computer Vision and Image Understanding,* **76**, 135–145.

Bhattacharya, A. 2008a. *Nonparametric statistics on manifolds with applications to shape spaces.* Ph.D. Thesis, University of Arizona.

Bhattacharya, A. 2008b. Statistical analysis on manifolds: a nonparametric approach for inference on shape spaces. *Sankhya,* **70-A**, part 2, 223–266.

Bhattacharya, A. and Bhattacharya, R. N. 2008a. Nonparametric statistics on manifolds with application to shape spaces. In *Pushing the Limits of Contemporary Statistics: Contributions in Honor of J.K. Ghosh.* IMS Collections, **3**, 282–301.

Bhattacharya, A. and Bhattacharya, R. N. 2008b. Statistical on Riemannian manifolds: asymptotic distribution and curvature. *Proceedings of the American Mathematical Society,* **136**, 2957–2967.

Bhattacharya, A. and Bhattacharya, R. N. 2009. Statistical on manifolds with application to shape spaces. In *Perspectives in Mathematical Sciences I: Probability and Mathematics*, edited by N. S. Narasimha Sastry, T. S. S. R. K. Rao, M. Delampady and B. Rajeev. Indian Statistical Institute, Bangalore, 41–70.

Bhattacharya, A. and Dunson, D. 2010a. Nonparametric Bayesian density estimation on manifolds with applications to planar shapes. *Biometrika*, **97**, 851–865.

Bhattacharya, A. and Dunson, D. 2010b. Nonparametric Bayes classification and testing on manifolds with applications on hypersphere. Discussion paper, Department of Statistical Science, Duke University.

Bhattacharya, A. and Dunson, D. 2011. Strong consistency of nonparametric Bayes density estimation on compact metric spaces. *Ann. Institute of Statistical Mathematics*, **63**.

Bhattacharya, R. N. and Ghosh, J.K. 1978. On the validity of the formal Edgeworth expansion. *Ann. Statist.*, **6**, 434–451.

Bhattacharya, R. N. and Patrangenaru, V. 2003. Large sample theory of intrinsic and extrinsic sample means on manifolds. *Ann. Statist.*, **31**, 1–29.

Bhattacharya, R. N. and Patrangenaru, V. 2005. Large sample theory of intrinsic and extrinsic sample means on manifolds—II. *Ann. Statist.*, **33**, 1225–1259.

Bhattacharya, R. N. and Patrangenaru, V. 2012. *A Course in Mathematical Statistics and Large Sample Theory*. Springer Series in Statistics. To appear.

Bhattacharya, R. N. and Qumsiyeh, M. 1989. Second order and L^p-comparisons between the bootstrap and empirical Edgeworth expansion methodologies. *Ann. Statist.*, **17**, 160–169.

Bhattacharya, R. N. and Waymire, E. 2007. *A Basic Course in Probability Theory*. Universitext, Springer, New York.

Bhattacharya, R. N. 1977. Refinements of the multidimensional central limit theorem and applications. *Ann. Probability*, **5**, 1–27.

Bhattacharya, R. N. 1987. Some aspects of Edgeworth expansions in statistics and probability. In *New Perspectives in Theoretical and Applied Statistics*, edited by M. Puri, J. Villaplana, and W. Wertz. Wiley, New York.

Bhattacharya, R. N. 2007. On the uniqueness of intrinsic mean. Unpublished manuscript.

Bhattacharya, R. N. and Denker, M. 1990. *Asymptotic Statistics*. Vol. 14. DMV Seminar, Birkhauser, Berlin.

Bhattacharya, R. N. and Ranga Rao, R. 2010. *Normal Approximation and Asymptotic Expansions*. SIAM, Philadelphia.

Bickel, P. J. and Doksum, K. A. 2001. *Mathematical Statistics, 2nd ed.* Prentice Hall, Upper Saddle River, NJ.

Bingham, C. 1974. An antipodally symmetric distribution on the sphere. *Ann. Statist.*, **2**, 1201–1225.

Bookstein, F. 1978. *The Measurement of Biological Shape and Shape Change*. Lecture Notes in Biomathematics, Springer, Berlin.

Bookstein, F. L. 1986. Size and shape spaces of landmark data (with discussion). *Statistical Science*, **1**, 181–242.

Bookstein, F. L. 1989. Principal warps: thin plate splines and the decomposition of deformations. *Pattern Analysis and Machine Intelligence*, **11**, 567–585.

Bookstein, F. L. 1991. *Morphometric Tools for Landmark data: Geometry and Biology*. Cambridge University Press, Cambridge.

Boothby, W. M. 1986. *An Introduction to Differentiable Manifolds and Riemannian Geometry, 2nd ed.* Academic Press, New York.

Burgoyne, C. F., Thompson H. W., Mercante, D. E., and Amin, R. E. 2000. Basic issues in the sensitive and specific detection in optic nerve head surface change within longitudinal LDT TopSS image: Introduction to the LSU experimental glaucoma (LEG) study. In *The Shape of Glaucoma, Quantitative Neural Imaging Techniques*, edited by H. G. Lemij and J. S. Shuman, 1–37. Kugler Publications, The Hague, The Netherlands.

Casella, G. and Berger, R. L. 2001. *Statistical Inference.* Duxbury Press, Pacific Grove, CA.

Chandra, T. and Ghosh, J. K. 1979. Valid asymptotic expansion for the likelihood ratio statistics and other pertubed chi-square variables. *Sankhya, Ser. A.*, **41**, 22–47.

Chikuse, Y. 2003. *Statistics on Special Manifolds.* Springer, New York.

Diaconis, P., Holmes, S. and Shakshahani, M. 2012. *Sampling from a manifold.* To appear.

Dieudonné, J. 1970. *Treatise on Analysis.* Vol. 2. Academic Press, New York.

Dimitric, I. 1996. A note on equivariant embeddings of Grassmannians. *Publ. Inst. Math (Beograd) (N.S.)*, **59**, 131–137.

Dimroth, E. 1963. Fortschritte der Gefugestatistik. *Neues Jahrbuch der Mineralogie*, Montashefte **13**, 186–192.

Do Carmo, M. 1992. *Riemannian Geometry.* Birkhäuser, Boston.

Dryden, I. L. and Mardia, K. V. 1992. Size and shape analysis of landmark data. *Biometrica*, **79**, 57–68.

Dryden, I. L. and Mardia, K. V. 1998. *Statistical Shape Analysis.* Wiley, New York.

Dryden, I. L., Faghihi, M. R., and Taylor, C. C. 1997. Procrustes shape analysis of spatial point patterns. *J. Roy. Statist. Soc. Ser. B*, **59**, 353–374.

Dryden, I. L., Kume A., Le, H., and Wood, A. T. A. 2008. A multi-dimensional scaling approach to shape analysis. *Biometrika*, **95** (4), 779–798.

Dunford, N. and Schwartz, J. 1958. *Linear Operators—I.* Wiley, New York.

Dunson, D. B. and Bhattacharya, A. 2010. Nonparametric Bayes regression and classification through mixtures of product kernels. *Bayesian Statistics*, **9**, 145–164.

Efron, B. 1979. Bootstrap methods: another look at jackknife. *Ann. Statist.*, **1**, 1–26.

Ellingson, L., Ruymgaart, F. H., and Patrangenaru, V. 2011. Nonparametric estimation for extrinsic mean shapes of planar contours. To appear.

Embleton, B. J. J. and McDonnell, K. L. 1980. Magnetostratigraphy in the Sydney Basin, SouthEastern Australia. *J. Geomag. Geoelectr.*, **32**, 304.

Escobar, M. D. and West, M. 1995. Bayesian density-estimation and inference using mixtures. *J. Am. Statist. Assoc.*, **90**, 577–588.

Ferguson, T. S. 1973. A Bayesian analysis of some nonparametric problems. *Ann. Statist.*, **1**, 209–230.

Ferguson, T. S. 1974. Prior distributions on spaces of probability measures. *Ann. Statist.*, **2**, 615–629.

Ferguson, T. S. 1996. *A Course in Large Sample Theory.* Texts in Statistical Science Series. Chapman & Hall, London.

Fisher, N. I., Hall, P., Jing, B., and Wood, A. T. A. 1996. Improved pivotal methods for constructing confidence regions with directional data. *J. Amer. Statist. Assoc*, **91**, 1062–1070.

Fisher, N. I. 1993. *Statistical Analysis of Circular Data*. Cambridge University Press, Cambridge.

Fisher, N. I., Lewis, T., and Embleton, B. J. J. 1987. *Statistical Analysis of Spherical Data*. Cambridge University Press, Cambridge.

Fisher, R. A. 1953. Dispersion on a sphere. *Proc. Roy. Soc. London Ser. A*, **217**, 295–305.

Fréchet, M. 1948. Lés élements aléatoires de nature quelconque dans un espace distancié. *Ann. Inst. H. Poincaré*, **10**, 215–310.

Gallot, S., Hulin, D., and Lafontaine, J. 1990. *Riemannian Geometry*. Universitext, Springer, Berlin.

Ghosh, J. K. and Ramamoorthi, R. 2003. *Bayesian Nonparametrics*. Springer, New York.

Goodall, C. R. 1991. Procrustes methods in the statistical analysis of shape (with discussion). *J. Roy. Statist, Ser. B*, **53**, 285–339.

Hall, P. 1992. *The Bootstrap and Edgeworth Expansion*. Springer, New York.

Hendriks, H. and Landsman, Z. 1996. Asymptotic tests for mean location on manifolds. *C.R. Acad. Sci. Paris Sr. I Math.*, **322**, 773–778.

Hendriks, H. and Landsman, Z. 1998. Mean location and sample mean location on manifolds: asymptotics, tests, confidence regions. *J. Multivariate Anal.*, **67**, 227–243.

Hjort, N., Holmes C., Muller, P., and Walker, S. G. 2010. *Bayesian Nonparametrics*. Cambridge University Press. Cambridge.

Hopf, H., and Rinow, W. 1931. Über den Begriff der vollständigen differentialgeometrischen Flache. *Comment. Math. Helv.*, **3**, 209–225.

Hopkins, J. W. 1966. Some considerations in multivariate allometry. *Biometrics*, **22**, 747–760.

Huckemann, S., Hotz, T., and Munk, A. 2010. Intrinsic shape analysis: geodesic PCA for Riemannian manifolds modulo isometric Lie group actions (with discussions). *Statist. Sinica*, **20**, 1–100.

Irving, E. 1963. Paleomagnetism of the Narrabeen Chocolate Shale and the Tasmanian Dolerite. *J. Geophys. Res.*, **68**, 2282–2287.

Irving, E. 1964. *Paleomagnetism and Its Application to Geological and Geographical Problems*. Wiley, New York.

Ishwaran, H. and Zarepour, M. 2002. Dirichlet prior sieves in finite normal mixtures. *Statistica Sinica*, **12**, 941–963.

Johnson, R. A. and Wehrly, T. 1977. Measures and models for angular correlation and angular-linear correlation. *J. Royal Stat. Soc. B*, **39**, 222–229.

Karcher, H. 1977. Riemannian center of mas and mollifier smoothing. *Comm. Pure Appl. Math.*, **30**, 509–554.

Kendall, D. G. 1977. The diffusion of shape. *Adv. Appl. Probab.*, **9**, 428–430.

Kendall, D. G. 1984. Shape manifolds, procrustean metrics, and complex projective spaces. *Bull. London Math. Soc.*, **16**, 81–121.

Kendall, D. G. 1989. A survey of the statistical theory of shape. *Statist. Sci.*, **4**, 87–120.

Kendall, D. G., Barden D., Carne, T. K., and Le, H. 1999. *Shape and Shape Theory*. Wiley, New York.

Kendall, W. S. 1990. Probability, convexity, and harmonic maps with small image I: uniqueness and fine existence. *Proc. London Math. Soc*, **61**, 371–406.

Kent, J. T. 1992. New directions in shape analysis. In *The Art of Statistical Science*, edited by K. V. Mardia. Wiley, Chichester.

Kent, J. T. 1994. The complex Bingham distribution and shape analysis. *J. Roy. Statist. Soc. Ser. B*, **56**, 285–299.

Kent, J. T. and Mardia, K. V. 1997. Consistency of Procustes estimators. *J. Roy. Statist. Soc. Ser. B*, **59**, 281–290.

Krim, H. and Yezzi, A., eds. 2006. *Statistics and Analysis of Shapes*. Birkhauser, Boston.

Lahiri, S. N. 1994. Two term Edgeworth expansion and bootstrap approximation for multivariate studentized M-estimators. *Sankhya, Ser. A.*, **56**, 201–226.

Le, H. 2001. Locating Fréchet means with application to shape spaces. *Adv. Appl.Prob.*, **33**, 324–338.

LeCam, L. 1973. Convergence of estimates under dimensionality restrictions. *Ann. Statist.*, **1**, 38–53.

Lee, J. and Ruymgaart, F. H. 1996. Nonparametric curve estimation on Stiefel manifolds. *J. Nonparametri. Statist.*, **6**, 57–68.

Lee, J. M. 1997. *Riemannian Manifolds: An Introduction to Curvature*. Springer, New York.

Lele, S. 1991. Some comments on coordinate free and scale invariant methods in morphometrics. *Amer. J. Physi. Anthropology*, **85**, 407–418.

Lele, S. 1993. Euclidean distance matrix analysis (EDMA): estimation of mean form and mean form difference. *Math. Geology*, **25**, 573–602.

Lele, S. and Cole, T. M. 1995. Euclidean distance matrix analysis: a statistical review. In *Current Issues in Statistical Shape Analysis*, edited by K. V. Mardia and C. A. Gill. University of Leeds Press, Leeds, 49–53.

Lewis, J. L., Lew, W. D., and Zimmerman, J. R. 1980. A non-homogeneous anthropometric scaling method based on finite element principles. *J. Biomech.*, **13**, 815–824.

Lo, A. Y. 1984. On a class of Bayesian nonparametric estimates. 1. Density estimates. *Ann. Statist.*, **12**, 351–357.

Mardia, K. V. and Jupp, P. E. 2000. *Directional Statistics*. Wiley, New York.

Mardia, K. V. and Patrangenaru, V. 2005. Directions and projective shapes. *Ann. Statist.*, **33**, 1666–1699.

Micheas, A. C., Dey, D. K., and Mardia, K. V. 2006. Complex elliptic distributions with applications to shape analysis. *J. Statist. Plan. and Inf.*, **136**, 2961–2982.

Millman, R. and Parker, G. 1977. *Elements of Differential Geometry*. Prentice-Hall, Upper Saddle River, NJ.

NOAA National Geophysical Data Center Volcano Location Database, 1994. http://www.ngdc.noaa.gov/nndc/struts/results?

Oller, J. M. and Corcuear, J. M. 1995. Intrinsic analysis of statistical estimation. *Ann. Statist.*, **23**, 1562–1581.

Parthasarathy, K. R. 1967. *Probability Measures on Metric Spaces*. Academic Press. New York.

Patrangenaru, V. 1998. *Asymptotic statistics on manifolds and their applications*. Ph.D. Thesis, Indiana University, Bloomington.

Patrangenaru, V., Liu, X., and Sagathadasa, S. 2010. A nonparametric approach to 3D shape analysis from digital camera images-I. *J. Multivariate Analysis*, **101**, 11–31.

Prentice, M. J. 1984. A distribution-free method of interval estimation for unsigned directional data. *Biometrica*, **71**, 147–154.

Prentice, M. J. and Mardia, K. V. 1995. Shape changes in the plane for landmark data. *Ann. Statist.*, **23**, 1960–1974.

References

Schwartz, L. 1965. On Bayes procedures. *Z. Wahrsch. Verw. Gebiete*, **4**, 10–26.

Sepiashvili, D., Moura, J. M. F., and Ha, V. H. S. 2003. Affine-permutation symmetry: invariance and shape space. *Proceedings of the 2003 Workshop on Statistical Signal Processing*, St. Louis, MO, 293–296.

Sethuraman, J. 1994. A constructive definition of Dirichlet priors. *Statist. Sinica*, **4**, 639–50.

Singh, K. 1981. On the asymptotic accuracy of Efron's bootstrap. *Ann. Statist.*, **9**, 1187–1195.

Small, C. G. 1996. *The Statistical Theory of Shape*. Springer, New York.

Sparr, G. 1992. Depth-computations from polihedral images. In *Proc. 2nd European Conf. on Computer Vision*, edited by G. Sandimi. Springer, New York, 378–386. Also in *Image and Vision Computing*, **10**, 683–688.

Sprent, P. 1972. The mathematics of size and shape. *Biometrics.*, **28**, 23–37.

Stoyan, D. 1990. Estimation of distances and variances in Bookstein's landmark model. *Biometrical J.*, **32**, 843–849.

Sugathadasa, S. 2006. *Affine and projective shape analysis with applications.* Ph.D. dissertation, Texas Tech University.

von Mises, R. 1918. Uber die "Ganzzaligkeit" der atomgewichte und verwandte fragen. *Phys. Z.*, **19**, 490–500.

Watson, G. S. 1965. Equatorial distributions on a sphere. *Biometrica*, **52**, 193–201.

Watson, G. S. 1983. *Statistics on Spheres*. Vol. 6. University Arkansas Lecture Notes in the Mathematical Sciences, Wiley, New York.

Wu, Y. and Ghosal, S. 2008. Kullback–Liebler property of kernel mixture priors in Bayesian density estimation. *Electronic J. of Statist.*, **2**, 298–331.

Wu, Y. and Ghosal, S. 2010. The L_1-consistency of Dirichlet mixtures in multivariate Bayesian density estimation on Bayes procedures. *J. Mutivar. Analysis*, **101**, 2411–2419.

Yau, C., Papaspiliopoulos, O., Roberts, G. O., and Holmes, C. 2011. Nonparametric Hidden Markov Models with application to the analysis of copy-number-variation in mammalian genomes. *J. Roy. Statist. Soc. B*, **73**, 37–57.

Ziezold, H. 1977. On expected figures and a strong law of large numbers for random elements in quasi-metric spaces. *Transactions of the Seventh Prague Conference on Information Theory, Statistical Functions, Random Processes and of the Eighth European Meeting of Statisticians*, **A**, 591–602. Tech. Univ. Prague, Prague.

Index

Printed in the United States
by Baker & Taylor Publisher Services